THE NATIONAL ACADEMIES
Advisers to the Nation on Science, Engineering, and Medicine

The **National Academy of Sciences** is a private, nonprofit, self-perpetuating society of distinguished scholars engaged in scientific and engineering research, dedicated to the furtherance of science and technology and to their use for the general welfare. Upon the authority of the charter granted to it by the Congress in 1863, the Academy has a mandate that requires it to advise the federal government on scientific and technical matters. Dr. Ralph J. Cicerone is president of the National Academy of Sciences.

The **National Academy of Engineering** was established in 1964, under the charter of the National Academy of Sciences, as a parallel organization of outstanding engineers. It is autonomous in its administration and in the selection of its members, sharing with the National Academy of Sciences the responsibility for advising the federal government. The National Academy of Engineering also sponsors engineering programs aimed at meeting national needs, encourages education and research, and recognizes the superior achievements of engineers. Dr. Charles M. Vest is president of the National Academy of Engineering.

The **Institute of Medicine** was established in 1970 by the National Academy of Sciences to secure the services of eminent members of appropriate professions in the examination of policy matters pertaining to the health of the public. The Institute acts under the responsibility given to the National Academy of Sciences by its congressional charter to be an adviser to the federal government and, upon its own initiative, to identify issues of medical care, research, and education. Dr. Harvey V. Fineberg is president of the Institute of Medicine.

The **National Research Council** was organized by the National Academy of Sciences in 1916 to associate the broad community of science and technology with the Academy's purposes of furthering knowledge and advising the federal government. Functioning in accordance with general policies determined by the Academy, the Council has become the principal operating agency of both the National Academy of Sciences and the National Academy of Engineering in providing services to the government, the public, and the scientific and engineering communities. The Council is administered jointly by both Academies and the Institute of Medicine. Dr. Ralph J. Cicerone and Dr. Charles M. Vest are chair and vice chair, respectively, of the National Research Council.

www.national-academies.org

COMMITTEE ON THE REVIEW OF THE TSUNAMI WARNING AND FORECAST SYSTEM AND OVERVIEW OF THE NATION'S TSUNAMI PREPAREDNESS

JOHN A. ORCUTT (*Chair*), University of California, San Diego, La Jolla
MARTHA R. GRABOWSKI (*Vice-chair*), Le Moyne College, Syracuse, New York, and Rensselaer Polytechnic Institute, Troy, New York
BRIAN F. ATWATER, U.S. Geological Survey, Seattle, Washington
ANN BOSTROM, University of Washington, Seattle
GEORGE CRAWFORD, Washington State Emergency Management Division (Retired), Camp Murray
RICHARD K. EISNER, Research Center for Disaster Reduction Systems, Disaster Prevention Research Institute, Kyoto University, Japan
JIAN LIN, Woods Hole Oceanographic Institution, Massachusetts
DOUGLAS S. LUTHER*, University of Hawaii at Manoa, Honolulu
HUGH B. MILBURN, National Oceanic and Atmospheric Administration (Retired), Lake Forest Park, Washington
DENNIS S. MILETI, University of Colorado (Retired), Rancho Mirage, California
EMILE A. OKAL*, Northwestern University, Evanston, Illinois
COSTAS E. SYNOLAKIS, University of Southern California, Los Angeles
NATHAN J. WOOD, U.S. Geological Survey, Vancouver, Washington
HARRY YEH, Oregon State University, Corvallis, Oregon

Staff

CLAUDIA MENGELT, Study Director
SUSAN PARK, Senior Program Officer (until December 2009)
DEBORAH GLICKSON, Program Officer (from January 2010)
PAMELA LEWIS, Administrative Coordinator
SHERRIE FORREST, Research Associate
JEREMY JUSTICE, Senior Program Assistant
HELENA ANTOUN, Christine Mirzayan Science and Technology Policy Fellow (from September to December 2009)
IAN BROSNAN, Christine Mirzayan Science and Technology Policy Fellow (from January to April 2010)

---

* Resigned from the committee.

Acknowledgments

This report was greatly enhanced by the participants of the meetings held as part of this study. To begin with, the committee would like to thank Jenifer Rhoades (National Oceanic and Atmospheric Administration [NOAA]) for her tremendous assistance with providing supporting documentation from across NOAA and her presentation. The committee would like to acknowledge the efforts of those who gave presentations at meetings: Susan Asturias (San Diego County Office of Emergency Services), Rodey Batiza (National Science Foundation), Eddie Bernard (NOAA), Linda Bourque (University of California, Los Angeles), Gary Carver (Humboldt State University), Kwok Fai Cheung (University of Hawaii), Herb Dragert (Geological Survey of Canada), Paula Dunbar (NOAA), John Ferree (NOAA), James Goltz (California Emergency Management Agency), Roger Hansen (Geophysical Institute, University of Alaska, Fairbanks), Barry Hirshorn (NOAA), Paul Huang (NOAA), William Knight (NOAA), William Leith (U.S. Geological Survey), Michael Lindell (Texas A&M University), Michael Mahoney (Federal Emergency Management Agency), Chris Maier (NOAA), Charles McCreery (NOAA), Susan McLean (NOAA), Anu Mittal (Government Accountability Office), David Oppenheimer (U.S. Geological Survey), Ervin Petty (Alaska Division of Homeland Security and Emergency Management), George Priest (Oregon Department of Geology and Mineral Industries), Karlene Roberts (University of California, Berkeley), John Sorensen (Oak Ridge National Laboratory), Adam Stein (NOAA), Elena Suleimani (Geophysical Institute, University of Alaska, Fairbanks), Uri S. ten Brink (U.S. Geological Survey), Vasily Titov (NOAA), Louis Uccellini (NOAA), Christa von Hillebrandt (University of Puerto Rico, Mayagüez), Stuart Weinstein (NOAA), Paul Whitmore (NOAA), Gene Whitney (Office of Science and Technology Policy), Jay Wilson (Clackamas County Emergency Management), Rick Wilson (California Geological Survey). These talks helped set the stage for fruitful discussions in the closed sessions that followed.

The committee is also grateful to a number of people who provided important discussion and/or material for this report: Leighton Ah Cook (Hawaii State Civil Defense), Christine Brown (Rochester Institute of Technology), Laura Kong (International Tsunami Information Center), Aurelio Mercado-Irizarry (University of Puerto Rico), Vickie Nadolski (NOAA), Therese Pierce (NOAA), Kevin Richards (Hawaii State Civil Defense), John Schelling (Washington Emergency Management Division), Edward Teixeira (Hawaii State Civil Defense), Tim Walsh (Washington State Department of Natural Resources), Ray Willemann (Incorporated Research Institutions for Seismology), Brian Yanagi (International Tsunami Information Center), Joseph Zhang (Oregon Health and Science University).

This report has been reviewed in draft form by individuals chosen for their diverse perspectives and technical expertise, in accordance with procedures approved by the NRC's Report Review Committee. The purpose of this independent review is to provide candid and critical comments that will assist the institution in making its published report as sound as possible and to ensure that the report meets institutional standards for objectivity, evidence, and responsiveness to the study charge. The review comments and draft manuscript remain confidential to protect the integrity of the deliberative process. We wish to thank the following individuals for their participation in the review of this report:

MANI K. CHANDY, California Institute of Technology, Pasadena
LORI DENGLER, Humboldt State University, Arcata, California
ERIC GEIST, U.S. Geological Survey, Menlo Park, California
FRANK I. GONZÁLEZ, University of Washington, Seattle
EVE GRUNTFEST, University of Oklahoma, Norman
ARLEEN A. HILL, University of Memphis, Tennessee
PHILIP LI-FAN LIU, Cornell University, Ithaca, New York
CARL MAIDA, University of California, Los Angeles
FREDRIC RAICHLEN, California Institute of Technology, Pasadena
PETER M. SHEARER, University of California, San Diego

Although the reviewers listed above have provided many constructive comments and suggestions, they were not asked to endorse the conclusions or recommendations nor did they see the final draft of the report before its release. The review of this report was overseen by **Robin K. McGuire**, William Lettis and Associates, Inc., appointed by the Divison on Earth and Life Studies, and **Robert A. Dalrymple**, Johns Hopkins University, appointed by the Report Review Committee, who were responsible for making certain that an independent examination of this report was carried out in accordance with institutional procedures and that all review comments were carefully considered. Responsibility for the final content of this report rests entirely with the authoring committee and the institution.

Contents

Executive Summary

Many coastal areas of the United States are at risk from tsunamis. Since 1800, tsunamis have taken lives in Hawaii, Alaska, Puerto Rico, American Samoa, the Virgin Islands, California, and Oregon. Tsunamis happen rarely enough to allow a false sense of security, but when they do occur there may be just minutes or hours for people to reach a safer location.

The catastrophic 2004 tsunami in the Indian Ocean provided the impetus for legislation to expand U.S. tsunami warning capabilities (P.L. 109-13). This was followed by the Tsunami Warning and Education Act of 2006 (P.L. 109-424), which asked the National Oceanic and Atmospheric Administration (NOAA) and the National Tsunami Hazard Mitigation Program (NTHMP) to strengthen the nation's existing tsunami detection, warning, education, and preparedness efforts. In the 2006 law, Congress requested the National Academy of Sciences (NAS) to review progress toward the nation's ability to detect and forecast tsunamis. The NAS expanded the study's scope to include assessment of the nation's ability to reduce tsunami losses by educating and preparing the American public.

In this report, the study committee finds that the nation has made progress in several areas since 2004. At the federal level, NOAA has improved the ability to detect and forecast tsunamis by expanding the sensor network (specifically the Deep-ocean Assessment and Reporting of Tsunamis, or DART, buoys). Other federal and state activities to increase tsunami safety include: improvements to tsunami hazard and evacuation maps for many coastal communities; vulnerability assessments of some coastal populations in several states; and new efforts to increase public awareness of the hazard and how to respond.

Despite these advances, many coastal communities in the United States still face challenges in responding to a tsunami that arrives in less than an hour after the triggering event. If the source were so close to shore that only minutes were available before the tsunami reached the coast, the public would need to recognize natural cues—mainly, ground shaking from the tsunami-triggering earthquake—and know to evacuate even without official warnings. If the nearby source earthquake is weak, or if the tsunami takes more than an hour or two to reach a coast, technological tsunami detection and forecasting could give advance warning for evacuation of coastal areas but seamless coordination between the two Tsunami Warning Centers and clear communications to local officials and the public would be required for a timely and effective response.

The current organizational structure of the two Tsunami Warning Centers has not been optimized for coordinated, clear communication of tsunami warnings. The two centers have different areas of responsibilities; are managed by different regional offices; use different technology; have separate support and organizational cultures; and do not provide functional redundancy. As a result, the public could receive conflicting warning messages from the two centers. In addition, the content of the warning messages is inconsistent with social science findings on the composition and delivery of effective warning messages, especially with re-gard to the importance of delivering a consistent and clear message when there are multiple information sources. A range of remedies is discussed in the report, from harmonizing message content to changing the organizational structure, such as merging the centers.

Previous reports have called for a national tsunami risk assessment to allocate funding based on the number of people, their vulnerability, or economic assets at risk. The committee endorses this concept and finds that progress is slow toward the completion of such an assess-ment and is limited by several factors:

- incomplete knowledge of tsunami sources;
- inconsistent access to high-quality bathymetric and topographic data;
- differences in inundation modeling approaches and choice of source parameters for the same event between states; and
- lack of vulnerability assessments that inventory the number, type, awareness, levels of preparedness, and evacuation potential of populations in tsunami-prone areas.

This report recommends stronger NOAA and NTHMP leadership in assessing tsunami sources, developing national guidelines and metrics for creating consistent evacuation maps, identifying vulnerable populations, and inventorying and evaluating education and prepared-ness efforts. Also, it is important to design effective interagency exercises, use professional emergency-management standards to prepare communities, and prioritize funding based on tsunami risk.

In addition, the report describes areas of research and development that would improve tsunami education, preparation, and detection:

- metrics to assess progress in education and preparedness efforts;
- inundation and forecast models that include an open validation and accreditation process, as well as post-event data validation;
- improved reliability, station coverage, and operations of the newly deployed DART network;
- periodic and comprehensive vulnerability assessments;
- coordination for post-tsunami event reconnaissance; and
- new tsunami detection techniques and analysis.

Regular, independent scientific review of the various elements of the tsunami program would be valuable in identifying and addressing research needs and in ensuring the effective implementation of new technologies and protocols.

Minimizing future losses to the nation from tsunamis requires persistent progress across the broad spectrum of efforts the report reviews: risk assessment, public education, govern-ment coordination, detection and forecasting, and warning-center operations. Sustained efforts in all these areas will be needed for communities to prepare for an event that may occur years to decades in the future, but only affords minutes or hours for people to respond.

Summary

A tsunami is a series of waves that can move on shore rapidly, but last for several hours and flood coastal communities with little warning. Tsunamis can be triggered by a variety of geological processes such as earthquakes, landslides, volcanic eruptions, or meteorite impacts. Since modern record keeping began in 1800, they have taken many lives in Hawaii, Alaska, Puerto Rico, the Virgin Islands, California, Oregon, and American Samoa. The threat of a potentially catastrophic tsunami on U.S. soil looms in seismically active regions in the Pacific and Atlantic (Figure S.1). More recently, tsunamis generated by earthquakes in West Java (July 2006), Samoa (September 2009), and Chile (February 2010) have flooded some U.S. coastlines, highlighting the need for a focused and well-coordinated effort to minimize the loss of life and property.

In the wake of the catastrophic 2004 Indian Ocean tsunami, which caused more than 200,000 deaths and widespread destruction, Congress passed two laws intended to increase efforts to diminish the potential impact of a tsunami. The first P.L. 109-13 in 2005 was aimed at expanding the current tsunami detection system; and the second P.L. 109-424 in 2006 asked the National Oceanic and Atmospheric Administration (NOAA) and the National Tsunami Hazard Mitigation Program (NTHMP) to strengthen the nation's tsunami detection, warning, education, and preparedness efforts.

At the same time, Congress charged the National Academy of Sciences (NAS) to review the nation's progress toward the ability to detect and forecast tsunamis. In particular, the committee was asked to review how the expansion of the sea level sensor network has improved the ability to detect and forecast tsunamis; how the tsunami program could be improved; and how well it is coordinated with other efforts. The NAS expanded the scope of the study to also review the nation's ability to minimize the impact from future tsunamis by educating and preparing the American public. The complete statement of task is provided in Appendix B.

Overall, the committee found that the nation's tsunami efforts have improved in several ways since 2004. For example, the expansion of the Deep-ocean Assessment and Reporting of Tsunamis (DART) sensor network has improved the ability to detect and forecast the size of tsunamis, the number and quality of hazard and evacuation maps has increased, and several states have assessed the number and types of vulnerable individuals in tsunami-prone areas. In addition, numerous tsunami education and awareness efforts have been initiated.

However, current capabilities are still not sufficient to meet the challenge posed by a tsunami generated close to land (see Box S.1). Near-field tsunamis can reach the coast just minutes after the triggering event—leaving little time to disseminate official warning messages. Tsunami education and preparation is necessary to ensure people are aware of the tsunami risk in their community and know how to recognize natural cues, such as the tremors of a tsunami-triggering earthquake, even if they do not receive an official warning. Communities at a

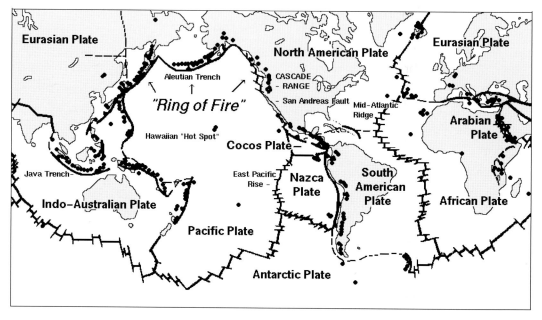

FIGURE S.1 Global map of active volcanoes and plate tectonics illustrating the "Ring of Fire" and depicting subduction zones; both areas associated with frequent seismic activity. SOURCE: http://vulcan.wr.usgs. gov/Imgs/Gif/PlateTectonics/Maps/map_plate_tectonics_world_bw.gif; USGS.

BOX S.1
What Is a Near-Field and a Far-Field Tsunami?

A tsunami generated close to the shoreline is known as a "near-field" tsunami; a tsunami generated by a source far from the point of impact is referred to as a "far-field" tsunami. In general, it takes a large earthquake (magnitude >7.0) to generate a damaging tsunami in the near-field and it takes a great earthquake (magnitude >8.0) to generate a damaging tsunami in the far-field. Near-field tsunamis pose a greater risk for coastal communities because the first waves can move on shore in minutes. Far-field tsunamis will not reach the coast for hours, and allow time to issue warning and evacuation notices. In some intermediate cases, a tsunami could hit the coast less than an hour after an event at a moderate distance from the coast. In this latter case, the earthquake may not be felt strongly, so warnings and evacuation notices will be essential for an effective response. The implications for detection and disaster warning and response are different for near-field and far-field tsunamis, and each scenario is considered in depth as part of this report.

greater distance from the triggering event might feel the ground shaking only weakly and not recognize the need to evacuate although a tsunami could arrive in as little as an hour. In this case, detection, forecast, and warning systems would need to operate efficiently, and decision makers at Tsunami Warning Centers and state and local emergency managers would need to coordinate closely to ensure that a clear, consistent message is delivered— a daunting challenge that will require major improvements in all parts of the end-to-end tsunami program.

Considering the many independent and disparate efforts currently ongoing as part of the nation's tsunami hazard mitigation efforts, the committee concluded that the best way to improve the current end-to-end tsunami warning system would be to define the characteristics of an ideal system that detects and forecasts the threat and coordinates risk assessment, public education, and the response to minimize loss of life and property in the event of a tsunami. Each component (risk assessment, education, detection and forecasting, and warning management) of this idealized system could then be compared against current and planned efforts to identify areas that need improvement.

LAYING THE FOUNDATION:
A COMPREHENSIVE NATIONAL ASSESSMENT OF TSUNAMI RISK

Understanding the nation's tsunami risk[1] is the first step to building a comprehensive tsunami preparedness program. The nation is just beginning to define the hazards tsunamis pose, the populations and societal assets they threaten, and the readiness of individuals and communities to evacuate. Although much progress has been made, the nation remains far from understanding enough of its tsunami risk to set risk-based priorities for state and national efforts in tsunami preparedness, education, detection, and warning.

Assessing tsunami risk is challenging: there is a paucity of information about the frequency, source, and characteristics of past tsunamis and their reoccurrence intervals; vulnerability of coastal communities cannot be assessed without considering interdependencies from a larger economic and sociopolitical context; and the potential impacts of future events are uncertain. Although difficult to assess, societal risk from tsunamis is critical information in the development and prioritization of risk-reduction efforts including: education, preparedness planning, warning-system development, mitigation, and response strategies at the local, state, and federal level. The level of sophistication, accuracy, resolution, and format required for assessing societal risk to tsunamis will depend on the intended use of the information.

> **Recommendation:** NOAA and its NTHMP partners, in collaboration with researchers in social and physical sciences, should complete an initial national assessment of tsunami risk in the near term to guide prioritization of program elements.

Tsunami hazard assessments focus on the physical characteristics of future tsunamis, especially on those that can pose a threat to people and the things they value. These characteristics

---

[1] Tsunami risk is defined in this report as the likelihood and amount of death, injury, and economic loss that may result from a tsunami.

include the speed of onset, impact forces, currents, and the area that will be flooded. Understanding the hazard also requires an understanding of tsunami sources, for example, how often a coast is likely to have a tsunami and how large of a tsunami a source might generate. A typical tsunami hazard assessment accordingly includes:

- studies of the locations, sizes, and histories of tsunami sources, which are usually earthquakes but can also be landslides or volcanic eruptions;
- inundation models, which determine the areas most likely to be flooded;
- hazard maps, which portray inundation models on maps that show roads, elevation and buildings, and other critical infrastructure; and
- evacuation maps, which depict areas that need to be evacuated in the event of a tsunami and show evacuation routes to safe havens.

Modeling tsunami inundation begins at the tsunami source with estimates of the seafloor deformation that initiates the tsunami. The simulation also requires accurate information on the topography of the seafloor to understand the surface over which tsunami waves propagate, and a robust computational model to simulate the formation of tsunami waves.

Knowledge of the sizes and recurrence intervals of tsunami-triggering events is only now emerging and is improving the understanding of tsunami sources critical to producing a comprehensive tsunami hazard assessment; but currently, no formal procedures for periodic re-evaluation of tsunami risks exist. The U.S. Geological Survey's (USGS) National Earthquake Hazards Reduction program serves as an example of a successful and useful approach to periodic national hazard assessments that NOAA and its NTHMP partners could adapt. In particular, the USGS updates the U.S. National Seismic Hazard Maps at six-year intervals nationwide and at other intervals regionally.

Recommendation: NOAA and its NTHMP partners should institute a periodic assessment of the sources of tsunamis that threaten the United States.

The committee concludes that the accuracy and realism of tsunami inundation models is limited by scientific uncertainties in determining the source of tsunamis, limited spatial resolution of bathymetry, a lack of topography data, and difficulties in modeling the complexity of processes that take place when the tsunami wave interacts with buildings and natural features of the coast. Modeling efforts would greatly benefit from a rigorous vetting process, peer-review, and validation with field data.

Recommendation: To improve tsunami inundation modeling, the NTHMP should periodically review progress in hydrodynamic modeling.

Moreover, the committee found that the development and use of inundation models is not occurring in a coordinated or standardized fashion across the NTHMP. Instead, each member state independently selects the tsunami source, bathymetric and topographic data, and numerical code. Although state resources are used to leverage federal resources, this state-by-state based approach to tsunami inundation mapping, coupled with inadequate coordination and consensus among NTHMP modelers and no external peer review, has created significant disparities in the methods, criteria, and judgments employed in tsunami inundation modeling and the resulting hazard maps that are based on these models.

> **Recommendation:** The NTHMP should reduce unnecessary and costly disparities in inundation modeling approaches among states and territories. The NTHMP should conduct modeling efforts consistently across political boundaries and execute efforts through a cooperative partnership among NOAA, the USGS, and NTHMP members.

Evacuation maps are critical tools for preparing and educating the public about the hazard and the appropriate response before an impending tsunami. For most at-risk communities, the committee concludes that progress has been made toward generating improved evacuation maps. However, methods to produce evacuation maps vary greatly among NTHMP member states; that means at-risk populations must try to interpret different representations of tsunami risk. Due to the absence of uniform quality standards, evaluative metrics, or guidelines on effective approaches, the committee found it difficult to assess whether current evacuation maps are sufficient for enabling effective evacuations or preparing the public.

> **Recommendation:** The NTHMP Mapping and Modeling Subcommittee should develop guidelines on evacuation-map production that fosters consistency in format and quality across the United States, and a national, online repository for tsunami evacuation maps.

Tsunamis pose risks only if they have the potential to impact people or the things people value. Therefore, a first step in understanding vulnerability is to inventory the number and characteristics of individuals in tsunami hazard zones. In addition, emergency managers should assess their demographic characteristics, as these can affect an individual's ability to receive, understand, and respond to warning messages. For example, the very young and very old may need evacuation assistance and thus have higher sensitivity to tsunami hazards. Currently, there is no national assessment of population exposure and sensitivity to tsunamis, including the number and types of individuals in tsunami hazard zones. This lack of information limits abilities to assess national tsunami risk, develop realistic evacuation plans, and tailor education efforts to at-risk individuals.

> **Recommendation:** The NTHMP should periodically inventory the number and type of people in tsunami hazard zones, with special attention to groups whose heightened sensitivity to tsunamis could constrain their ability to prepare for and evacuate from future tsunamis. The NTHMP should provide guidelines on how to use this information to tailor evacuation planning and education efforts.

Many communities in the United States are threatened by near-source tsunamis, but few evacuation studies have been conducted to evaluate the ability of at-risk individuals to reach higher ground before tsunami waves arrive. For example, local earthquakes that generate near-source tsunamis have the potential to impact roads, infrastructure such as bridges, or facilities essential for response efforts. Preparedness efforts would greatly benefit from assessing how these earthquake damages impact the ability to evacuate. Without such information, emergency managers are not able to identify where targeted outreach and evacuation assistance will be needed.

Recommendation: For all communities with near-source tsunami threats, the NTHMP should conduct evacuation modeling studies to assess the likelihood of successful evacuations.

PREPARING IN ADVANCE THROUGH A CONSISTENT PUBLIC EDUCATION CAMPAIGN

Surviving a tsunami depends on the ability of an individual in the hazard zone to recognize warning signals, make correct decisions, and act quickly. For near-field tsunamis, waves will arrive within minutes after generation; therefore, at-risk individuals will need to recognize natural cues such as the ground shaking or the receding of the water line as the primary warning. Knowledge and readiness gained through pre-event education may save lives. For far-field tsunamis, waves will arrive several hours after generation, and individuals need to understand official warnings and follow instructions given by local agencies. Regardless of the tsunami source, integrated public education and preparedness planning are necessary to protect lives and to make tsunami knowledge commonplace and ingrained into local culture and folk wisdom.

Educating At-Risk Individuals

Tsunami education in U.S. coastal communities is a major challenge because it requires reaching hundreds of coastal communities with hundreds of thousands of residents, employees, and tourists. The NTHMP Mitigation and Education Subcommittee is charged with assessing tsunami education needs for the nation, addressing these needs through targeted products and activities, and then sharing these products with other at-risk coastal areas.

Tsunami outreach and educational programs can draw from a rich base of research on enhancing hazard education to motivate the public to prepare for future hazards. For example, research has shown that training campaigns and the dissemination of education products are more effective when tailored to the strengths and vulnerabilities of specific communities. A campaign designed for long-time residents would capitalize on familiarity of the surroundings, as well as emphasize household preparation strategies and the importance of creating community networks. In contrast, a program designed for tourists and other transient populations would focus on easily identifiable landmarks, would provide information via signs posted in prominent locations, and would train hotel and tourist services staff such as tour guides, life guards, and vendors to provide assistance to tourists.

The committee was requested to review the availability and adequacy of tsunami education and outreach. One obstacle to this task was that no systematic evaluation of U.S. tsunami education efforts has been conducted at a national scale. No compilation or inventory of

NTHMP-related tsunami efforts was available at the time of this review, although it is a current goal of the NTHMP. A second obstacle is a lack of pre- and post-outreach evaluations and post-event assessments. Because there are few studies that documented the perceptions, knowledge, and capacity to prepare at-risk populations, there are no consistent baselines or metrics to gauge the effectiveness of education programs.

The committee concludes that current tsunami education efforts are not sufficiently coordinated and run the danger of communicating inconsistent and potentially confusing messages.

Recommendation: To increase the effectiveness of tsunami education, the NTHMP should

(1)   develop consistent education efforts among its members using evidence-based approaches,
(2)   tailor tsunami education to local circumstances,
(3)   create and maintain an online repository of education efforts,
(4)   develop and implement an evaluation program of the effectiveness of education efforts, and
(5)   leverage hazard-education efforts and expertise of other NOAA entities.

Because pre-event education is critical to saving lives during a near-field tsunami, the committee concludes that tracking progress in education and outreach efforts in communities threatened by near-field tsunamis is a high priority.

Recommendation: The NTHMP should prioritize systematic, coordinated perception and preparedness studies of communities with near-field tsunami sources to determine whether at-risk individuals are able to recognize natural cues of tsunamis and to take self-protective actions.

Preparing Communities

Because of the breadth and diversity of actions that could be taken to increase preparedness, the committee restricted its review of community preparedness to NOAA's TsunamiReady Program, which has emerged in recent years as a framework for improving tsunami preparedness in coastal jurisdictions. TsunamiReady is a voluntary program that aims to help communities reduce the potential impacts from tsunami-related disasters through redundant and reliable warning communications, better preparation through community education, and official readiness through formal planning and exercises. It sets minimum guidelines, such as having the ability to communicate warnings to the local population, and encourages consistency in educational materials. Traditionally, the program has measured its success by the number of communities recognized as TsunamiReady annually. However, the committee questions the effectiveness of the program and its success criteria, because the program lacks the following elements:

(1) a professional standard to guide its development,
(2) metrics to assess baseline readiness and community needs,
(3) evaluative criteria to assess community performance during a tsunami,
(4) accountability measures to ensure recognized communities meet and continue to meet mandatory requirements,
(5) local points of contact with training in community preparedness, and
(6) criteria and guidance on what constitutes effective public outreach and preparedness efforts.

The Emergency Management Accreditation Program (EMAP) aims to improve community preparedness for natural hazards. Unlike TsunamiReady, EMAP is more broadly geared to all-hazards mitigation. EMAP is the nationally recognized standard for emergency management and provides criteria to assess current programs or to develop, implement, and maintain a program to mitigate, prepare for, respond to, and recover from disasters and emergencies. Its process for accreditation is transparent and applied through peer review. Because TsunamiReady's current requirements are not well structured and do not fit the concept, terminology, and format of a standard, the program could improve by drawing on EMAP's well-established standard, process, and experience with the emergency management community.

Recommendation: The NOAA Tsunami Program could strengthen the TsunamiReady Program by modeling it after the Emergency Management Accreditation Program.

Developing and Delivering Effective Warning Messages

The likelihood of individuals responding to tsunami warnings depends on the quality, clarity, and accuracy of the official warning messages they receive from the two Tsunami Warning Centers (TWCs) and/or local and state emergency management agencies. An effective message contains the necessary information to motivate individuals to take self-protective action and must reach at-risk people in a timely fashion. It is critical that warning messages: are accurate and consistent; use language that allows a person to visualize the proper response; make clear when recommended actions should begin and finish; identify who needs to evacuate and who does not; and explain how taking the protective action will reduce the pending consequences of not taking action at all.

Currently, both the TWCs issue a tsunami warning, advisory, watch, and information statement through multiple official channels following detection of a tsunami-triggering event. It then becomes the responsibility of local or state officials to take the appropriate actions and issue their own messages and evacuation orders to individuals in tsunami-prone areas. The generation of two different tsunami warning messages has created confusion among the media, some local officials, and the general public, and will likely continue to do so unless message content is improved or a single message is issued.

Recommendation: If distinct messages are to be produced by the two TWCs, then the messages should be consistent. Ideally, the committee recommends that one message be released by the two TWCs that includes information for all areas under their responsibility.

Recommendation: The NOAA/National Weather Service (NWS) should better integrate the TWC warning functions with the state, county, and city warning functions with regard to message content and dissemination methods for the public by developing formal TWC outreach plans and assessing needs and priorities of TWC customers.

Coordinating Across All Levels of Government

State working groups, regional groups, and the NTHMP facilitate the coordination and planning across jurisdictional boundaries, including the coordination of educational efforts and opportunities to provide feedback to the TWCs on warning messages. These efforts are valuable in contributing to pre-event planning and coordination, but they could be strengthened through additional exercises and drills focused on improving evacuation procedures during an event.

Current efforts to practice evacuation procedures and protocols include community-led evacuation drills, live code "end-to-end" tests, table-top exercises among emergency management agencies, and functional exercises to test interagency communication and coordination. The committee concludes that the importance of these approaches vary based on local conditions and tsunami threat, and include specific conclusions for both far-field and near-field tsunamis.

- *Far-field tsunami threats:* Evacuations will be managed by multiple agencies over many hours; therefore, exercises are important to engage agencies to discuss and test coordination and communication. However, the committee concludes that evacuation drills are not advisable because of the risks associated with such drills, especially in larger communities.
- *Near-field tsunami threats:* Initial evacuations will be self-directing after at-risk individuals recognize natural cues. The committee concludes, however, that table-top and functional exercises are still important because of the significant response and relief operations after the initial tsunami wave arrives. The committee concludes that these community-led, voluntary drills may be useful in promoting tsunami awareness, providing social cues, and building social networks but only in small communities that have limited vertical-evacuation options and may have less than 30 minutes to evacuate.

Recommendation: The NTHMP should actively encourage member states to develop and maintain active tsunami working groups to help facilitate and coordinate tsunami education, preparedness, and warning dissemination.

Recommendation: To ensure that managed evacuations for far-field tsunamis are effective and minimize societal and economic interruptions, the NTHMP should develop guidelines on the design of effective exercises for use by emergency management agencies.

Considering their multiple responsibilities and limited resources, the TWCs should be commended for their commitment to establishing connections with external groups and coordinating their efforts. However, relatively few staff resources have been dedicated to maintaining partnerships with customers, and existing efforts are secondary to the technical aspects of the warning centers. There are no formal outreach plans for media training or working with emergency management and response personnel, no formal training interaction for TWC watchstanders and state emergency management officials, and no formal standard operating procedures for evaluating the effectiveness of warning message content or channels. Retrospective reviews could enhance the effectiveness of the ongoing educational efforts and guide further improvements in community preparedness and coordination among decision makers.

Recommendation: After a significant tsunami warning is issued to U.S. communities (e.g., the 2010 Chilean event), the NOAA/NWS should initiate an independent review of TWC actions and its integration with its partners and customers through an external science review board and make findings public.

DETECTING AND FORECASTING TSUNAMIS

The two separate Tsunami Warning Centers monitor seismic activity to assess the potential for tsunami threats from earthquakes. The content of the first tsunami information statement, advisory, watch, or warning from the TWC is decided solely on seismic parameters and the historical record, if any, of past tsunamis generated in the area of the earthquake. Based on their own data analysis, the TWCs independently decide whether to issue alerts to the emergency managers in their respective areas of responsibility. This initial statement can be issued within 5-10 minutes after the earthquake is detected and might provide communities near the source with the only alert, if waves reach shore within minutes. However, this indirect seismic method has limited accuracy in its estimates of the strength of the tsunami. Because forecasters must err on the side of caution when human lives are at stake, the TWCs use conservative criteria for issuing advisories, watches, or warnings, which can lead to unwarranted evacuations costing millions of dollars.

Data from coastal sea level gauges and the open ocean DART network provide the only means to verify the existence of a tsunami and to forecast the height of the tsunami waves as they spread from the source. This information is used to adjust or cancel warnings, watches, and advisories. Coastal and open ocean sea level sensor networks can also detect tsunamis from sources that fail to generate seismic waves. Although the detection of the open ocean signal can occur within the first hour after the tsunami, forecasts might take longer. Thus, only communities farther from the tsunami source will benefit from these refined warning messages.

The committee concludes that the global networks that monitor seismic activity and coastal and open-ocean sea level variations remain essential to the tsunami warning process. The current global seismic network is adequate and sufficiently reliable for the purposes of detecting likely tsunami-triggering earthquakes. However, because the majority of the seismic stations are not operated by the TWCs, the availability of this critical data stream is vulnerable to changes outside of NOAA's control.

The complex seismic processing algorithms used by the TWCs, given the availability of seismic data, quickly yield adequate estimates of earthquake location, depth, and magnitude for the purpose of tsunami warning. However, the methodologies are inexact, in part because of the physically variable nature of tsunamis, and in part because of the need for rapid determination of earthquake parameters that may not be definitive until the entire rupture process is complete (potentially tens of minutes). In the case of a very large earthquake the initial seismological assessment, although adequate for most medium-sized earthquakes, can underestimate the earthquake magnitude and lead to errors in assessing tsunami potential. In parallel to their own analyses, staff at the TWCs could avail themselves of earthquake locations and magnitudes that are estimated within minutes of an event from the USGS National Earthquake Information Center (NEIC). An interagency agreement could be established to make these initial estimates available on secure lines between the USGS and NOAA.

> **Recommendation:** NOAA and the USGS could jointly prioritize the seismic stations needed for tsunami warnings to advocate for the upgrade and maintenance of the most critical stations over the long term.

> **Recommendation:** The TWCs should work jointly with NEIC to test the utility of the W-phase algorithm in the tsunami warning process, using both a sufficient dataset of synthetic seismograms and a set of waveforms from past great earthquakes, paying particular attention to the algorithm's performance during "tsunami earthquakes" and to the assessment of a lower-magnitude bound for its domain of applicability.

Detection of Tsunamis with Sea Level Sensors

A majority of the funds authorized by the Tsunami Warning and Education Act (P.L. 109-424) have been used to manufacture, deploy, and maintain an array of 39 DART stations, establish 16 new coastal tide gauges, and upgrade 33 existing water level stations. These new and upgraded sea level stations have closed significant gaps in the sea level sensor network that had left many U.S. coastal communities subject to uncertain tsunami warnings. The availability of these open-ocean DART stations makes it possible to forecast the height of tsunamis once waves are detected by the buoy and before they arrive onshore.

However, some fundamental issues remain. For example, gaps in coverage exist in the Caribbean region or off the South America coast. Most concerning is the committee's finding that as much as 30 percent of DART stations are inoperable at any given time. These high numbers for buoy outages jeopardize the ability of the TWCs to forecast tsunamis. Because the value of individual components in the network and the risk to the warning capability due to individual component failures has not been evaluated, the National Data Buoy Center (NDBC) has no guidance for high-priority buoy repairs.

Recommendation: The reliability of the DART network should be improved to allow the TWCs to fully utilize the capability this technology offers.

Recommendation: In order to bring NDBC into compliance with P.L. 109-424, the center should engage in a vigorous effort to improve the reliability of the DART stations and minimize the gaps caused by outages.

Recommendation: NOAA should regularly assess the appropriate spatial coverage of the coastal and DART sea level sensor network (U.S. and international).

Recommendation: NOAA should prioritize the sea level stations (both U.S. and international) according to their value to tsunami detection and forecasting for the areas of responsibility of the TWCs.

Recommendation: NOAA should assess on a regular basis the vulnerabilities to, and quality of, the data streams from all elements of the sea level sensor networks, beginning with the highest priority sites determined per the recommendations above.

Recommendation: NOAA should encourage access to the DART platform by other observational programs, because the platform presents an opportunity to acquire a long time series of oceanographic and meteorological variables.

Recommendation: NOAA should establish a "Tsunami Sea Level Observation Network Coordination and Oversight Committee" to oversee and review the implementation of the recommendations provided above.

Tsunami Forecasting

The expansion of the coastal and open-ocean sea level network has made it possible to forecast tsunamis in near-real time. It provides emergency managers with critical information about the time the first wave might arrive, the duration during which waves will continue to arrive (which can be for many hours), and the size of these waves. Currently, one forecast model is fully operational and one is being used by the TWCs as an additional source of forecast model output. Although these models have been relatively successful in forecasting recent events, a more open and transparent process is needed to evaluate model performance and how the results of the two models can be used to improve the results. For example, the

National Hurricane Center (NHC) runs ensemble models to take advantage of several model outputs to create a single product for the forecast. Tsunami forecasting would benefit from the development of a process that identifies (1) benchmarks to evaluate model performance, (2) how different solutions can be used to create a single forecast, and (3) how field data can be used to validate and improve the models.

> **Recommendation:** Both the TWCs and the NOAA Center for Tsunami Research should continue to work together to advance current forecasting methodologies and bring all available methodologies into full operational use.

Future Research and Technology Development

Currently, given the distance of the source to the closest coastal or open-ocean sea level sensor, it can take up to an hour or more to confirm a tsunami forecast and potentially even longer to forecast the size. This is not only of concern in the case of near-field tsunamis, but also in the event of a tsunami earthquake or underwater landslide, events that generate only a small amount of ground shaking but could trigger a tsunami of much greater amplitude than would be expected. In this circumstance, official warnings may be the only way to notify people. An example comes from the Meiji Sanriku tsunami of 1896 in northeast Japan. The earthquake was large (magnitude 7.2) but generated such weak ground shaking that few people were concerned about the potential for a tsunami. More than 22,000 people perished in the huge tsunami that followed.

To detect a tsunami earthquake or underwater landslide, direct measurements of the water-surface variations and currents are required in real time. This rapidly sampled data could also help issue warnings to communities a little farther away from the source. Infrastructure such as sea level sensor networks and communications equipment could be destroyed by the initial impact of the tsunami wave, leading to a lack of official warnings.

One way to accomplish real-time measurements is to collect data using cabled seafloor observatories. These comprise various sensors connected to each other and to shore by a sea-floor communications cable that serves both to deliver power to the sensors and to transmit data from the sensors back to onshore data servers. Several types of instruments are useful for tsunami detection, including bottom pressure sensors, seismometers, current meters, and accelerometers. Observatories currently in operation include the North-East Pacific Time-Series Underwater Networked Experiments (NEPTUNE) Canada, off the coast of British Columbia and the Monterey Accelerated Research System (MARS) in Monterey Bay, California. Another large U.S. observatory, the Ocean Observatories Initiative (OOI), has been funded by the National Science Foundation (NSF) for deployment across Oregon's continental shelf, slope, and the Cascadia Subduction Zone, over the Juan de Fuca plate, and on the Juan de Fuca Ridge. The committee concluded that tsunami detection, warning, and preparedness activities for near- and mid-field tsunamis could benefit from the expansion of existing alternative technologies for real-time detection.

Recommendation: To develop more rapid and accurate warnings of local tsunamis, the TWCs should coordinate with the NEPTUNE-Canada and OOI observatory managers to obtain access to their seismic and bottom pressure data in near-real time. Data interpretation tool(s), jointly applied to the seismic and bottom pressure data, should be developed to realize the most rapid tsunami detection possible.

Another promising area of research that could improve the ability to more rapidly predict the magnitude of a tsunami comes from Global Positioning System (GPS) measurements. In combination with seismic data, continuous GPS measurements have proven to be powerful in studying continental earthquakes such as illuminating the processes of earthquake after-slip. Continuous GPS can provide a map of the three-dimensional deformation of the earth's surface as a result of the earthquake rupture. This information can be used to help predict tsunami generation and provide accurate forecasts of wave heights.

The use of GPS holds great promise for extending the current seismic networks to include capabilities for measuring displacements in the coastal environments for the large earthquakes that can be underestimated using seismic techniques alone. Displacements onshore could potentially be used to infer offshore displacements in times as short as five minutes in an area such as the Cascadia fault zone.

Recommendation: NOAA should explore further the operational integration of GPS data into TWC operations from existing and planned GPS geodetic stations along portions of the coast of the U.S. potentially susceptible to near-field tsunami generation including Alaska, Cascadia, the Caribbean, and Hawaii. Where GPS geodetic coverage is not adequate, NOAA should work with NSF and the states to extend coverage, including the long-term operation and maintenance of the stations.

The report identifies several other areas of research and technology development that have the potential to improve the nation's ability to detect, forecast, and provide timely and accurate warnings regarding tsunamis. For example, the seismic analysis could be improved using high-frequency P-waves. In addition, satellite altimetry or island seismometers might be used to detect tsunami waves, or GPS satellites could detect tsunami waves in the atmosphere. These research avenues hold promise but are far from becoming operational in the tsunami detection and warning process.

Regular, independent scientific review of the various elements of the tsunami warning system would be valuable in identifying and addressing research needs and in ensuring the effective implementation of new technologies and protocols. Science needs to be brought to bear more systematically across the spectrum of tsunami preparation, education, detection, and warning systems. The establishment of an external science advisory panel consisting of physical and social scientists, and practitioners in emergency management, is one option to provide advice and oversight across the spectrum of tsunami efforts.

IMPROVING TSUNAMI WARNING CENTERS' RELIABILITY
AND SUSTAINABILITY OF OPERATIONS

The TWCs provide services to a wide community that includes emergency managers, the scientific community, and the public. They are responsible for gathering information from sensor and observational systems, detecting tsunamigenic earthquakes, developing decision support information, and providing and disseminating warnings to the public and other entities. Although operational procedures for earthquake detection are similar at both TWCs, the technologies used are considerably different, with different hardware platforms, software suites, processes, and interfaces to the public and their users. The committee concludes that these differences lead to technological incompatibilities and limited capabilities for back up, redundancy, and checks and balances, which are important mission capabilities for the tsunami warning system.

The success of the TWC mission is critically dependent on technical infrastructure and human capital, both of which the committee assessed to be insufficiently supported. Several issues have been identified that are associated with the reliance on dated software technology hampering easy interfacing with current network and mobile data structures. Addressing these problems is difficult for the TWCs, as most scientific personnel and watchstanders have training in the geophysical sciences and not in software engineering.

To harmonize software and hardware suites, NOAA developed an information technology (IT) Convergence Plan, with the goal of creating a single, platform-independent technology architecture to be deployed at each TWC and a shared tsunami portal. The committee believes that the NOAA's IT Convergence Plan is well-motivated, but it notes that the plan is a single project effort and is not part of a systematic, comprehensive IT system plan and enterprise architecture.

Recommendation: The NOAA/NWS should harmonize and standardize checklists, tsunami warning products, and decision support tools, and it should use standard TWC software tools and applications.

Recommendation: Given the importance of IT and the rapid evolution of IT, the tsunami warning program should undertake a comprehensive, enterprise-wide long-range technology planning effort, consistent with international technology process and product standards, in order to develop both an enterprise-wide technology architecture for TWC operations and the accompanying enterprise-wide technology support processes.
As part of this long-range planning effort NOAA/NWS should:

- consider providing the TWCs with stronger IT commitment and leadership, and greater resources for software and hardware personnel, planning, development, operations, maintenance, and continuous process and product improvement.
- provide sufficient IT staff to the TWCs so that IT hardware and software design, development, and maintenance are not a collateral duty of a watchstanding scientist, as is the case presently.

- adopt national, and where applicable, international, standards, best practices, and lessons learned for all functions, technology, processes, and products.
- regularly and systematically apply continuing process and product improvement models for hardware and software planning, development, operations, and maintenance and for organizational processes; and develop a learning organizational culture.

The committee reviewed the pre-written messages delivered by the TWCs and found that many documented principles for effective warning messages have not been applied. The committee concluded that the centers' warning products would be much more effective if their content and delivery incorporated the latest social science on composing effective warning messages and were compatible with current software, hardware, and social media.

Recommendation: The TWCs should consider alternative warning message composition software and should improve protocols by undertaking an external review by IT specialists in the area of communication technology to identify the latest technology in message composition software and formats to ensure compatibility with current and next generation information and communication technology for message dissemination.

The committee also found inconsistencies between the warning products of the TWCs and those of the NWS. For example, a watch means that an event has an 80 percent chance of becoming a warning in the NWS, but this is not the case with the TWCs whose watches rarely become warnings. Another current inconsistency is how the TWCs and the NWS deal with "all-clears." The TWCs cancel a bulletin, which could be read by the public as a signal that it is safe to return, which is not the same as an "all-clear" issued by the NWS. The NWS will soon move from using the "alert bulletin system" to an "impact based system," which will introduce another inconsistency with the TWCs.

Recommendation: Current and future adjustments of TWCs and NWS warning products should be made in a consistent fashion. A mechanism should be put in place so that future changes in warning products are quickly reviewed for inconsistencies, which are then addressed, so that products from the TWCs and the NWS match.

Human Resources

Each TWC relies on a dedicated staff, including nine science duty officers that perform watchstanding duties in addition to research and development. The watchstander has a critical role in tsunami decision support by maintaining situational awareness and issuing correct notification and warning products. Although visualization software assists by monitoring seismic and sea level data and mapping event locations, it is the watchstanders' training, experience, and expert judgment that are essential in making the appropriate decisions when creating warning products. To enhance the effectiveness of TWC decision making and the TWC staff's ability to inform decision making processes, regular and varied types of training are needed. In addition, because of the importance of technical and scientific know-how within the TWCs, opportunities for interactions between TWC staff and external scientific and professional communities are important and need to be encouraged and institutionalized within the tsunami program.

> **Recommendation:** Because of the importance of technical and scientific expertise to the TWC's function, TWC human capital requirements, recruiting, training, re-training, development, mentoring, and professional exchange should be included, reassessed, and updated as part of the NOAA/NWS enterprise-wide technology planning effort.

Organizational Structure

The goal of having two geographically distributed TWCs in Hawaii and Alaska with distinct areas of responsibility is to provide the system with backup in the case of critical failure at the other center. However, there are significant differences in IT architecture and software suites that thwart this backup function. Also, inconsistencies in warning products issued by the two TWCs have caused confusion. Because clear communication and consistency in message content are two key principles to effective warning message composition, the TWCs' warning products are less effective in eliciting the appropriate response. Because the TWCs are managed by two different regional NWS offices, use different analytical software and hardware, and appear to have distinct organizational cultures, the committee concludes that they do not function as redundant systems. Significant organizational changes will be needed to allow them to truly function as redundant systems that provide true backup capabilities. The committee considered the following options to address shortcomings of the current structure of the TWCs: harmonizing the two TWCs' operations, merging the two TWCs into a single center, or co-locating one or more TWCs with other research or forecasting units.

> **Recommendation:** Organizational structures for the two TWCs should be evaluated and fully described as part of an enterprise-wide technology planning effort. Whether there should be a single or multiple TWCs, or whether the TWC operations should be consolidated in a different location, should be addressed during the enterprise-wide long-range planning effort.

Conclusions

The numerous distributed efforts in tsunami detection, warning, and preparedness are linked together to reduce loss of life and economic assets from a tsunami. In the event of a tsunami, all these distributed efforts must come together in less than a day to produce an effective, adaptable response and function like a single organization. A challenge for tsunami warning preparedness efforts is to develop effective organizational structures that provide reliable and sustainable operations in non-tsunami periods as well as during catastrophic incidents, especially given the short time available to respond in a crisis.

The committee found that personnel responsible for these efforts (e.g., at the TWCs and Tsunami Program, state emergency managers, etc.) are highly committed to serving each of the program's functions, from detection through education and community outreach. However, the committee found many shortcomings of the TWCs in terms of function, technology, human capital, and organizational structures, and many opportunities for significant improvements in center operations. Improvements will depend upon an organizational culture change within the NOAA/NWS Tsunami Program that supports and celebrates operational excellence, adopts national and international standards, processes, best practices, and lessons learned for all functions, technologies, processes, and products, and continuously seeks process improvements.

Recommendation: Tsunami warning system processes and products should reflect industry best practices, as well as lessons learned from other operational real-time, large-scale, mission-critical distributed systems, and should comply with international information technology and software engineering product and process standards.

Recommendation: NOAA/NWS and the TWCs should undertake ongoing, joint or NOAA-wide, continuous process improvement activities for their functional, technological, organizational, and human capital initiatives.

CHAPTER ONE

Introduction

THE TSUNAMI THREAT IN THE UNITED STATES

The 2004 Indian Ocean tsunami resulted in catastrophic losses of life and property and demonstrated how destructive tsunamis can be. More than 200,000 people died, with most occurring in Indonesia, which was near the tsunami source, but deaths were also reported in countries as far away as Somalia. Recently, the Samoan (September 2009) and Chilean (February 2010) tsunamis reminded the world of how quickly a tsunami can move onshore and destroy lives. In comparison to extreme weather—such as floods, hurricanes, or tornadoes—tsunamis have caused comparatively few fatalities in the United States over the past 200 years. Modern records kept since 1800 tally less than 800 lives lost due to tsunamis in the United States and territories.[1] In 1960, a magnitude 9.5 Chilean earthquake generated tsunami waves that killed 61 people and caused $24 million in property damage in Hilo, Hawaii (Eaton et al., 1961). The 1964 Good Friday earthquake in Alaska generated a tsunami that devastated local Alaskan communities and inundated distant communities as far south as Crescent City, California.

Earlier tsunamis—yet to be repeated in modern times—include tsunami waves of North American origin in the year 1700 that caused flooding and damage as far away as Japan. Paleo-records indicate that the Cascadia subduction zone off the Washington, Oregon, and northern California coasts has repeatedly generated potentially catastrophic tsunamis (Atwater et al., 2005). Because of the relative infrequency of catastrophic tsunamis in recent U.S. history, mobilizing the required resources to maintain the nation's warning and preparedness capabilities is challenging.

Tsunamis are caused by a variety of geological processes, such as earthquakes, subaerial and submarine landslides, volcanic eruptions, or very rarely from meteorite impacts (Box 1.1). However, it takes a large event (e.g., typically an earthquake of magnitude greater than 7.0) to generate a damaging tsunami. Therefore, determining the likelihood of future tsunamis for U.S. coastal communities requires an understanding of the likelihood of reoccurrence of such geological processes, the likely magnitude of such events, and the location of the sources (see Chapter 3 for additional details). Because most tsunamis result from earthquakes, the tsunami hazard is high along U.S. shores that adjoin boundaries between tectonic plates, particularly along the subduction zones of Alaska, the Pacific Northwest, the Caribbean, and the Marianas (Figure 1.1). However, U.S. shores are also exposed to tsunamis generated far from them. For example, Hawaii has been struck by tsunamis that have been generated by earthquakes off the coasts of South America, Russia, and Alaska (Cox and Mink, 1963). Submarine landslides,

---

[1] http://www.ngdc.noaa.gov.

BOX 1.1
What Is a Tsunami?

A tsunami is a series of waves generated by an impulsive disturbance in a body of water such as a lake or ocean. The disturbance is typically a fault displacement. Less commonly, the displacement results from a landslide (into or below the water surface), volcanic activity, or rarely from a meteorite impact. It typically takes a large earthquake (magnitude >7.0) to generate a damaging tsunami in the near-field and a great earthquake (magnitude >8.0) to generate a tsunami in the far-field. The height of a tsunami ranges from centimeters to more than tens of meters and depends on the initial disturbance and the bathymetry of the seafloor over which the waves travel. The waves are very small in deep water, but grow in height once they move into shallow water. The velocity at which the tsunami travels away from the source depends on the water depth. A tsunami in deep water (≥4 km) is about as fast as a jetliner (700-800 km/hr) and slows to the speed of a car (60 km/hr) in shallow water. For example, a tsunami originating in the central Aleutian Islands near Alaska would arrive on the Hawaiian shores in about five hours. Such tsunamis, originating far from where it runs ashore, are commonly referred to as "far-field" tsunamis, which allow at-risk communities several hours to evacuate if the warning is received promptly. However, a tsunami triggered by an event close to shore (e.g., an earthquake associated with the Cascadia subduction zone), would reach communities in low-lying coastal areas in a few minutes, allowing for little time to warn and evacuate. When the tsunami originates close to the at-risk community, it is termed a "near-field" tsunami. For the near-field tsunami, people need to recognize the triggering earthquake as their warning to evacuate. Tsunami arrival times can range from minutes to many hours. The report mainly discusses the near- and far-field tsunamis (at either extreme end of the arrival time range) because of the unique challenges they pose to the warning and preparedness efforts. However, a tsunami can be triggered by an earthquake that is only weakly felt onshore, yet may arrive in under an hour. Such events are also considered and mentioned because they heighten the challenges associated with preparing and warning the public.

---

probably triggered by earthquakes, account for much of the known tsunami hazard along the U.S. Atlantic and Gulf coasts, and in southern California (Dunbar and Weaver, 2008). Seismically active faults and the potential for landslides in the Caribbean pose a significant tsunami risk for that region (Dunbar and Weaver, 2008).

Tsunami hazard zones of U.S. coastal communities contain thousands of residents, employees, and tourists, and represent significant economic components of these coastal communities (Wood, 2007; Wood et al., 2007; Wood and Soulard, 2008). The economic and social risks from tsunamis grow with increasing population density along the coasts. To reduce societal risks posed by tsunamis, the nation needs a clear understanding of the nature of the tsunami hazard (e.g., source, inundation area, speed of onset) and the societal characteristics of coastal communities (e.g., the number of people, buildings, infrastructure, and economic activities)

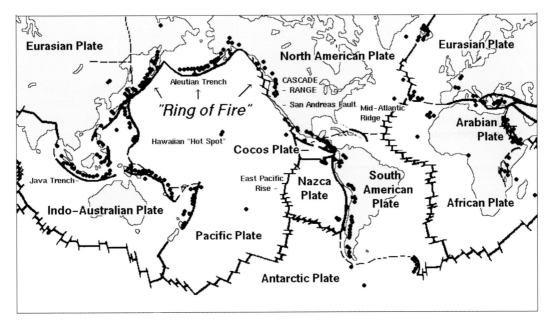

FIGURE 1.1 Global map of active volcanoes and plate tectonics illustrating the "Ring of Fire" and depicting subduction zones; both areas associated with frequent seismic activity. SOURCE: http://vulcan.wr.usgs.gov/Imgs/Gif/PlateTectonics/Maps/map_plate_tectonics_world_bw.gif; USGS.

that make them vulnerable to future tsunamis. With a clear understanding of the tsunami hazards and social vulnerability that comprise tsunami risk, officials and the general public can then prepare for future events and hopefully reduce this risk.[2]

When assessing tsunami hazard and developing risk reduction measures, it is important to consider the distance between a coastal community and potential tsunami sources as well as the probability of occurrence. Near-field tsunamis (see Box 1.1) pose a greater threat to human life than far-field tsunamis because of the short time between generation and flooding; because the extent of flooding is likely greater; and because the flooded area may be reeling from an earthquake (National Science and Technology Council, 2005). Near-field tsunamis account for most U.S. tsunami deaths outside of Hawaii, but even Hawaii has suffered losses from near-field tsunamis. Because it takes a very large earthquake to impact the far-field, more triggering events have the potential to impact communities that are within an hour or less from the source. For example, an earthquake generated within the Cascadia fault zone along the northern California, Oregon, and Washington coasts will allow only minutes for evacuation of

---

[2] Risk is a concept used to give meaning to things, forces, or circumstances that pose danger to people or what they value. Risk descriptions are typically stated in terms of the likelihood of harm or loss of a vulnerable thing or process (e.g., health of human beings or an ecosystem, personal property, quality of life, ability to carry on an economic activity) due to a physical event (i.e., hazard).

the coastal communities after the earthquake is felt. In addition, tsunami observations demonstrate an increase in wave height with proximity to the source, resulting in extensive coastal flooding by a near-field tsunami. Consequences of a near-field tsunami are far greater for any given location.

Far-field tsunamis afford hours of advance notice for evacuation and are likely to have smaller wave heights than those in the tsunami's near field. However, the farther a coastal community from the earthquake source the less likely it is to have felt the earthquake and the more dependent it is on an instrumental detection system to provide warnings. Timely and accurate warnings are required to implement orderly evacuations and to avoid frequent unnecessary evacuations, which can be costly. The National Science and Technology Council (NSTC) report (2005) concludes that "the challenge is to design a tsunami hazard mitigation program to protect life and property from two very different types of tsunami events."

GOALS AND SCOPE OF THIS REPORT

The 2004 Indian Ocean tsunami, spurred two congressional acts intended to reduce losses of life and property from future tsunamis. The Emergency Supplemental Appropriations Act for Defense, the Global War on Terror, and Tsunami Relief, 2005 (P.L. 109-13), included $24 million to improve tsunami warnings by expanding tsunami detection and earthquake monitoring capabilities. This Act was followed in 2006 by the Tsunami Warning and Education Act (P.L. 109-424), which directs the National Oceanic and Atmospheric Administration (NOAA) to strengthen the nation's tsunami warning system (TWS), work with federal and state partners toward the mitigation of tsunami hazards, establish and maintain a tsunami research program, and assist with efforts to provide tsunami warnings and tsunami education overseas.

Section 4(j) of the Tsunami Warning and Education Act calls upon the National Academy of Sciences (NAS) "to review the tsunami detection, forecast, and warning program established under this Act to assess further modernization and coverage needs, as well as long-term operational reliability issues." In response, NOAA asked the NAS to assess options to improve all aspects of the tsunami program. This request is reflected in the first part of the committee's charge (see Appendix B) and accordingly focuses on efforts on tsunami detection, forecasting, and warning dissemination.

The NAS, in accepting this charge and in consultation with NOAA, broadened the review's scope to include an assessment of progress toward additional preparedness efforts to reduce loss of life and property from tsunamis in the United States as part of the National Tsunami Hazard Mitigation Program (NTHMP). The main rationale for this broadened scope was to address Section 5(a) in P.L. 109-424, which called for "a community-based tsunami hazard mitigation program to improve tsunami preparedness of at-risk areas in the United States and its territories." Such a tsunami hazard mitigation program requires partnership among federal, state, tribal, and local governments. Its strategies include identifying and defining tsunami hazards, making inventories of the people and property in tsunami hazard zones, and providing the public with knowledge and infrastructure for evacuation, particularly for near-field tsunamis that come ashore in a few minutes. The broadened scope aims at encompassing the range of national tsunami warning and preparedness efforts.

The Range of Options Available for Tsunami Hazard Mitigation

As the scope of the study was broadened to include aspects of tsunami hazard mitigation, the committee recognized the need to define the term "mitigation" and set some boundaries for the study, because the full suite of mitigation options exceeds the purview and capacity of this particular study. The definition of hazard mitigation and the actions it includes differ among various hazard communities. Some members of the academic community consider the full range of hazard mitigation options to include three classes of actions (White and Haas, 1975): (1) modifying the natural causes of hazards, (2) modifying society's vulnerability (e.g., levees, wind- and seismic-resistant houses), and (3) redistributing the losses that occur (e.g., insurance, emergency response). In contrast, natural hazard practitioners consider the range of human adjustment to natural hazards to fall into two major classes of actions: (1) mitigation of potential losses through interventions in the constructed world in ways that lessen potential losses from nature's extremes (e.g., land-use management, control and protection works, building codes), and (2) preparedness for, response to, and recovery from specific events and their associated losses (Mileti, 1999).

Focus on Warning and Preparedness

Although land-use planning and adjusting building codes is important in mitigating the impacts of tsunamis, the charge to the committee is focused primarily on the detection, forecast, and warning for near- and far-field tsunamis and issues directly related to the effective implementation of those warnings. To be responsive to its charge, the report focuses on the second class of mitigation actions, which generally includes pre-event planning to develop preparedness plans, appropriate organizational arrangements, training and exercises for issuing event-specific public warnings, an adequate emergency response, and plans for recovery and reconstruction. These types of adjustment are based on the notion that the adequacy of pre-event planning determines the effectiveness of event-specific response. This view also places insurance in the preparedness class.

THE NATION'S TSUNAMI WARNING AND PREPAREDNESS EFFORTS

Only very recently has there been a national interest in tsunami warning and preparedness. Before 2004, most efforts were spearheaded by local, state, or regional initiative operating on very limited budgets. Integrating these existing individual efforts into a national tsunami program has led to a very different type of program than that of a national tsunami warning program designed from the outset. The history of tsunami warning and preparedness efforts can be traced back to two of the six destructive tsunamis that caused causalities on U.S. soil.

These efforts were originally part of the National Geodetic Survey, which developed the two tsunami warning centers (TWCs) in Hawaii and Alaska after the 1946 Aleutian tsunami (Unimak Island, AK) and the 1964 Alaskan tsunami (Prince William Sound, AK) (Figure 1.2). These centers eventually became part of NOAA's National Weather Service (NWS), but each is located in different NWS regions and is managed independently.

Concern about tsunamis in Washington, Oregon, and California increased in the late 1980s and early 1990s when several new scientific studies revealed their near-field tsunami threat from the Cascadia subduction zone (Atwater, 1987; Heaton and Hartzell, 1987). California was reminded of its potential tsunami threat by an earthquake near Cape Mendocino in 1992, which generated a small tsunami that arrived in Eureka only minutes after the earthquake occurred. These and other developments prompted a more urgent call to produce comprehensive assessments of tsunami risk and preparedness at the state and federal level.

Congress responded to this call in a 1995 Senate Appropriations Committee request to NOAA to develop a plan for reducing tsunami risk to coastal communities. NOAA suggested the formation of a national committee to address tsunami threat, leading to the establishment of the NTHMP that same year. The NTHMP is tasked with coordinating the various federal, state, territorial, and commonwealth tsunami efforts. NOAA's Tsunami Program was established in 2005 to incorporate all the current tsunami efforts at NOAA (see below). To respond to the committee's charge (see Appendix B) and assess progress made toward improved tsunami warning and preparedness, the committee begins its evaluation with an inventory of the elements of the NTHMP and NOAA's Tsunami Program.

National Tsunami Hazard Mitigation Program

The NTHMP has a Coordinating Committee (steering committee) that works to collaborate on the tsunami mitigation efforts of the NTHMP and three subcommittees: a Mapping and Modeling Subcommittee, a Warning Coordination Subcommittee, and a Mitigation and Education Subcommittee.[3] In addition to coordinating individual efforts, the NTHMP provides guidance to NOAA's TWSs. Federal partners include NOAA, the U.S. Geological Survey (USGS), and the Federal Emergency Management Agency (FEMA). State partners originally included Hawaii, Alaska, Washington, Oregon, and California, and now include all 29 U.S. coastal states and territories.

The USGS contributes to the seismic network that the TWCs use through operating and maintaining their respective seismic networks and to the tsunami research and risk assessments and conducts an independent seismic analysis of potential tsunamigenic earthquakes at its National Earthquake Information Center (NEIC). The USGS and NOAA both support the Global Seismographic Network (GSN), which provides high-quality seismic data to assist earthquake detection (including tsunamigenic earthquakes). Both agencies also support earthquake and seismic studies to improve tsunami warning efforts and tsunami disaster response and hazards assessments. FEMA is responsible for hazard mitigation and emergency response; as

---

[3] nthmp.tsunami.gov.

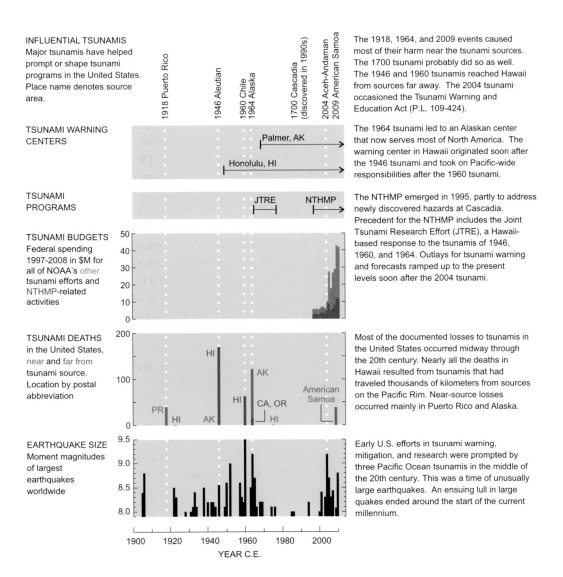

The 1918, 1964, and 2009 events caused most of their harm near the tsunami sources. The 1700 tsunami probably did so as well. The 1946 and 1960 tsunamis reached Hawaii from sources far away. The 2004 tsunami occasioned the Tsunami Warning and Education Act (P.L. 109-424).

The 1964 tsunami led to an Alaskan center that now serves most of North America. The warning center in Hawaii originated soon after the 1946 tsunami and took on Pacific-wide responsibilities after the 1960 tsunami.

The NTHMP emerged in 1995, partly to address newly discovered hazards at Cascadia. Precedent for the NTHMP includes the Joint Tsunami Research Effort (JTRE), a Hawaii-based response to the tsunamis of 1946, 1960, and 1964. Outlays for tsunami warning and forecasts ramped up to the present levels soon after the 2004 tsunami.

Most of the documented losses to tsunamis in the United States occurred midway through the 20th century. Nearly all the deaths in Hawaii resulted from tsunamis that had traveled thousands of kilometers from sources on the Pacific Rim. Near-source losses occurred mainly in Puerto Rico and Alaska.

Early U.S. efforts in tsunami warning, mitigation, and research were prompted by three Pacific Ocean tsunamis in the middle of the 20th century. This was a time of unusually large earthquakes. An ensuing lull in large quakes ended around the start of the current millennium.

FIGURE 1.2 Timelines for U.S. tsunami warning centers, programs, tsunami budget, deaths from tsunamis in the United States and its territories, and earthquakes of magnitude 8.0 or larger worldwide since the year 1900. Sources of data for this figure include: NOAA (federal spending); http://www.ngdc.noaa.gov/hazard/tsu_db.shtml (tsunami fatalities); http://earthquake.usgs.gov/earthquakes/eqarchives/ (great earthquake history). SOURCE: Committee member.

part of its mitigation efforts it has issued *Guidelines for Design of Structures for Vertical Evacuation from Tsunamis* (Federal Emergency Management Agency, 2008). FEMA becomes the lead federal agency in managing the emergency response once a tsunami has caused damage to U.S. coastlines.

The National Science Foundation (NSF) used to be a partner of the NTHMP, but as its involvement decreased the decision was made in 2009 to remove it from the NTHMP. Its primary function is to provide research funding and to partner with other federal agencies in research and development. NSF provides funding for the GSN. NSF has also been actively involved with investments regarding tsunami research infrastructure, such as the Network for Earthquake and Engineering Simulation (NEES), Earthquake Engineering and Research Centers (EERCs), and the Southern California Earthquake Center (SCEC) (Bement, 2005). Because it is not part of the NTHMP and its funding decisions are primarily driven by the demand in the research community, this report does not include an explicit discussion of NSF's role but rather discusses the role of the broader research community in the nation's tsunami efforts.

NOAA has been carrying most of the responsibility and obtains most of the funding to provide tsunami warnings, maintain observing networks (including seismic networks not funded by the USGS in Alaska and Hawaii), manage and archive data, and conduct research (further discussed in the next section).

The coastal states, U.S. territories, and commonwealths contribute their own initiatives and resources to the nation's preparedness and education efforts; these vary in extent and approach from state to state. In particular, states are responsible for providing communities with inundation maps that allow municipalities to produce evacuation maps and guidance, and to educate the public about the hazard and appropriate responses. Local officials in turn are responsible for transmitting tsunami alerts throughout their respective jurisdictions, issuing evacuation orders, managing evacuations, and declaring all-clears.

NOAA's Tsunami Program

In 2006, the Tsunami Warning and Education Act (P.L. 109-424) charged NOAA with addressing the nation's priorities in tsunami detection, warning, and mitigation. NOAA's Tsunami Program assumed the responsibilities to plan and execute NOAA's tsunami efforts, primarily the program's budget, strategic plan, and the coordination of activities among its NOAA organizational components and external partners, including the NTHMP. NOAA's Tsunami Program advocates an end-to-end TWS, which includes detection, warnings and forecasts, message dissemination, outreach and education, and research.

NOAA's Tsunami Program is supported by five line offices (Table 1.1): NWS; the Office of Marine and Aviation Offices (OMAO); the National Ocean Service (NOS); Oceanic and Atmospheric Research (OAR); and the National Environmental Satellite, Data, and Information Service (NESDIS). The NWS, as the administrator for NOAA's Tsunami Program, is primarily responsible for helping community leaders and emergency managers in strengthening their local tsunami

TABLE 1.1 Tsunami Program Matrix

	Tsunami Program Capabilities	Line Offices				
		NWS	**OAR**	**NOS**	**NESDIS**	**OMAO**
Weather and Water Goals	Hazard Assessment	Modeling and Mapping (tsunami source, propagation, coastal elevations, and inundation) Assessments (exposure and vulnerability, historical event records, social and economic impacts)				
	Warning Guidance	Earthquake Detection		Sea level Monitoring, Inundation Detection	Geophysical and Oceanographic Data Management and Long-Term Archive	Detections System O and M Support
		Bottom Pressure (Tsunami) Detection		Water Current and Geodetic Reference		Aerial and Ship Coastal Survey
		Data Management and Communication		Ocean-Data Management and Communications		
		Warning and Forecast Analysis and Notification	Warning and Forecast Development			
	Mitigation and Preparedness	Coordination (Federal-State-Local-Tribal) for Community-Based Education, Outreach, and Planning				
		Capacity Building; Standards and Guidelines; Coastal Resilience				
		Hazard Awareness and Integrated Coastal Area Management				
	Research	Technology Transfer, Transition, and Training				
		Tsunami Science, Modeling and Mapping, Sensors, Detection and Prediction, Ocean Exploration				
	International Coordination	Inter-governmental (IOC, WMO, ISDR, GEO…):[a] Hazard Mapping, Technology Transfer and Education, Data/Information Exchange, Observations and Communication, Standards and Policy				
		Interagency (USGS, FEMA, NSF, DOS/USAID):[b] Hazard and Risk Assessment, Coastal-Ocean Observations, Data Management, Direct Foreign Assistance, Capacity Building, and Foreign Policy				

[a] IOC: Intergovernmental Oceanographic Commission; WMO: World Meteorological Office; ISDR: International Strategy for Disaster Reduction; GEO: Group on Earth Observations.
[b] USGS: U.S. Geological Survey; FEMA: Federal Emergency Management Agency; DOS: Department of State; USAID: U.S. Agency for International Development.
SOURCE: Data from National Oceanic and Atmospheric Administration, 2008a.

warning and preparedness programs through its TsunamiReady program as well as operating the TWCs.

The Pacific Region's Pacific Tsunami Warning Center (PTWC) and the Alaska Region's West Coast/Alaska Tsunami Warning Center (WC/ATWC) are administered within the NWS, although the two TWCs report to their respective regional NWS offices. The two TWCs have distinct areas of responsibility as described in Chapter 5. The NWS also houses the National Data Buoy Center (NDBC), which operates and maintains the Deep-ocean Assessment and Reporting of Tsunamis (DART) buoys. These buoys monitor and alert the TWCs of sea level changes associated with a tsunami. OMAO collaborates by providing detection system maintenance support and conducting coastal surveys. NOS provides state and local coastal emergency managers with hazard-related information such as training and assessment tools, and also operates coastal tide stations and sea level gauges that monitor changes in sea level. OAR comprises a research network involving internal research laboratories, grant programs, and collaborative efforts between NOAA and academic institutions. Pacific Marine Environmentla Laboratory (PMEL), within OAR, focuses on designing optimal tsunami monitoring networks, improving forecast modeling, and improving impact assessment on coastal communities. NESDIS provides access to global environmental data; such as climate, geophysical, and oceanographic data. The National Geophysical Data Center (NGDC), housed within NESDIS, manages a database for historic tsunami events, maps, and DART and tide gauge records. Some negative consequences arising from this distribution of tsunami detection, forecast, warning, and planning functions across different parts of NOAA and across different NTHMP partners is discussed in greater detail in Chapters 3 and 5.

ASSESSING THE NATION'S EFFORTS

Because tsunami warning and preparedness efforts are distributed across federal and state agencies and were historically conducted without a federal coordination mechanism, the committee faced a number of challenges in assessing progress in the nation's ability to warn and prepare for the threat of tsunamis. The first challenge results from the need to assess many individual activities. Secondly, it is difficult to extrapolate from these individual activities to assess whether all the distributed efforts can function coherently during a tsunami to warn and evacuate people in a timely fashion. To help address these challenges, the committee began its analysis by sketching the required functions and elements of an idealized integrated warning and preparedness effort based on available research findings in the hazards and high-reliability organizations (HRO) literature (see section below). The committee then sought to compare its vision of an idealized system with the evolving status quo.

An ideal integrated TWS comprises multiple technologies, systems, individuals, and organizations. A comprehensive view of the elements therefore includes technical, organizational, social, and human components. The ideal system incorporates risk assessment, public education, tsunami detection, warning management, and public response (Figure 1.3).

Protecting and warning the public begins with an understanding of the tsunami risk envi-

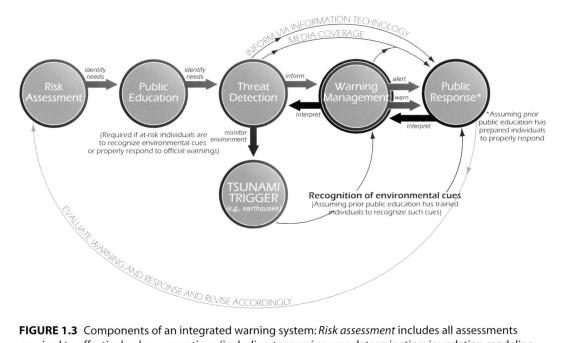

FIGURE 1.3 Components of an integrated warning system: *Risk assessment* includes all assessments required to effectively plan evacuations (including tsunami source determination, inundation modeling, and evacuation mapping) and prepare the communities to evacuate in the event a warning is issued or received. Risk assessments identify needs for public education. *Public education* aims to ensure maximum preparedness and a public that knows what to do when it receives a warning or feels the ground shaking in the case of near-field tsunamis. *Threat detection* comprises the continuous monitoring of the natural and technological environments that could create an emergency; it informs the warning management and public response component using threshold criteria and communication technology. *Warning management* interfaces the threat detection component with the public response component and is responsible for tsunami alerts, warnings, and evacuations; in consultation with the threat detection component it will alert and warn the public. *Public response* is the ultimate outcome of the integrated warning system, and it integrates public education, threat detection, natural cues from tsunami triggers, and warning management. SOURCE: Committee member; design by Jennifer Matthews, University of California, San Diego.

ronment. This must be done before a tsunami is generated in order to design the threat detection system, the education and awareness campaigns, and the evacuation and response plans. To understand the risk environment, both hazards (the physical characteristics of tsunamis and the inundation area) and vulnerabilities (the people and properties in harm's way) need to be characterized (National Research Council, 2006). Pre-event public education is required to enable at-risk populations to correctly interpret: (1) natural cues from the environment (e.g., ground shaking from the earthquake) or (2) warnings from a technical detection system as a signal to evacuate to higher ground in a timely fashion. The threat detection component monitors the environment for threshold events using cues from natural and technical systems (Mileti, 1999; Mileti and Sorenson, 1990).

Once a significant tsunami is detected, the warning process needs to be managed. Tsunami information needs to be analyzed and decisions have to be made about the extent of the warning. Managers and decision makers issue warnings directly to the public. Ideally, officials managing the response also maintain situational awareness and information flow between the technical detection system and the public to update warnings and messages with the required protective actions to be taken. Because of the dominance of real-time communications, the Internet, and social networking, both the general public and media will increasingly access tsunami information directly from real-time information sources (e.g., the TWCs, seismometers, and water-level gauges) before local officials are able to respond. The public's real-time access to different information sources, such as social media and networking systems, underscores the importance of public education to prepare both the public and the press for proper interpretation of information and response to detected hazards. An effective warning system monitors the public's response and reactions in order to improve its processes for effective, understandable, actionable, reliable, and accurate warnings of impending danger. In the following chapters, the report covers the system components and compares the idealized system with current and/or planned efforts.

An integrated TWS has an impact on large populations and on a wide range of resources and, in the event of failure, has the potential to cause enormous economic, social, organizational, technological, and political losses. Although often seen as mainly comprising technical and technological elements, a warning system must, out of necessity, include the human dimension, such as people's behavior, policies, procedures, and organizations. However, it is the human dimension that poses a significant challenge:

> This involves the setting and running of national services (people), and the implementation of complex emergency-preparedness and awareness plans at the national and local levels to immediately inform every person of the threat. In the building of any early warning system, this is the difficult part. (Intergovernmental Oceanographic Commission, International Strategy for Disaster Reduction, and World Meteorological Organization, 2005).

CHALLENGES TO REDUCING THE NATION'S VULNERABILITY TO TSUNAMIS

Reducing the vulnerability of coastal settlements and infrastructure to tsunami risk poses some unique challenges. Although tsunamis can be devastating, as was seen during the 2004 Indian Ocean event, catastrophic tsunamis are relatively infrequent. This infrequency makes it more challenging to sustain the capacity to educate, warn, and prepare for this particular hazard. As discussed above, the history of tsunami warning and mitigation efforts in the United States shows that significant new funding is often made available only after a tsunami has devastated a coastal community and caused casualties. High funding levels and commitment to tsunami mitigation dissipate over time, leading to difficulties in maintaining efforts, knowledge, and lessons learned over time. Another challenge is the need to relay warnings from the federal government to state and local officials in just minutes (in the case of a near-field tsunami) or hours (in the case of a far-field tsunami). Sustaining the organizational preparedness and coordination across many jurisdictional boundaries presents a daunting challenge.

The committee recognizes that the nation's tsunami detection, warning, and preparedness efforts originated in many diverse efforts distributed across several coastal states, and that attempts to integrate these distributed components into a coherent program have only recently begun. In particular, because tsunamis are rapid onset events, there is very little margin for error in the system before failure becomes catastrophic. An organization that operates in a low probability, high-risk environment, allowing few errors, is called an HRO (Roberts, 1990). HROs manifest a number of common properties: flexible and adaptable organizational structures, continually reinforced organizational learning, decision making that is both flexible and mobile, a strongly reinforced organizational culture, constant and effective communication, and trust among members of the system, particularly across organizations (Grabowski and Roberts, 1999; Grabowski et al., 2007). Because the committee identified the need for high-reliability operations in TWSs, the committee draws from the research literature on HROs (Roberts, 1990) and resilient systems (Hollnagel et al., 2008) to highlight particular characteristics that reduce the risks of failure in an idealized end-to-end warning system:

- **Situational Awareness in an Emergency:** Because tsunamis are events that allow only minutes to hours for evacuation, a keen sense of situational awareness and the ability to respond quickly and effectively is required (Weick, 1990, 1993, 2003). HROs require decision making that is adaptable to change and surprise, and that is able to continually reassess needs across distributed organizations (Weick, 1993, 1998; Weick et al., 1999). Such is the case with the nation's tsunami warning and preparedness efforts, where the TWCs, the state and local offices, and emergency managers and the affected public are geographically dispersed and often lack face-to-face contact. The dispersed and decentralized nature of the end-to-end tsunami warning and preparedness efforts make it a significant challenge to maintain awareness of the evolving situation during a crisis.

- **Learning and Training:** To maintain situational awareness under changing conditions requires training. Therefore, an effective TWS requires that watchstanders, emergency managers, regulators, the public, and the media learn together, and engage in learning that enhances sense-making and developing alertness to small incidents that may cascade into much larger disasters (Weick, 1993; Farber et al., 2006). Because of the low frequency of tsunamis (e.g., California is issued an alert bulletin on average once every three years; Dengler, 2009), a TWS has few opportunities to learn from an event and therefore needs to learn from exercising the system through drills. Trial and error can be disastrous not only because disasters are rare, but also because in the absence of a major catastrophe to focus attention in the system, lessons learned from previous events may be forgotten or misapplied (March et al., 1991; Levitt and March, 1988; De Holan and Phillips, 2004). Learning in a high-reliability organization needs to be systematic, continually reinforced, measured, and made part of the system's core values.

- **Fluid Organizational Structures:** HRO structures are often adaptable and fluid, allowing the system to expand or contract in response to its environment (Roberts, 1990). TWSs with flexible organizational structures would be able to expand and contract resources in response to shifts and changes in environmental demands, disasters, or periods of slack resources. In the event of a tsunami, TWS managers need to grow effective, functioning response organizations in a period of less than 24 hours, and then adjust the organizational structures to the needs of the response (Tuler, 1988; Bigley and Roberts, 2001). The ability to provide varied organizational structures in response to environmental demands may be critical to the success of TWS organizations, similar to the way fire and emergency organizations expand and contract in response to fire demands (Grabowski and Roberts, 1999). Distributed information technology that connects the system responders can provide the technological glue that ties HRO members together, and fluid organizational structures can allow the organization to grow, expand, contract, and respond to changes in a dynamic, high tempo environment (Bigley and Roberts, 2001). Similar requirements for members and organizations in TWSs can be envisioned as tsunami conditions unfold.
- **Strong Organizational Culture:** Schein (1992, 1996) defines "culture" as a set of basic tacit assumptions, that a group of people share, about how the world is and ought to be; it determines their perceptions, thoughts, feelings, and to some degree, their overt behavior. In many organizations, shared assumptions typically form around the functional units of the organization and are often based on members' similar educational backgrounds or experiences (Grabowski and Roberts, 1996, 1997). HROs are characterized by strong cultures and norms that reinforce the organization's mission and goals and that focus attention on procedures, policies, and reward structures consistent with the organization's mission and safety (LaPorte and Consolini, 1991). HROs have cultures attentive to errors; cultures where closely held ideas about the organization, its mission, and member roles in reliability enhancement are articulated; cultures that encourage learning; and cultures where safe areas—for decision making, communication, and the like—are created as buffers (Weick, 1993). Constructs such as oversight and checks and balances reinforce the strong cultural norms of the HRO. Melding the varied cultures that integrate the system into a cohesive whole can be extremely difficult in distributed systems that are connected by linkages that can dissolve and wane as requirements, organizational structures, and political will change (Weick, 1987; Weick and Roberts, 1993; Grabowski and Roberts, 1999).

Managing decision making across organizations that report to different management structures is a challenge for highly dispersed efforts; this is certainly the case with U.S. tsunami detection, warning, and preparedness efforts. A particular challenge is that the federal government has responsibility to forecast and warn about potential hazards, yet local governments order evacuations. Failure to consider distributed decision making within groups and across multiple units can lead to lack of readiness for the next large-scale catastrophe; e.g., Hurricane Katrina (Roberts et al., 2005; Farber et al., 2006). Building good communication and trust aid in effective decision making and can increase the likelihood of success in geographically distributed organizations. Trust can be built by common training; opportunities for scientific and operational exchange; and workshops, conferences, exercises, and simulations that build community and coherence across distributed organizations.

TYING IT ALL TOGETHER: REPORT ROADMAP

In the following chapters, the committee assesses progress in the nation's distributed tsunami preparedness, detection, and warning efforts and compares it to its vision of an idealized warning system (Figure 1.3). Chapter 2 evaluates progress in hazard and vulnerability assessments and identifies potential improvements that could guide the nation's tsunami risk-assessment efforts. Chapter 3 discusses education and outreach efforts and evaluates pre-event community and organizational preparedness and the coordination between the various entities at the local, state, and federal levels. Chapter 4 examines the technical hazard detection system, including the seismic and sea level sensor networks. Chapter 5 examines the TWCs' operations and how technology and human capital are used to provide their functions. Appendices present supporting data on tsunami sources, hazard and evacuation maps, educational efforts, seismological methods, and several case-study tsunamis.

CHAPTER TWO

Aligning Priorities with Societal Risks from Tsunamis

SUMMARY

This chapter reviews progress toward understanding the nation's tsunami risk, which is the first step in a comprehensive tsunami program. The knowledge of the hazards tsunamis pose is evolving. The nation is just beginning to understand the populations and social assets that tsunamis threaten, the readiness of individuals and communities to evacuate, and the losses of life and property they may cause. Although much has been learned about the nation's tsunami risk, the nation remains far from understanding enough of its tsunami risk to set priorities and allocate resources for tsunami mitigation efforts based on risk.

The chapter evaluates progress and identifies opportunities for the National Oceanic and Atmospheric Administration (NOAA) and the National Tsunami Hazard Mitigation Program (NTHMP) to advance the goal of hazard and evacuation mapping and a comprehensive national risk assessment. Among other things, hazard modeling serves as the basis to produce evacuation maps, which are a critical tool in educating and preparing the public. The committee concludes that it is unclear whether current evacuation maps are sufficient for enabling effective evacuations or preparing the public due to the absence of uniform quality standards, evaluative metrics, or guidelines on what constitutes effective mapping approaches.

The advice, directed mainly at NTHMP partners and listed here in summary form, includes:

- Completion of an initial, national tsunami risk assessment in the near term to inform program prioritization.
- Periodic reappraisal of tsunami sources and modeling codes, achieved in part through workshops and peer review.
- Greater consistency, across state boundaries, in the methods, criteria, and judgments employed in modeling of tsunami inundation, achieved in part through collaboration among federal and state partners and through external review of inundation maps.
- Evaluations of the effectiveness of hazard maps, leading to standards that increase the overall quality and consistency of these maps.
- Recurring inventories of the number and kinds of people in tsunami hazard zones, with special attention to high-risk groups including children, the infirm, and tourists.
- Identification of areas where successful evacuation from a tsunami would require buildings or engineered berms.

INTRODUCTION

Understanding risk is a prerequisite to effectively preparing and warning endangered people of potential and imminent tsunamis. To that effect, officials must understand in advance (1) where and to what extent tsunami inundation may occur and the amount of time before waves arrive, (2) the characteristics of the population and communities in tsunami-prone areas, and (3) how prepared individuals and communities are for evacuation.

Risk, as used in this report, is a concept used to give meaning to things, forces, or circumstances that pose danger to people or what they value (see also Box 2.1). It takes into account

---

BOX 2.1
Definitions

Risk is a concept used to give meaning to things, forces, or circumstances that pose danger to people or what they value. Risk descriptions are typically stated in terms of the likelihood of harm or loss of a vulnerable thing or process (e.g., health of human beings or an ecosystem, personal property, quality of life, ability to carry on an economic activity) due to a physical event (i.e., hazard) (National Research Council, 1996a). Some researchers have used the term risk to quantify the likelihood of future tsunamis, while others have defined it as a product of the probability of tsunami-attributable social damage (e.g., buildings, lives, businesses) and the magnitude of that damage. Research shows that managers, policy makers, and members of the public rarely define risk as an objective calculation; instead, perceptions vary according to differences in awareness, experiences, and social context (Fischhoff et al., 1984; Weichselgartner, 2001).

Hazard is the physical characteristics of an event (e.g., tsunami: speed of onset, impact forces, currents, inundation area) that can pose a threat to people and the things they value.

Vulnerability is the personal or situational conditions that increase the susceptibility of people or resources to harm from the hazard.

Inundation refers to the process of coastal flooding due to tsunamis or storm waves regardless of the impact to human activities.

Run-up height is the vertical elevation of the most landward penetration of the tsunami wave with respect to the initial sea level (figure opposite page). Run-up is a vertical distance, while inundation is a horizontal distance.

Inundation models determine the areas likely to be flooded by a tsunami and involve numerical computations of tsunami evolution for specific tsunami scenario or consider an ensemble of tsunami scenarios that might affect the map area.

Hazard maps depict inundation areas on base maps that typically include contours, imagery, buildings, roads, and/or critical infrastructure and take into account local geologic knowledge.

Evacuation maps depict areas that need to be evacuated in the event of a tsunami and to show evacuation routes to safe havens. Evacuation maps are based on the same inundation zones in hazard

the likelihood and physical characteristics of tsunamis (i.e., the hazard), the personal or situational conditions that increase the susceptibility of people or resources to harm from the tsunami (i.e., vulnerability), and associated uncertainties.

This chapter first introduces tsunami risk assessment by identifying roles it may play in reducing the losses of life and property to tsunamis and by summarizing broad approaches to risk assessment. The rest of the chapter reviews progress in assessing tsunami hazard and tsunami vulnerability. Aspects of perceptions, knowledge, and preparedness levels that influence individual resilience are discussed in the following chapter.

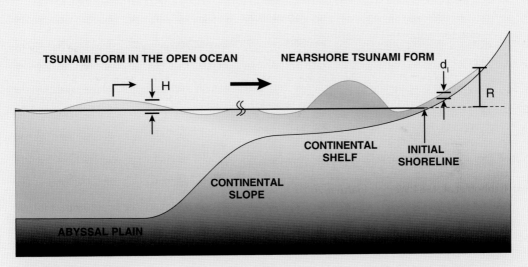

H = tsunami wave height
R = run-up height
d_I = inundation depth
The depth in the continental shelf is typically less than 200 m.
The average depth in the abyssal plain in the Pacific is about 4,000 m.

The definition of tsunami run-up height. SOURCE: Committee member.

maps but these zones are typically enlarged to allow easier identification of landmarks, such as major intersections or highways.

Evacuation modeling refers to the process of simulating how people evacuate any given area and is used to quantitatively evaluate whether high ground can be reached in time.

COMPREHENSIVE RISK ASSESSMENT

Tsunami risk assessment is fundamental to the nation's tsunami programs, because it can help support and guide risk reduction efforts, including tsunami education, preparedness planning, and warning system development. In particular, it can help allocate resources according to greatest risk and thus ensure that efforts are prioritized to protect the most people. The following three examples illustrate the benefits of such an assessment:

- *Priorities among states.* A comprehensive tsunami risk assessment could improve resource allocations among states, territories, and commonwealths. In 2006, the Government Accountability Office (GAO) observed that "with the likely expansion of the NTHMP from 5 state participants to potentially 28 state and territorial participants in 2006, it will be difficult for NOAA to ensure that the most threatened states receive the resources they need to continue and to complete key mitigation activities without an updated, risk-based strategic plan" (Government Accountability Office, 2006). To this date, NTHMP resources are being allocated on the basis of estimated hazards, without regard for vulnerability (National Tsunami Hazard Mitigation Program, 2009a).
- *Priorities among program elements.* A comprehensive tsunami risk assessment coupled with cost-benefit analyses could also help NOAA weigh the value of warning and forecasting on the one hand against that of education and community preparedness on the other. The GAO's 2006 report noted that

  > NOAA's initial strengthening efforts emphasize detection and warning for distant tsunamis, while the greater risk to most locations in the United States— according to NOAA data as well as the National Science and Technology Council's (NSTC) December 2005 report on tsunami risk reduction—[is] likely to be posed by local tsunamis. For example, the deployment of Deep-ocean Assessment and Reporting of Tsunamis (DART) stations and warning center enhancements will not reduce the local tsunami risk as directly as other strategies such as educating vulnerable populations to immediately head for high ground when the earth shakes near the coast (Government Accountability Office, 2006).

- *Program development.* Understanding tsunami risk helps averts surprises. In the 1980s, the Cascadia subduction zone became known as a source of catastrophic tsunamis, first from geophysical clues that it might produce such waves and then through geological signs that it had (Atwater et al., 2005). These discoveries about the tsunami hazard were the first indication of the need for Washington, Oregon, and California to take steps to prepare to build tsunami-resilient communities before the next great Cascadia tsunami strikes. These discoveries also helped create the NTHMP itself. When the NTHMP originated in the mid-1990s, it was founded in part on then-recent discoveries about tsunami risk. Oregon's concern about Cascadia tsunami hazards played a central role in the NTHMP's establishment, according to committee members' interviews with founding members of the NTHMP steering group.

Tsunami risk assessments are challenging due to (1) the paucity of information about the frequency, sources, and characteristics of past tsunamis, (2) the complex and interdependent nature of coastal communities in larger economic or sociopolitical systems, (3) the poorly defined mix of near-field and far-field tsunami hazards that coastal communities face, and (4) the uncertainty of potential impacts of future events. Although societal risk from tsunamis is challenging to assess, it is critical information for the development and prioritization or risk-reduction efforts, such as education, preparedness planning, warning-system development, mitigation planning, and response strategies.

An effective risk characterization should be undertaken with local decision makers in mind and should be directed toward informing specific choices of public officials and affected individuals (National Research Council, 1996a). When developing the analysis for decision tools (e.g., inundation maps, vulnerability assessments, evacuation maps), scientists need to engage and deliberate with decision makers to assess their information needs and create useful decision tools (e.g., Fischhoff et al., 1978; National Research Council, 1996a, 2007; Gregory and Wellman, 2001; Bostrom et al., 2008; Renn, 2008). Risk analysis involves the systematic collection and interpretation of quantitative or qualitative data to better understand hazards and vulnerable communities (National Research Council, 1996a). Risk deliberation is any communicative process in which technical experts, public officials, and affected parties collectively consider risk issues to ensure that decision-relevant knowledge and diverse perspectives are included in any risk reduction process (National Research Council, 1996a).

There is no single method to properly characterize all aspects of risk from tsunamis; different techniques are needed to address different aspects (e.g., demographic sensitivity, structural fragility, financial exposure) and their potential risk-reducing adjustments (e.g., education programs, structural mitigation, insurance). However, best practices suggest including assessments of possible exposure sources and pathways (including geospatially specific susceptibilities to tsunamis), potential consequences, the effects of feasible risk reduction options, and the probabilities and uncertainties of exposures and ensuing effects. Probabilistic assessment can be a tool to provide a basis for cost-benefit analysis and design considerations for tsunami mitigation efforts (e.g., design criteria of an evacuation structure) and a transparent basis for prioritizing resources. Geographic information system (GIS)-based analyses of socioeconomic exposure to tsunamis are useful for identifying demographic sensitivities within a community that could impact evacuations. Assessments of adaptive capacity and resilience can identify educational needs and pre-event preparedness levels.

The level of sophistication, accuracy, resolution, and format required for assessing societal risk to tsunamis will depend on the intended use of the information. For example, education efforts designed to raise hazard awareness, such as teaching people to recognize natural cues, may only require oral histories from tsunami survivors (Dudley, 1999) or coarse delineations of run-ups from past events (Theilen-Willige, 2006) to indicate tsunami-prone areas (see Box 2.1 for definitions). By contrast, evacuation planning usually requires computer modeling of various tsunami evacuation scenarios (Katada et al., 2006; Yeh et al., 2009). Urban planning or detailed assessments of economic impacts require even higher resolution calculations (Borrero et al., 2005).

So far, no comprehensive risk assessments have been undertaken that could guide the setting of priorities at the state or national level. Existing risk assessments are uneven and are typically isolated efforts for specific sites (e.g., González et al., 2006). The only national assessment to date is an evaluation of tsunami hazards, based on the written historical records available at the NOAA/National Geophysical Data Center (NGDC) (Dunbar and Weaver, 2008) with the rationale being that "the state of geologic knowledge [of tsunami sources] does not permit the calculation of meaningful probabilities of occurrence."

Although hazard researchers may be uncomfortable with the lack of information about occurrence, probabilistic risk assessment can be an indicator for the relevance and import of existing evidence, despite the uncertainties. Probabilistic risk assessment for nuclear power plants and for earthquakes show how risk assessments can be constructed for rare, high-consequence events in order to inform planning efforts and designs (e.g., McGuire and Becker, 2004; McGuire, 2008). Where data are sparse, expert judgment can be used to qualify the available data appropriately. For example, tsunami planning has been so far based on worst-case scenarios, with inconsistent choices made about specific scenarios and little understanding of consequences or expected losses. Where extremely rare, high-consequence events are the worst case, communities that are much more likely to see smaller yet damaging tsunamis may not be prioritized for funding. Such tradeoffs deserve to be explicitly considered and can be incorporated into quantitative and qualitative risk assessments using expert judgment in a deliberative process (National Research Council, 1996a).

Conclusion: The United States lacks a national tsunami risk assessment that characterizes the hazards posed by tsunamis, inventories the populations and social assets threatened by tsunamis, measures the preparedness and ability of individuals and communities for successful tsunami evacuations, and forecasts expected losses. This information is needed to help spur and prioritize investments in preparedness, education, detection, and warning efforts and for developing long-term strategic planning at the local, state, and federal level.

Recommendation: NOAA and its NTHMP partners, in collaboration with researchers in social and physical sciences, should complete an initial national assessment of tsunami risk in the near term to guide prioritization of program elements. The national tsunami risk assessment should (1) incorporate the best possible relevant science (social and behavioral science, geography, economics, engineering, oceanography, and geophysics) and (2) include broad stakeholder and scientific participation to ensure that efforts are responsive to the needs of at-risk communities and decision makers.

TSUNAMI HAZARD ASSESSMENT

Tsunami hazard assessments focus on characterizing and visualizing the physical characteristics of future tsunamis (e.g., speed of onset, impact forces, currents, inundation area) that can pose a threat to people and the things they value. Understanding where tsunami inundation is likely and how much time at-risk individuals have to evacuate frame the discussion of societal risk to tsunamis and are the foundations upon which education, preparedness plans, response plans, and mitigation strategies are developed. Tsunami hazard assessments typically entail three elements: (1) inundation models to determine the areas likely to be flooded, (2) hazard maps that portray inundation-model outputs on community base maps (e.g., roads, elevation, structures), and (3) evacuation maps that depict areas that may need to be evacuated in the event of tsunamis. The purpose of this section is to describe each of these elements, progress in the development and implementation of each element, and areas for improvement.

Before discussing each individually, it is important to distinguish the differences between the three elements. For example, although hazard and evacuation maps are both used for reducing tsunami risk, the two types of tsunami maps are developed in different ways and for different purposes (Figure 2.1). Inundation model outputs simply denote the physical characteristics of tsunami generation, propagation, and inundation areas and do not recognize political boundaries. Tsunami hazard maps portray inundation model outputs with some modification to reflect local knowledge of land conditions, are organized by communities, and include basic societal assets (e.g., roads, major structures). Tsunami evacuation maps incorporate the same inundation areas as hazard maps, but typically denote larger zones to accommodate local risk tolerance and to allow for easier identification of landmarks (e.g., at-risk individuals can identify major roads easier than a specific elevation contour). Evacuation maps also differ from hazard maps in that they are educational tools designed to be easily understood by non-scientists and typically identify evacuation shelters and assembly sites (both natural and manmade), suggested routes, locations of warning loudspeakers, and transportation infrastructure to facilitate evacuations (e.g., streets, bridges). Inundation models and hazard maps are developed by scientists to identify and communicate maximum inundation areas and flow conditions, whereas evacuation maps are designed by scientists in collaboration with local, state, and federal emergency management and public safety agencies to identify areas of public safety concern.

Inundation Modeling

Credible tsunami inundation modeling requires three elements: (1) an understanding of the tsunami source that generates the tsunami, usually through estimation of seafloor displacement; (2) accurate and precise bathymetric and topographic data to understand the surface over which the waves propagate; and (3) a robust hydrodynamic computational model to simulate tsunami evolution. Each of these three elements are treated below in turn and offer two sets of conclusions and recommendations—one on ways of reducing uncertainties about tsunami sources and the other on hydrodynamic modeling.

(A)

(B)

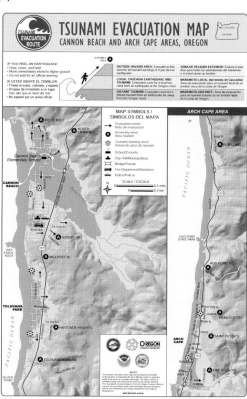

FIGURE 2.1 Tsunami maps for Cannon Beach, Oregon, including (A) a tsunami hazard map developed for a Cascadia subduction zone earthquake scenario and (B) an evacuation map that includes evacuation zones for a Cascadia-related tsunami (in yellow) and a far-field tsunami (in orange). SOURCE: http:www.oregongeology.org/sub/default.htm; image courtesy of DOGAMI.

Tsunami Sources

The societal value of inundation models depends largely on estimates of tsunami sources (Synolakis et al., 1997). If a modeler underestimates a tsunami source, a real tsunami may inundate places the modeling had deemed safe and lives could be unnecessarily lost as a result. If instead the modeler overestimates the tsunami source, risk reduction efforts may be cost-prohibitive and more people and businesses may be unnecessarily affected in future evacuations.

Large uncertainties remain concerning the sources of tsunamis that could inundate U.S. shorelines. These sources include subduction zones of the Pacific Rim and Caribbean, underwater landslides off the Atlantic and Gulf Coasts and off southern California, and volcanoes in

Alaska (Figure 2.2, Appendix A). The next six paragraphs illustrate unknowns that are spelled out more fully as questions in Appendix A.

Far-field heights of tsunamis from circum-Pacific sources. For hazard and evacuation mapping, simulations of far-field tsunami heights on U.S. Pacific shores require estimates about tsunami sources at subduction zones on the Ring of Fire (Figure 2.2). The simulations are sensitive to earthquake size because far-field tsunami height increases substantially with earthquake magnitude (Abe, 1979). For example, to make worst-case simulations for Pearl Harbor, a NOAA group used 18 tsunami sources, together spanning all subduction zones on the Pacific Rim (Tang et al., 2006). For each source, the group used an earthquake of magnitude 9.3 (McCaffrey, 2008). For some of the subduction zones, the assumed magnitude 9.3 differs from the largest known earthquake magnitude of the past, which falls short of 8.0 or 8.5 (blue or brown, respectively, in Figure 2.2). This difference in assumed source magnitude results in the corresponding worst-case tsunami on distant U.S. shores to be larger by a factor of about 10 from the tsunamis of the past.

Far-field hazards from Alaskan sources. Earthquake-generated tsunamis from sources along the Aleutian-Alaskan subduction zone pose far-field tsunami hazards in Washington, Oregon, California, and Hawaii. If a hazard map is to give each of these sources appropriate weight, it is necessary to estimate how large the earthquakes can get and how often they happen (Geist and Parsons, 2006; González et al., 2006). For tsunamis from most of the Aleutian-Alaskan subduction zone, currently there is almost no basis for such weighting except for geophysical estimates of how much plate motion gets spent on earthquakes of various sizes (González et al., 2006; Wesson et al., 2008). Little is known about how often the tsunamis actually recur except for the source of the oceanwide 1964 Alaskan tsunami (Carver and Plafker, 2008). Even for the 1964 source, it is unclear whether the next large tsunami is expected to recur sooner than average because the 1964 earthquake ended a recurrence interval, which was 300 years longer than the previous 600-year average.

Near-field hazards from Cascadia tsunamis. Although documented geological histories of great Cascadia earthquakes extend thousands of years into the past, persistent uncertainties about them have yielded wide-ranging estimates of tsunami hazards. A probabilistic analysis of tsunami hazards in Seaside, Oregon, showed the hazard to be sensitive to variability in earthquake size and recurrence (González et al., 2006). A subsequent hazard map for nearby Cannon Beach doubled the maximum tsunami height relative to previous estimates (Priest et al., 2009).

Caribbean hazards. Islands of the Caribbean are threatened by tsunamis mainly from local tectonic sources and the subsequent potential for submarine landslide. One tsunami source, near Puerto Rico, may even threaten coastlines as far away as Massachusetts. Local tsunami sources caused loss of life in the Virgin Islands in 1867 and in western Puerto Rico in 1918 (O'Loughlin and Lander, 2003). The 1867 tsunami began during an earthquake (Reid and Taber, 1920; McCann, 1985), as did the 1918 tsunami (Reid and Taber, 1919). However, the 1918

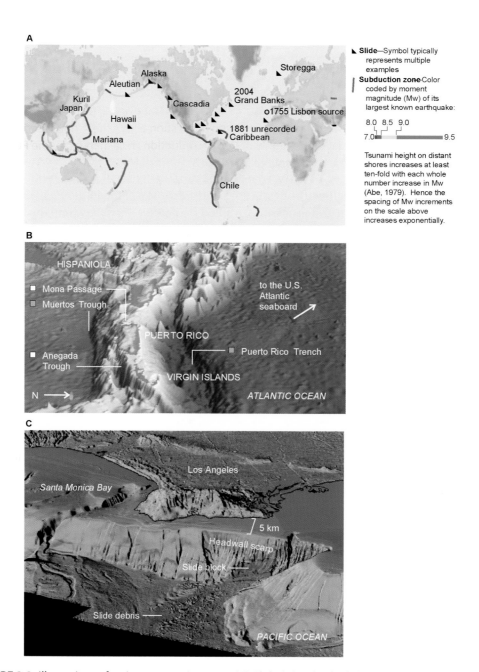

FIGURE 2.2 Illustrations of various tsunami sources. (A) Global sketch of subduction zones and landslides known or inferred to pose tsunami hazards. The blue, brown, and yellow lines do not necessarily represent maximum earthquake size, as discussed in the text and further illustrated in Appendix A. The depiction of landslides emphasizes those regarded as posing a tsunami threat to the United States and its territories. SOURCE: Committee member. (B) The offshore area of Puerto Rico. SOURCE: Image courtesy of Uri ten Brink. (C) The offshore area near Los Angeles. SOURCE: Normark et al., 2004; with permission from Elsevier.

tsunami may owe most of its size to a submarine landslide in the earthquake's focal region (López-Venegas et al., 2008). The 1867 and 1918 disasters are probably just the tip of an iceberg; they represent but a sample, during the geological instant of the past 150 years, of Caribbean tsunami sources that can be inferred from the region's active tectonics (McCann, 1985; Grindlay et al., 2005; Mercado-Irizarry and Liu, 2006) and from its abundance of steep submarine slopes (ten Brink et al., 2004, 2006). Probably the biggest open question about these many Caribbean sources is the tsunami potential of the highly oblique subduction zone marked by the Puerto Rico Trench. This hypothetical tsunami source faces the low-lying metropolis of San Juan (pop. 0.5 million) and, farther afield, may threaten the Atlantic seaboard from the Carolinas to Massachusetts (Geist and Parsons, 2009). In addition to having all these tsunami sources of its own, the Caribbean bore the brunt of the documented far-field effects of the 1755 Lisbon tsunami (Barkan et al., 2009; Muir-Wood and Mignan, 2009). The tsunami did not appear to have a significant effect on San Juan, based on the absence of documentation in the extensive Spanish-language records from that part of the 18th century (McCann et al., undated).

Near-field hazards from slides off U.S. coasts. Submarine slides abound off the Atlantic coast, particularly in the Caribbean (above) and off New England and the Middle Atlantic states (Twichell et al., 2009). Submarine slides are also present beneath the Gulf of Mexico (Trabant et al., 2001) and off southern California (Lee, 2009). The probabilistic tsunami hazard the slides pose is poorly known. It may be low because most of the sliding appears to have occurred soon after the last glaciation, at a time when sediment supply and sea levels greatly differed from today's (Lee, 2009). Much remains to be learned about slide size, speed, and duration (Locat et al., 2009), all of which affect a slide's efficiency in setting off a tsunami (Geist et al., 2009).

Tsunami sources that have escaped notice. That such sources remain undiscovered can be inferred from the recent identification of tsunami threats that had previously gone unrecognized—from great earthquakes on the Cascadia subduction zone (Atwater et al., 2005), faults and landslides beneath Puget Sound (Bucknam et al., 1992), outsize subduction earthquakes off northeast Japan (Nanayama et al., 2003), and landslides off Norway (Haflidason et al., 2005), Puerto Rico (ten Brink et al., 2006), the U.S. Atlantic coast (ten Brink, 2009), and southern California (Lee, 2009).

Determining worst-case source scenarios. Decisions about worst-case tsunami sources for the purpose of inundation modeling (see below) vary among NTHMP members. Inundation modeling in Alaska uses historical events (e.g., the 1964 Great Alaskan Tsunami) as well as a set of hypothetical tsunami scenarios unique for each local community for the tsunami sources. Inundation modeling in Hawaii is also based on historical distant tsunamis (1946 Aleutian, 1952 Kamchatka, 1957 Alaska, 1960 Chile, and 1964 Alaska tsunamis). Inundation modeling in California is based on 6 to 15 local and distant sources (depending on map location) that result in a single maximum tsunami inundation scenario. The primary subduction-type fault threat for northern California is the Cascadia scenario, and other potentially important tsunami sources include distant tsunamis (e.g., earthquakes near Alaska or Japan) and submarine landslides.

Like northern California, the Oregon and open-ocean Washington coastlines are threatened by tsunamis generated by the Cascadia subduction zone earthquakes, and the different modeling approaches taken by each state illustrate the disparities in tsunami source determinations. Inundation modeling for the open-ocean coast of Washington related to a Cascadia scenario is based on tsunami source determinations conducted over 10 years ago in Oregon (Priest, 1997). Modeling efforts for the Oregon coast related to Cascadia scenarios are based on newer methods and source determinations (González et al., 2006; Priest et al., 2009), but those in the state of Washington were not. Input received by the committee from NTHMP representatives from Oregon and Washington revealed differences in opinions on the costs and benefits of newer modeling approaches. Oregon representatives believed improvements in inundation modeling approaches are essential for hazards assessments, whereas Washington representatives consider the existing maps to be sufficient and instead focus NTHMP funds on education and other preparedness efforts. The committee cannot fully comment on which philosophy is more appropriate because there has not been adequate discussion within the NTHMP on the relative costs, benefits, and utility of higher order models compared to existing methods. Until these discussions occur, individual states will likely continue their independent approaches. In addition, not all these source determinations are subject to rigorous peer-review. One consequence of these independent efforts is that different tsunami-source determinations in two states for the same tsunami scenario (e.g., Cascadia) complicate national efforts to compare societal risk from tsunamis.

With so many unknowns, what overall strategies are likely to define the sources of tsunamis that threaten the United States the most? One strategy is to focus on sources for near-field tsunamis. This focus is suitable for Alaska, the Cascadia coast of Washington, Oregon, and northern California, and the Caribbean. It can be justified by existing simulations of tsunami inundation that show greater flooding from near-field tsunamis than from far-field tsunamis in Alaska, Oregon, and northern California (Appendix C). It can also be justified by the importance of public education about tsunami hazards as a means of saving lives from near-field tsunamis (Chapter 3). Yet another reason to focus on near-field tsunamis is their sensitivity to local properties of their sources. These include the coastwise extent, depth range, and localized concentration of slip on a fault plane, and the orientation and speed of a landslide.

A second overall strategy is to look further back in time. A tsunami hazard assessment based on U.S. written history alone would overlook nearly all the tsunami hazards from earthquakes on the Cascadia subduction zone and the Seattle fault, and from most of the landslides off southern California and the U.S. Atlantic coast. Time intervals, or recurrence intervals, between earthquakes causing catastrophic tsunamis can be as long as centuries or millennia. The 2004 Indian Ocean tsunami, for example, seemingly lacked historical precedent because its most recent predecessor occurred 550-700 years ago (Jankaew et al., 2008). The preliminary hazard assessment by Dunbar and Weaver (2008) explicitly recognizes this issue and addresses it partially by drawing on national probabilistic seismic hazard mapping that incorporates earthquake histories, inevitably incomplete, that are drawn from geological records that span centuries or even millennia.

A third and final strategy is to employ tsunami source research. The uncertainties noted above, and in Appendix A, make clear that wide-ranging research is needed to advance the understanding of tsunami sources. The challenge is how to keep abreast of, and selectively encourage, research on the myriad sources of tsunamis that threaten U.S. shores. Nearly all this research is supported outside NOAA's Tsunami Programs, through the National Science Foundation (NSF), the U.S. Geological Survey (USGS), and other organizations. The research engages physical scientists and engineers worldwide, illustrated by the research papers cited in Appendix A. Two elements of the National Earthquake Hazards Reduction Program (NEHRP) illustrate how the nation's tsunami programs might harness their efforts to help define the subset of tsunami sources that are located near U.S. shores. The first of these elements is the periodic assessment of the nation's seismic hazards. Every six years, the USGS produces, for use in the seismic provisions of building codes, a set of probabilistic National Seismic Hazard Maps (http://pubs.usgs.gov/fs/2008/3017/). These are supplemented by probabilistic seismic hazard maps of regional scope—Alaska, the Pacific Northwest, California. Updates to the earthquake maps involve a series of workshops (held in various parts of the country) that bring together scientists from government, academia, and consulting companies. The workshops provide a venue for new findings to be shared and vetted, earlier findings and assumptions to be checked, and new collaborations and research directions to be inspired. Much of this research bears directly on tsunami sources that adjoin U.S. shores. A second NEHRP element that might be adapted by the nation's tsunami programs is an external grants program. NEHRP external grants administered by the USGS support research that commonly complements the earthquake research carried out internally by the USGS. A parallel external research program through NOAA might be used, in part, to help reduce uncertainties about key tsunami sources.

> **Conclusion:** Because knowledge about the sizes and the recurrence intervals of tsunami sources are only emerging, improving the understanding of the tsunami sources is critical to producing comprehensive tsunami risk assessment. Currently, no formal procedures for periodic re-evaluation of tsunami risks exist. The NEHRP serves as an example of a successful and useful approach to periodic national assessments.

> **Recommendation:** NOAA and its NTHMP should institute a periodic assessment of the sources of tsunamis that threaten the United States, focusing mainly on earthquakes, but also on landslides and volcanoes.

These assessments could be modeled after, and made jointly with, the assessments for National Seismic Hazard Maps. Like those maps, the appraisals could be updated every six years, but with regional assessments as needed. Key unknowns emphasized through this process could then become priority areas for an external research program that NOAA could model on the external grants program of the NEHRP in the USGS.

Bathymetry and Coastal Topography Data

Because of the wavelength of tsunamis being very long relative to the depth (typically on the order of several tens to hundreds of kilometers in deep water), tsunami propagation from source area to coastlines is strongly affected by the ocean bathymetry and coastal topography over which it travels. Therefore, inundation modeling requires integrated bathymetry data over the entire ocean basin and coastal topographic data. Requirements for horizontal resolution of bathymetric data are not constant and vary depending on depth. The General Bathymetric Chart of the Oceans (GEBCO) group recommends grid spacing of no more than 1 arc-minute ($\approx$ 2 km) in a 4,000 m deep open ocean, 10 arc-second ($\approx$ 300 m) in a 100 m deep continental shelf, 3 arc-second ($\approx$ 90 m) in 10 m deep near-shore waters, and even smaller resolutions for modeling onshore run-up motions (Intergovernmental Oceanographic Commission and International Hydrographic Organization, 2005).

These recommendations for bathymetric data resolution are largely being met for the majority of the U.S. coastline. Gridded global bathymetry data with 30 arc-second ($\approx$ 1 km) resolution recently became available from GEBCO (http://www.gebco.net/), and data were generated by the combination of ship depth soundings with interpolation between sounding points guided by satellite-derived gravity data from radar altimetry (Sandwell and Smith, 2009). High-resolution digital elevation models (DEMs) for U.S. coastal regions are being developed by NOAA's NGDC (http://www.ngdc.noaa.gov/mgg/inundation/tsunami/inundation.html). Coastal bathymetric, topographic, and shoreline data are combined and integrated to yield the grid size ranging from 1/3 arc-second (~10 m) to 36 arc-seconds (~1 km) based on data compiled from various data sources (e.g., NGDC, National Ocean Service (NOS), USGS, Federal Emergency Management Agency (FEMA), U.S. Army Corps of Engineers (USACE), and other federal, state, and local government agencies, academic institutions, and private companies). These DEMs were developed specifically to support tsunami modeling efforts with NOAA's numerical code Method of Splitting Tsunami (MOST) operated at NOAA/Pacific Marine Environmental Laboratory (PMEL) but are publicly available (http://www.ngdc.noaa.gov/mgg/inundation/tsunami/inundation.html). Although efforts by GEBCO and NOAA/NGDC have accelerated the development and access of precise bathymetric data, certain regions of the United States still lack adequate data to perform inundation modeling. For example, in coastal Alaska, only 3 sets of 1/3 arc-second data are currently available, whereas 16 sets are available for the Washington, Oregon, and California coasts.

Hydrodynamic Computational Models

Hydrodynamic modeling of tsunami generation, propagation, and run-up is challenging because of its complexity and multi-scale nature. It is complex because it involves multi-phase (water, air, solid) interactions in a three-dimensional domain where some hydrodynamic fundamentals (e.g., turbulence) remain unsolved. The starting point for all inundation models is the source and how the bottom motion, in the case of tectonic displacements at the seafloor, translates to the water-surface displacement. This requires advancing the understanding of the source and its rupture mechanisms given the current uncertainties around many of the potential sources. In addition to the source parameters, there is inherent uncertainty in the models and the accuracy of topographic and bathymetric data that precludes the possibility of a completely accurate and precise tsunami inundation model. It is multi-scale because of differences in the wavelength of tsunamis (on the order of hundreds of kilometers) and the effects of their inundation (to be described at scales of a few meters or less), and as such, numerical simulations must use nested or adapted grids (e.g., large grids for the propagation across the abyssal plain, smaller grids across the continental shelf, and the smallest grids for onshore run-up motions). The current state of knowledge regarding hydrodynamic modeling at various stages of tsunami propagation is discussed in the following section.

Tsunami propagation is usually computed based on the shallow water wave theory. The theory comprises conservations of fluid volume and linear momentum with the assumptions of hydrostatic pressure field, uniform horizontal velocities over depth, and water being incompressible. The shallow water wave theory can be justified because tsunamis from a seismic source are very long and the depth of ocean is relatively shallow (on the order of 4 km). Although tsunamis contain a wide range of spectral components at the source, most of the energy is contained in the long wave components, and shorter-length (higher frequency) waves are soon left behind and disperse. For tsunami propagation in the open ocean, the nonlinearity effect may be insignificant because the wave amplitude (less than a few meters) is much smaller than the depth; hence the linear shallow water wave theory with large spatial grid size (but less than 1 minute = 2 km) may be adequate for the propagation computations (Scientific Committee on Oceanic Research, 2001). Given that the actual resolution of the seafloor data is generally much poorer than 2 km, however, the impact of potential, realistic variability (e.g., assuming bathymetry obeys a power law) should be studied.

When the tsunami reaches the continental slope, a portion of incident tsunami energy could reflect back to the ocean, depending on the abruptness of the depth change. For the tsunami that travels onto the continental shelf, the amplitude will increase due to the shoaling effect; hence, the nonlinearity effect (i.e., measured by the ratio of wave amplitude to the depth) increases. At the same time, the dispersion effect (i.e., measured by the ratio of water depth to the wave length) could become important depending on the length of the incoming tsunami and the width of the continental shelf. When the continental shelf is sufficiently wide in comparison to the tsunami wavelength, a single pulse of the incoming tsunami could be transformed to a series of shorter waves. This phenomenon is often called "fission" (Madsen and Mei, 1969) when the incident wave is specifically a solitary wave—a stable permanent-form wave in shallow water. Although co-seismically generated tsunamis do not evolve to the "exact" form of solitary waves because of the insufficient distance to evolve in any oceans (Hammack and Segur, 1978), it is anticipated that the fission-like phenomenon must take place. When the width of the continental shelf is much smaller than the incoming tsunami wavelength, the intruding tsunami does not have sufficient time to split itself into a series of shorter waves; hence the tsunami could reach shore without significant dispersion. In the former case (i.e., dispersion effects are important), the model based on the Boussinesq approximation (weakly nonlinear and weakly dispersive model) may be appropriate (Yeh et al., 1996; Liu et al., 2008), and in the latter case (i.e., nondispersion), it is appropriate to use the fully nonlinear shallow water wave theory to model tsunami propagation. Because the tsunami wavelength becomes shorter on the shelf than in the abyssal plain, and the bathymetry becomes more complex, a finer grid size is required for the simulation model in the continental shelf (Scientific Committee on Oceanic Research, 2001).

Once the tsunami commences the run-up onto the beach, hydrodynamic models need to consider and parameterize natural and human configurations (e.g., buildings, trees, mounds, roads) by assigning proper values to the friction factors. More sophisticated numerical models (e.g., based on 3-D Navier-Stokes model) may need to be implemented to assess the detailed effect of tsunami forces on structures (Yeh et al., 1996). Although modelers acknowledge the importance of natural and human configurations in influencing tsunami run-up, there is still significant uncertainty of their effect on tsunami propagation and inundation models.

More than 10 years ago, NOAA/PMEL attempted to initiate a tsunami community modeling activity to develop a functionality to coordinate and improve hydrodynamic models specifically used for tsunami prediction practice (Titov et al., 1999). The activity was intended to provide sharing of community models and databases, which would promote accelerating improvement of the models and applications. Because of lack of support this activity did not materialize at PMEL. However, similar concepts have been carried on elsewhere. The Arctic Regional Super-computing Center (ARSC) and Northwest Alliance for Computational Science & Engineering (NACSE) have developed the "Tsunami Computational Portal" (https://tsunamiportal.arsc.edu/). The portal provides a common interface for running and comparing the results of differ-ent hydrodynamic models for tsunamis and provides uniform access to the topographic and bathymetry data. Another instance is a model repository within one of NSF's EVOs (Engineering Virtual Organization, Inundation Science & Engineering Cooperative (ISEC): http://isec.nacse. org/). The source codes of hydrodynamic models and full documentation are available in the repository, and the site includes a discussion forum. There appears to be a trend toward making hydrodynamic models, once proprietary codes are developed by individuals or organizations, more transparent and more easily shared, facilitating further improvement and applications.

Conclusion: Rapid improvements in the global and regional bathymetry datasets are substantial and provide adequate data for inundation modeling efforts in most places throughout the nation. However, some regions, such as coastal Alaska, still lack bathymetric data of sufficient quality and quantity, which impedes their hazard and evacuation mapping efforts. Regarding hydrodynamic models, the accuracy and precision of run-up models are hampered by uncertainties associated with the effects of coastal bathymetry, topography, vegetation, and structures.

Recommendations: To improve tsunami inundation modeling, the NTHMP, through workshops convened with the USGS and academic institutions, should periodically review progress in hydrodynamic models, including peer-reviews of modeling codes and discussions on how to better incorporate variations in coastal bathymetry, topography, vegetation, and buildings into inundation models.

Hazard Maps

The development of tsunami hazard maps in the United States has had three distinct historical phases: (1) before the establishment of the NTHMP in 1995; (2) between the establishment of the NTHMP in 1995 and the Tsunami Warning and Education Act in 2006; and (3) after the Tsunami Warning and Education Act of 2006. Prior to the development of the NTHMP in 1995, tsunami maps were only available in Hawaii and Alaska. With the creation of the NTHMP, funding became available for tsunami inundation modeling and mapping in the five Pacific states. Efforts were implemented within each state, with the intention of building local modeling capacity, improving tsunami awareness, leveraging state funds, and creating local stakeholders that could sustain tsunami efforts through periods of low federal funding. Progress in hazard mapping efforts as of 2001 was significant but not uniform across the states (Table 2.1; González et al., 2001).

Following the 2004 Indian Ocean tsunami, demand by coastal communities for hazard and evacuation maps increased, and several NTHMP members began to produce next generation maps at much higher resolution (Barberopoulou et al., 2009; Priest et al., 2009). The most recent tabulation of completed mapping efforts provided by the NTHMP (Table 2.2) yields little detailed information and no ability to compare the current inventory to the 2001 inventory; hence, it is difficult to assess progress since 2001 with regard to the numbers of coastal communities for which hazard mapping efforts had been completed. In general, however, it appears there has been some progress (yet not uniform as was the case in 2001) in mapping efforts across the NTHMP. Many NTHMP members cite a lack of resources as one of the key challenges to making progress in mapping inundation zones. Despite an increase in demand for tsunami hazard maps in the wake of the 2004 Indian Ocean tsunami, federal support to undertake mapping efforts did not increase and, in contrast, some states actually received fewer financial resources thereafter (Figure 2.3).

The 2006 Tsunami Warning and Education Act (P.L. 109-424) represents a new phase in tsunami inundation and hazard mapping because of mandates and recommendations in the public law. In response to language in the 2006 Act and a report by the Government Accountability Office (2006) to strengthen mapping efforts, the NTHMP formed a Mapping and Modeling Subcommittee (MMS) to coordinate its efforts. Additional language in the 2006 Act calls on the NOAA Tsunami Program to "provide tsunami forecasting capability based on models and measurements, including tsunami inundation models and maps" (P.L. 109-424, Section 4(b)(4)) and for the NTHMP to "use inundation models that meet a standard of accuracy defined by the Administration (i.e., NOAA) to improve the quality and extent of inundation mapping ... in a coordinated and standardized fashion to maximize resources and the utility of data collected" (P.L. 109-424, Section 5(c)). The remainder of this sub-section discusses current efforts in relation to this new language.

Tsunami generation, propagation, and inundation are solely physical problems and do not recognize political borders. However, the committee found (contrary to the 2006 Tsunami Warning and Education Act) that the development and use of inundation modeling is not occurring in a standardized fashion across the NTHMP and that each NTHMP state member

TABLE 2.1 Status of Inundation Mapping

	Pre-NTHMP	NTHMP				
	Completed	Completed	In Progress	Planned	Remaining	Total
ALASKA						
Maps		1	1	20	[30]	52
Communities		5	13	40	89	147
Pop. At-Risk		9,608	7,192	60,247	45,103	122,150
CALIFORNIA						
Maps	2	5	1	1	[11]	18
Communities	7	42	8	8	97	155
Pop. At-Risk	28,052	857,915	37,433	89,539	963,926	1,948,813
HAWAII[a]						
Maps	66	2	4	2	[13]	21
Communities	69	9	13	4	37	63
Pop. At-Risk	348,524	66,916	133,199	6,287	176,878	383,280
OREGON						
Maps	58[b]	6	1	6	[15]	28
Communities	31	7	1	11	12	31
Pop. At-Risk	95,094	41,743	2,220	23,371	27,760	95,094
WASHINGTON						
Maps		5	1	5	[8]	19
Communities		25	3	46	30	104
Pop. At-Risk		44,383	11,056	241,297	601,500	898,236
TOTALS						
Maps	124	19	8	34	[77]	138
Communities	100	88	38	109	265	500
Pop. At-Risk	443,618	1,020,565	191,100	420,741	1,815,167	3,447,573

[a] Not listed for Hawaii are two additional maps in progress for the Hilo and Kahului area funded by NASA's Solid Earth and Natural Hazards Program.
[b] Many of these Oregon maps cover populated regions that lack Census 2000 designated communities.
SOURCE: Reprinted from González et al., 2001.

continues to independently conduct its own inundation modeling efforts. In doing so, each NTHMP state member uses its own choice of tsunami source, bathymetry, topographic data, and numerical code for the hydrodynamic simulations (Table 2.3).

Under the recently adopted NTHMP strategic plan (National Tsunami Hazard Mitigation Program, 2009a), the MMS plans to address these issues through developing of national guidelines and an approval process for inundation mapping, as well as establishing priorities of areas to be mapped. However, at the time of this committee's evaluation (2010), little

TABLE 2.2 NTHMP Inventory of State Inundation/Evacuation Products (last updated April 2009)

State/Region	Product	Communities	Comments
AK	Report and maps of tsunami inundation	3 areas completed: Kodiak and vicinity, Homer and Seldovia, Seward	Significant challenge in obtaining recent bathymetric and topographic data of sufficient quality to generate accurate high-resolution DEMs for inundation mapping
CA	Report and maps of tsunami inundation	Inundations maps (131) for emergency response covering almost all risk, populated areas within California (+100 cities/towns; 15 coastal and 5 interior bay area counties) completed at 90 m grid resolution, enhanced by 3 to 10 m DEMs on shore	Inundation maps for several remaining communities are planned. Evaluation of 10 m DEM modeling will determine if another generation of maps are needed. Additional work will include evacuation and possible creation of tsunami hazard maps for land-use planning.
HI	Report and maps of tsunami inundation	Evacuation maps from 1-D modeling for state completed; Updating 28 communities with 2-D modeling; Oahu modeling/inundation mapping complete and Hawaii in progress; Expected completion Dec 2009	Initial 725 of 1,500 km of coastlines completed with 1-D modeling; Credible worst-case scenarios used; 2-D numerical modeling of five major trans-Pacific tsunamis during the past century under way; Validation with historical run-up records; Identified 10 locations of the current evacuation maps that may need modification
OR	Report, maps, GIS projects	Detailed hazard maps for 10 communities completed; Tsunami evacuation zone maps for 23 communities completed	Distinguish between a tsunami caused by an undersea earthquake near the Oregon coast (local tsunami) and an undersea earthquake far away from the coast (distant tsunami)
Puerto Rico	Report and maps of tsunami inundation	Puerto Rico inundations maps completed; U.S. Virgin Islands pending	
WA	Report and maps of tsunami inundation		
U.S. East Coast	Working on inventory of products and methods		Recent appointment, no established tsunami mapping practice common to all region; No combined inventory available yet
Gulf of Mexico	Working on inventory of products and methods		Recent appointment, no established tsunami mapping practice common to all region; No combined inventory available yet

SOURCE: NTHMP Mapping and Modeling Subcommittee; courtesy of R. Witter and S. McLean, co-chairs.

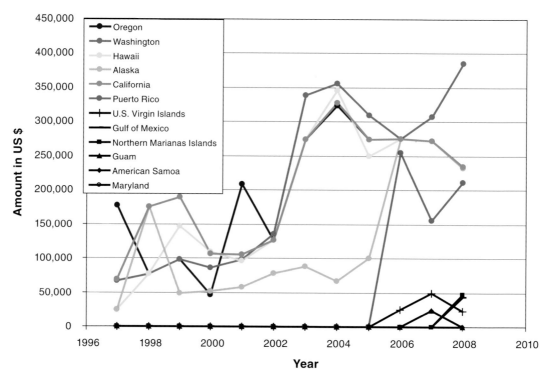

FIGURE 2.3 Federal funding for state efforts received from the NTHMP. SOURCE: Based on data from NOAA.

progress had been made. Potential tsunami sources to be examined collectively as regions, as opposed to current state-centric perspectives on segments of a source, could include (a) an Aleutian-Alaska tsunami along the Alaskan coasts, (b) a Cascadia tsunami along the Washington, Oregon, and northern California coasts, (c) an East-Caribbean tsunami for Puerto Rico, and (d) distant-source tsunamis around the Pacific Basin with special attention to Hawaii. With limited resources to devote to tsunami modeling, initial priorities could focus on sources near U.S. shores, because nearby sources pose the greater tsunami hazard, provide the least amount of time for at-risk individuals to react and evacuate, and details of slip distribution on a fault plane, or of the orientation of a landslide block, tend to have their greatest effects on tsunami heights nearby.

In addition to tsunami sources, several different numerical codes are being used by NTHMP members, including MOST, Cornell Multi-grid Coupled Tsunami Model (COMCOT), Non-hydrostatic Evolution of Ocean WAVE (NEOWAVE), University of Alaska, Fairbanks (UAF), Semi-implicit Eulerian-Langrangian Finite (SELFE), and TSUNAMI-N2. Some numerical codes are freely available in the public domain, such as COMCOT (http://ceeserver.cee.cornell.edu/pll-group/comcot.htm), for running through the Tsunami Computational Portal (https://tsunamiportal.

TABLE 2.3 Current Practice for Inundation Mapping

State/Region	Organization Conducting Inundation Modeling	Description
Alaska	Alaska Earthquake Information Center, University of Alaska	Alaskan inundation models are based on the UAF tsunami code that involves nonlinear shallow water wave theory. Historical events (e.g., the 1964 Great Alaskan Tsunami) as well as a set of hypothetical tsunami scenarios unique for each local community are used for the tsunami sources. Alaskan inundation modeling is challenging because of large tidal fluctuations (more than 7 m in some localities) and the lack of accurate bathymetry data.
Atlantic and Gulf Coasts	University of Delaware and Texas A&M University	It is unclear who is in charge of inundation modeling. Modelers at the University of Puerto Rico plan to create maps for some selected areas in the Gulf and East Coast. Modelers at the University of Delaware and Texas A&M University may also be responsible for the Gulf and East Coast, respectively.
California	University of Southern California and California Geological Survey (CGS)	Inundation maps were released in 2009 and were based on the MOST numerical code. CGS has modified the inundation maps using GIS smoothing and other techniques. Each map uses 6 to 15 local and distant sources resulting in a single maximum ("worst-case") tsunami inundation. The Cascadia subduction zone is the primary local tsunami threat for northern California, but the state lacks clear tsunami sources for the area south of Mendocino. Other potential but important tsunami sources would be distant tsunamis and those generated from submarine landslides.
Hawaii	University of Hawaii	Hawaiian inundation maps have used the COMCOT numerical code (http://ceeserver.cee.cornell.edu/pll-group/comcot.htm), which is a model based on nonlinear shallow water wave theory. Current efforts are transitioning to the non-hydrostatic model called NEOWAVE (Yamazaki et al., 2009). The maximum flow depth, maximum inundation, and maximum wave elevation are obtained from the simulations. Hawaii's inundation model is based on five historical distant tsunamis (1946 Aleutian, 1952 Kamchatka, 1957 Alaska, 1960 Chile, and 1964 Alaska tsunamis). After checking other hypothetical distant tsunamis as well as local sources from the west flank of the island of Hawaii, it was confirmed that those five historical events are the worst cases for Hawaii.

continued

TABLE 2.3 Continued

State/Region	Organization Conducting Inundation Modeling	Description
Oregon	Oregon Health and Science University	Oregon inundation maps are based on the SELFE numerical code, which is a semi-implicit finite-element Eulerian-Lagrangian algorithm to solve the shallow water wave equations using an unstructured grid. Modelers use Cascadia source scenarios that are an extension from their previous study (Priest, 1997) but different from the scenarios suggested by the USGS (Geist, 2005). The modeling utilizes information on the geological paleo-tsunami deposit data (both offshore and onshore deposits). Oregon produces two-tier inundation maps (for local and distant tsunami scenarios, e.g., see Figure 2.1), a GIS database with metadata, maximum wave elevation, maximum wave velocities, wave height time series, and 2- and 3-D animations.
Puerto Rico	University of Puerto Rico	Inundation maps were produced in 2003 using the numerical code called TSUNAMI-N2. The code is based on a finite-difference, shallow water wave model that was developed at Tohoku University, Japan. There are plans to compare results from TSUNAMI-N2 and MOST using the latest Light Detection and Ranging (LIDAR) topography/bathymetry data. Modelers here are inclined to use MOST instead of TSUNAMI-N2, although no reason for their choice has been given.
Washington	NOAA/PMEL	Washington inundation maps are based on the MOST numerical code (from NOAA/PMEL), which involves nonlinear shallow water wave theory. The MOST code selects the grid sizes and the time increments so that wave dispersion behaviors can be compensated for by the numerical dispersion inherent in the finite difference scheme. For Pacific coast sites, the Cascadia source scenarios recommended by the state of Oregon (Priest, 1997) are used for open-ocean sites and for sites along the Strait of Juan de Fuca. For Puget Sound sites, the Seattle fault deformation model that simulates the 900–930 AD event is used as a credible worst-case scenario.

SOURCE: Committee member.

arsc.edu/), while others, such as MOST, which has received substantial support from NOAA, are available to NTHMP partners upon request. Lastly, states also vary in their approach to post-processing of the inundation modeling output.

Similar to tsunami source determinations, inundation model outputs and the numerical codes that produced them have not been universally peer-reviewed or subject to quality control standards. A recent NOAA report is the first attempt at a general framework for achieving quality control of numerical codes for inundation modeling (Synolakis et al., 2007). Only in the fall of 2008 did the NTHMP MMS adopt this technical memorandum expressing its criteria for model benchmarking. The memorandum recommends that, at a minimum, inundation models should be examined by testing basic hydrodynamics and examining errors in mass conservation, and further describes 10 benchmark problems that should be used for model validation. The benchmarking comprises analytical, laboratory, and field data (e.g., comparisons to historical datasets of tsunami observations) with allowable error criteria of no more than 5 percent in mass conservation, 5 percent or 10 percent in analytical and laboratory, and 20-25 percent in the run-up accuracy for field data. NTHMP participants agreed to use the Office of Oceanic and Atmospheric Research (OAR) PMEL-135 for its benchmarking process, to share information, and to develop a process for model validation. Although this process is a good start toward achieving and maintaining quality control for different modeling practices performed by NTHMP member states, the OAR PMEL-135 has not been used consistently in the production of inundation maps among the states, and the outcomes have not been peer-reviewed. Therefore, progress is difficult to assess, and it is not clear whether current maps suffice to plan safe evacuations or prepare the public, given that they may have been produced with outdated technology. Use of benchmarked models is undoubtedly needed for improved hazard assessments and defensible evacuation and hazard maps.

Although the goal of the NTHMP MMS is to provide the guidance for accuracy and to coordinate and standardize the mapping efforts as called for by Congress, this mechanism for coordination has not resulted in more standardized maps. Coordination and standardization of modeling approaches may prove to be difficult within the current MMS structure, given the potential conflicts of interest of the state tsunami modelers that typically comprise the subcommittee.

Because of the MMS difficulty in developing consistent mapping approaches or standards, the committee discussed several options to reaching the goal of benchmarking models and meet agreed upon standards. Instead of developing a standardized approach through the MMS, the committee discussed the benefits of NOAA/PMEL providing the results of its hydrodynamic inundation modeling efforts to states. Because of the technical resources available at PMEL, such an approach might provide the necessary consistency in modeling methodology and input parameters across the states. This approach may also alleviate strains on state budgets and the lack of financial resources, which were cited by many NTHMP members as one of the key challenges to making progress in producing the next generation of inundation maps. This approach follows language in the 2006 Tsunami Warning and Education Act, which calls for the NOAA Tsunami Program to provide "tsunami inundation models and maps for use in increasing the preparedness of communities" (Section 4(b)(4)) and for NTHMP members to "use inundation models that meet a standard of accuracy defined by the Administration (NOAA) to improve the quality and extent of inundation mapping, including assessment of vulnerable inner coastal and nearshore areas" (Section 5(c)(1)).

During the NTHMP meeting of 2008, the option of PMEL providing the inundation modeling was indeed discussed, and PMEL offered the modeling output, but states did not express interest in using these results due to a lack in transparency regarding the choice of source parameters. Given the somewhat unstable history of federal support and funding, states may also be reluctant to rely on federally produced inundation efforts instead of in-house expertise that has been developed over the past decades. The committee acknowledges the foregoing states' concerns, in addition to their long-term investments in inundation model development and tsunami source research.

Alternatively, a model accreditation system might be designed to ensure greater adherence to guidelines, formal criteria and procedures, and minimum quality standards developed by the MMS (e.g., OAR PMEL-135). This model accreditation system could be administered by a body that uses peer-review or another objective and transparent process to judge whether modeling efforts meet these aforementioned guidelines, criteria, and minimum quality standards. Ideally, NTHMP would only make funding available to states that have been officially accredited. This body administering the accreditation system needs to be free of conflicts of interest.

> **Conclusion:** The development and use of inundation models is not occurring in a coordinated or standardized fashion across the NTHMP. Each NTHMP member state independently selects the tsunami source, bathymetric and topographic data, and numerical code when modeling inundation to create tsunami hazard maps. This state-based approach to tsunami inundation modeling, coupled with inadequate coordination and consensus among NTHMP modelers and no external peer-review, has created significant disparities in the methods, criteria, and judgments employed in tsunami inundation modeling and resulting tsunami hazard maps. There is also inadequate understanding and discussion on the cost, benefits, and utility of employing new higher order models over existing models. Disparities in modeling approaches and source determination also hamper the nation's ability to compare societal risk to tsunami hazards. Solving these issues may be difficult within the current MMS structure, given the potential conflicts of interest of the state tsunami modelers that comprise the subcommittee. An alternative approach might need to be identified, such as an accreditation system.

> **Recommendation:** The NTHMP should seek ways to reduce needless disparities among states and territories in their approaches to inundation modeling. Inundation from tsunami threats that transcend political boundaries (e.g., Cascadia subduction zone) should be modeled consistently across state lines, instead of the current state-centric approaches. To eliminate these unnecessary and costly differences in inundation modeling approaches, the committee recommends that inundation modeling be conducted consistently across state lines and be executed through a cooperative partnership among NOAA, the USGS, and NTHMP members, and not by the individual NTHMP members as is the current practice. The committee proposes the following strategy:

> - First, NOAA could take a lead to organize a workshop to establish community hydrodynamic models used for tsunami inundation modeling: one set for co-seismic tsunami sources and another set for landslide tsunami sources. To ensure continuing community support and improvement, the models—including full documentation and source code—should be accessible in a virtual repository that will be maintained by NOAA. The repository should host a discussion forum.

- Second, the USGS could take a lead to organize a series of workshops to determine the design tsunami sources. Tsunami sources should be determined for the regional base without state boundaries. (For example, the Cascadia scenario could be studied for (overlapped) subregions: north, central, and southern parts of the subduction zones.) The actual inundation modeling could be conducted by NTHMP members, and the outcomes of the inundation modeling must be transparent and accessible in the foregoing virtual repository.
- Alternatively, NOAA could undertake inundation modeling using a community of models; again, the outcomes would be accessible in the virtual repository. Throughout the task of inundation modeling, the committee emphasizes that NTHMP members would continue to be responsible for producing hazard maps that reflect local conditions and needs.

Evacuation Maps

Evacuation maps depict areas that may need to be evacuated in the event of a tsunami and are designed to be understood and used by at-risk individuals and by local emergency managers in their evacuation-planning efforts. The committee found it difficult to review progress made in evacuation mapping since 2006 with regard to the number of maps completed as well as the quality and the level of coordination of these efforts because of: (1) the lack of a comprehensive and continuous inventory of available evacuation maps relative to the number of at-risk communities and (2) inconsistencies in product quality. As mentioned in the previous section, the most recent tabulation of completed efforts provided by the NTHMP (Table 2.2) provides no information on the number of communities with evacuation maps relative to the number of communities with evacuation maps in 2001 or the total number of at-risk communities. Therefore, the committee cannot accurately assess progress with regard to the numbers of coastal communities for which evacuation-mapping efforts have been completed.

Although the production of evacuation maps is essential for emergency preparedness and evacuation planning, the committee observed that efforts to produce such maps vary substantially nationwide. This observation is based on input provided by the NOAA Tsunami Program and NTHMP members during the course of the committee's review, as well as information gathered via an online search for tsunami evacuation maps (see Appendix D for a list of evacuation maps). Oregon and Hawaii have tsunami inundation zones for their entire coastlines and evacuation maps for all coastal communities, and are currently updating their evacuation maps in select communities. Although most evacuation maps currently available only delineate a single tsunami scenario, the Oregon Department of Geology and Mineral Industries (DOGAMI) has recently created a new type of evacuation map that includes near- and far-field tsunami inundation scenarios (Figure 2.1). Washington lacks a statewide tsunami inundation zone but has evacuation maps for all communities on the open-ocean coast (e.g., Clallam, Jefferson, Pacific, and Grays Harbor counties) and for several Puget Sound communities (e.g., Whatcom County) that are threatened by tsunamis generated by the Cascadia subduction zone and earthquake faults in Puget Sound (e.g., the Seattle Fault, Tacoma Fault). California has recently completed new tsunami inundation mapping and is currently producing tsunami evacuation maps for all populated areas of the coastline. In other states and territories (Alaska, Puerto Rico, Guam), tsunami evacuation maps are available only for certain communities, and few are available online. No tsunami evacuation maps currently exist for communities in the Gulf of Mexico or the eastern U.S. coastline. As outlined in its strategic plan (National Tsunami Hazard Mitigation Program, 2009a), the NTHMP plans to inventory all tsunami evacuation maps by 2010 and to thereafter increase the number of maps annually by 10 percent.

The majority of evacuation maps across the United States are available as digital PDFs or printed brochures (Table 2.4). Public access to digital copies of evacuation maps is not consistent or intuitive across the states. For example, local evacuation maps are available online at the state geology department in Oregon and at a university-based seismic network organization in Puerto Rico, while in most other states, they are provided by the state emergency management or civil defense agency (Table 2.4). Hawaii was the first state to create a dynamic online map-

TABLE 2.4 Availability and Format of Tsunami Evacuation Maps in Several U.S. States and Commonwealths

State/Commonwealth	Online Location for Tsunami Evacuation Maps	Format of Available Maps
Alaska	• Alaska Department of Natural Resources	• Hazard maps for selected communities (pdf)
California	• California Emergency Management Agency	• Brochures (available online as PDF) • Online mapping application— "My Hazards"
Guam	• Guam Office of Civil Defense	• Text-based descriptions of evacuation plans
Hawaii	• Hawaii State Civil Defense	• Stand-alone evacuation maps (PDF) • Online mapping application (Google-enabled)
Oregon	• Oregon Department of Geology and Mineral Industries • Oregon Coastal Atlas	• Brochures (available online as PDF)
Puerto Rico	• Puerto Rico Seismic Network, based at the University of Puerto Rico-Mayaguez	• Brochures (available online as PDF)
Washington	• Washington State Emergency Management Division • Washington Dept. of Natural Resources	• Brochure (available online as PDF)

SOURCE: Committee member.

ping application (based on Google Maps) that allows individuals to find specific addresses and see tsunami evacuation zones relative to more identifiable landmarks (Appendix D). Because it is critical for individuals to understand the extent of tsunami evacuation zones in their communities, the committee commends the NTHMP for acknowledging a deficiency in map availability and citing the need for guidelines for the approval and distribution of maps in its 2009-2013 draft strategic plan. Efforts to centralize the availability of evacuation maps, or at the very least to develop a NTHMP portal that guides individuals to disparate state archives, will enable individuals to more easily find evacuation materials and prepare for future events.

In addition to inconsistencies in map access, tsunami evacuation maps are inconsistent with regard to formats, colors, and noted landmarks (e.g., bridges, assembly areas, hospitals, hotels to help tourists, instructions on whether to evacuate by foot or car). The formats of currently available tsunami evacuation maps vary and include community-based maps that emphasize landmarks (see Appendix D; Alaska), single tsunami evacuation zones (see Appendix D; Washington, California, Oregon, Puerto Rico), and multiple tsunami evacuation zones to differentiate between local and distant tsunamis (see Appendix D; Cannon Beach, Oregon; Figure 2.1). Recent updates to the Cannon Beach map that show two tsunami evacuation zones (one for a distant tsunami and another for a near-field event; see Figure 2.2) may be the most scientifically justified, but no studies have been conducted to determine whether at-risk individuals in this community understand these differences and would know whether to evacuate certain areas given live drills of the two scenarios. In general, there is no rigorous evaluation of how people respond to or interpret maps. Evacuation maps in Oregon, Washington, California, and Puerto Rico use yellow to denote tsunami evacuation zones, while in Hawaii the static maps found online denote the hazard zones as gray and the maps in phone books in red. Variations also exist among tsunami evacuation maps with regard to accompanying text on the map products that explains how to use the map and to prepare for future tsunamis (Table 2.5).

To date, state mapping efforts have largely relied on their own state advisory groups to guide evacuation map development. Because tsunami scenarios that form the basis for evacuation maps will vary among states (e.g., distant versus local events or worst-case versus most likely), map content will always need to be tailored to the special facilities and populations in tsunami-prone areas. However, the preparation and presentation of this information (e.g., symbols, colors) in a consistent way across the United States helps create a consistent voice in public education and is encouraged (see Chapter 3 for additional details). For example, a resident of the Oregon coast who works in a neighboring coastal town in Washington or is vacationing in Hawaii should be able to recognize and understand tsunami evacuation maps with little interpretation. State agencies have not evaluated the effectiveness of various map formats (either based on surveys and interviews or by testing their utility during evacuation drills) in promoting individuals to take protective action, and there are no NTHMP guidelines for evacuation map preparation.

Although consistency in evacuation map preparation does not currently exist, the committee commends the NTHMP for noting the need for guidelines on evacuation map preparation (including assistance to non-English speaking communities and criteria for defining evacuation routes and sites) in its 2009-2013 strategic plan (National Tsunami Hazard Mitigation Program,

TABLE 2.5 Availability of Explanatory Text with Tsunami Evacuation Maps in Various NTHMP States and Commonwealths

Text Included in Tsunami Evacuation Maps	Hawaii[a]		Oregon	California	Puerto Rico[b]	Washington
	Static Map[b]	On-line Mapping Application				
Tsunami explanation		X	X	X		X
Instructions to evacuate tsunami zone		X	X	X		X
Distant vs. local tsunamis distinction		X	X	X		X
Risk-reduction strategies (e.g., family plan, emergency kit)		X	X	X		X
Natural cues described		X	X	X		X
Evacuation signage described			X	X		X
Vehicle- vs. foot-based evacuation discussed		X	X	X		X
Where to evacuate to (e.g., assembly areas)		X	X	X		X
Directions for more information		1	X	X		X
Checklist for immediate action after warning		X	X	X		X
Post-tsunami actions to take			X	X		X
Mention of special-needs populations			X	X		

[a] Three of the four links for more information at county civil defense departments did not work. The fourth link (to Hawaii County Civil Defense) was active, but led to the static PDFs of evacuation maps for the island.
[b] For the static evacuation maps in Hawaii and in Puerto Rico, no accompanying documentation was included in the map or could be found.
SOURCE: Committee member.

2009a). The need for best practices or standards for evacuation map preparation was also noted in the 2007 NTHMP five-year review. These guidelines or best practices need to be based on an evaluation of how people process the maps and what conventions are most effective.

Conclusion: Evacuation maps are critical tools for understanding and communicating population vulnerability to tsunamis. Most at-risk communities in the Pacific states (Hawaii, Alaska, Washington, Oregon, and California) and Puerto Rico have produced evacuation maps with a single line indicating a worst-case scenario, except for Cannon Beach, Oregon, where the map includes evacuations zones for far- and near-field tsunamis. Approaches to evacuation map production vary greatly among NTHMP members (e.g., format, choice of scenarios); therefore, at-risk populations are expected to interpret different state-developed representations of tsunami risk. It is unclear whether current evacuation maps are sufficient for enabling effective evacuations or preparing the public due to the absence of uniform quality standards, evaluative metrics, or guidelines on what constitutes effective mapping approaches.

Recommendation: The NTHMP Mapping and Modeling Subcommittee should develop guidelines on evacuation map production that foster consistency in format and quality across the nation and that are based on sound cartographic principles, although map content must be tailored to the relevant facilities, populations, and characteristics of the local communities. To improve public access to evacuation maps, the NTHMP should develop a national, online repository for tsunami evacuation maps and host a consistent online mapping application for all tsunami evacuation zones across the United States. The NTHMP should annually update the inventory of evacuation maps relative to the number of at-risk communities.

VULNERABILITY ASSESSMENTS

Societal vulnerability to tsunamis refers to the physical, social, economic, and environmental conditions or processes that increase the potential for individuals or communities to incur losses or damages from future tsunamis (International Strategy for Disaster Reduction, 2004). Common elements of vulnerability within the natural hazard literature include exposure, sensitivity, and resilience (Dow, 1992; Hewitt, 1997; Cutter, 2003; Turner et al., 2003). Exposure refers to hazard proximity, sensitivity refers to differential degrees of potential harm given similar exposure (e.g., different building types), and resilience addresses the coping and adaptive capacities of an individual or community during and after an extreme event. Understanding societal vulnerability to tsunamis provides emergency managers with the required information to protect their communities and to determine whether individuals have the capacity to take protective actions.

Although this information is considered critical to reducing tsunami risk, until recently relatively little has been written about societal vulnerability to tsunamis compared to the amount of literature devoted to the physical characteristics of tsunamis (Keating, 2006; National Research Council, 2007). One possible reason is that vulnerability is not an integral part of the NTHMP and is not addressed by its subcommittees (Warning Coordination, Mapping and Modeling, Mitigation and Education). The NTHMP Mitigation and Education Subcommittee notes the importance of understanding vulnerability to tsunamis in the development of mitigation and education strategies but has not dedicated resources to doing such assessments or providing guidelines on how to do so. Efforts to describe societal vulnerability to tsunamis have come from other federal agencies (e.g., the USGS) and academic institutions. As the tsunami research community increases its efforts into assessments of population exposure and other elements of vulnerability assessment, it would benefit from leveraging ongoing efforts for other hazards in the various agencies.

Individual and community vulnerability to tsunami hazards are dynamic processes that require monitoring due to changing coastal populations, risk perceptions, and use of tsunami-prone areas. Methods of characterizing vulnerability vary depending on the intended use of the results (e.g., evacuation planning, land-use planning, infrastructure siting, and mitigation projects). Due to the committee's focus on national preparedness to tsunamis, we limit our discussion of vulnerability in this chapter to issues that relate to an individual's ability to evacuate tsunami-prone areas, including (1) population exposure and sensitivity to tsunamis and (2) evacuation potential for at-risk individuals in tsunami-prone areas. The purpose of this section is to briefly describe each element, progress in the nation's understanding of vulnerability, and areas for improvement.

Population Exposure and Sensitivity

Tsunamis pose risks only if they have the potential to impact humans or things they value. Therefore, a first step in understanding vulnerability is to inventory the number and types of individuals in tsunami hazard zones. Population exposure can be estimated for small geographic areas (e.g., a single coastal community) via building inventories in tsunami-prone areas (Morgan, 1984; Papathoma et al., 2003; Wood and Good, 2004; Dall'Osso et al., 2006) or community workshops that leverage local knowledge (Wood et al., 2002). For large geographic areas (e.g., counties, states), decadal population data gathered by national census agencies (e.g., U.S. Census Bureau) or business databases gathered by private companies can be integrated with tsunami hazard data using GIS tools to determine the number of individuals in tsunami-prone areas. Other regional approaches to estimating population exposure to tsunamis include global population models (Balk et al., 2005), and landcover data (Wood, 2009).

In addition to determining the number of individuals in tsunami-prone areas, it is important for emergency managers to assess their demographic characteristics, as these can amplify an individual's potential for losses and affect their ability to receive and understand warning messages (Mileti and Sorenson, 1990; Miller et al., 1999; Morrow, 1999; Cutter, 2003). People in tsunami-prone areas will vary in their hazard awareness, risk perception and tolerance, and ability to prepare or respond to an extreme event. If officials are to effectively motivate people to take protective measures in response to a tsunami warning, then they need to understand who they are trying to motivate and their capacity to respond. Assessing the types of people in tsunami-prone areas helps officials determine placement of warning signage and technology (e.g., sirens), tailor the format and delivery of education efforts to reach different populations (see Chapter 3 for additional discussion on education), and identify those who may need special assistance during an evacuation (e.g., elderly populations).

Characterizing the demographic attributes of individuals in tsunami-prone areas does not imply that all individuals of a certain demographic group will exhibit identical behavior during or after a tsunami since multiple demographic characteristics of an individual or neighborhood interact and likely amplify each other. Variations in local cultures and situations, as well as in individual and community resilience, will influence the extent of these demographic sensitivities. Individuals have multiple demographic characteristics (e.g., age, gender, economic status) and the interaction of factors may heighten or reduce their sensitivity to tsunamis (Wood et al., 2010). The importance of demographic sensitivities will also be influenced by characteristics of the hazard; for example, a high fraction of elderly people in a tsunami-prone area will be a larger issue if the likely warning time before inundation is 30 minutes (e.g., near-field tsunami) compared to several hours (e.g., far-field tsunami). With these caveats in mind, the following demographic groups may have higher sensitivity to tsunami hazards:

- the very young and the very old (Balaban, 2006; McGuire et al., 2007);
- households of racial and ethnic minorities becuase of historical societal inequalities (Laska and Morrow, 2006) and to potential exclusion from disaster preparedness efforts (Morrow, 1999);
- renters, who are less likely than homeowners to prepare for catastrophic events and have more limited exposure to hazard information (Morrow, 1999; Burby et al., 2003);
- individuals with pre-existing socioeconomic issues (e.g., homeless, living in poverty, low literacy levels, inability to speak the primary language of an area) that may inhibit their ability to prepare for future events (Wisner et al., 2004);
- individuals at hospitals, psychiatric facilities, adult residential care centers, daycare centers, schools, and correctional facilities that may have difficulty (e.g., hospital patient) or be incapable (e.g., correctional-facility inmate) of evacuating on their own and will require external assistance to evacuate; and
- employees or tourists who may have low or no exposure to awareness efforts or evacuation drills (Wood and Good, 2004; Johnston et al., 2007).

The committee found that by 2010, reports attempting to inventory the number and types of people in tsunami hazard zones in coastal communities have been completed for the tsunami-prone areas of Hawaii (Wood et al., 2007), Oregon (Wood, 2007), and the open-ocean coast of Washington (Wood and Soulard, 2008). In each of these reports, census block data from the 2000 U.S. Census and national business data were merged with tsunami hazard data to identify, by community, the number and types of residents, employees, and facilities that attract tourists and house special-needs populations in tsunami-prone areas. As a first approximation of economic exposure, these reports also include inventories of the amount and percentage of tax-parcel values, employee distributions, and business sales volume in tsunami hazard zones of each community and county. In reviewing the tsunami and risk/vulnerability assessment sections of several FEMA-approved state mitigation plans, the committee found little information related to the diversity of populations in tsunami-prone areas of the various coastal states. Some state plans estimate the number of individuals in tsunami hazard zones of each county, some simply list the communities with populations with tsunami risk, and some only discuss general tsunami-related population issues. Many plans, however, did include detailed information on financial, structural, and critical facility exposure, suggesting that plans are written to help document potential long-term economic impacts and not preparedness issues to save lives.

Conclusion: There is no national assessment of population exposure and sensitivity to tsunamis, including the number and types of individuals in tsunami hazard zones. The NTHMP's current subcommittee structure (Mapping and Modeling, Warning Coordination, and Mitigation and Education) does not coordinate mapping the exposure of at-risk individuals and communities and therefore does not provide leadership on this topic. The absence of this information and the leadership to attain it impacts national preparedness to tsunamis in several ways. First, the NTHMP will be unable to reach its stated goal of a national tsunami risk assessment (see earlier conclusion in this chapter on this topic). Second, efforts to develop realistic evacuation plans are compromised in communities threatened by far-field tsunamis if the magnitude of the at-risk population is not known. Third, communities threatened by near-field tsunamis cannot develop public awareness and education efforts that are tailored to local conditions and needs (discussed in the following chapter) without an understanding of the types of at-risk individuals.

Recommendation: The NTHMP should collaborate with state and federal agencies (e.g., the USGS, the Census Bureau) to periodically inventory the number and types of people in tsunami hazard zones at intervals no less frequently than that of the U.S. Census, with special attention to children, the infirm, tourists, and other groups whose heightened sensitivity to tsunamis could constrain their ability to prepare for and evacuate from future tsunamis. The NTHMP should expand Mapping and Modeling Subcommittee efforts to explicitly include community vulnerability. The NTHMP should establish a Science Advisory Committee to help develop guidelines on consistent approaches for identifying and mapping populations in tsunami-prone areas. The NTHMP should also provide guidelines on how to use this information to tailor evacuation planning and education efforts.

Evacuation Potential

In addition to population exposure and sensitivity, an individual's capacity to learn from past disasters, implement risk reduction measures, adapt during an event, and persevere after an event is another significant factor in understanding societal vulnerability to tsunamis (National Science and Technology Council, 2005). These resilience factors influence the ability of an individual to prepare for future tsunamis and to take self-protective measures when a tsunami occurs. This sub-section on evacuation potential focuses on assessing the ability of individuals to evacuate tsunami-prone areas based on physical characteristics of the tsunami hazard zone (e.g., distance to higher ground, integrity of egress routes, island with no high ground), while the role of perceptions, knowledge, and preparedness levels in increasing resilience is discussed in the following chapter on education, preparedness, and evacuation coordination.

Evacuation modeling estimates the amount of time necessary for people to reach safe havens from various locations in tsunami hazard zones (Post et al., 2008), which is especially critical for U.S. coastal communities that are threatened by tsunamis reaching shores in an hour or less. Such modeling efforts have been used to study the influence of congestion due to crowds and road bottlenecks on the ability of individuals to evacuate tsunami hazard zones (e.g., Lammel et al., 2008); the effectiveness of official routes in managing the typical number of people who will need to evacuate (Ismail et al., 2008); the likelihood of casualties (Koshimura et al., 2006); or the need for vertical evacuation structures (e.g., buildings, engineered berms) in places where time is not available to reach naturally occurring higher ground (Yeh et al., 2005; Federal Emergency Management Agency, 2008). Such analysis can answer fundamental questions that might ultimately determine the survival of people: Under what circumstances is it most effective to evacuate on foot instead of car due to congestion or earthquake-damaged infrastructure? Under what circumstances is it best to use vertical evacuation (e.g., tall buildings, engineered berms) because ground-based evacuations may not be practical in communities that must evacuate thousands of people from a large tsunami-prone area?

Although there is basic research focused on tsunami evacuation modeling, the committee found little applied evacuation modeling research (e.g., Yeh et al., 2009) to examine specific U.S. coastal communities that may only have minutes to an hour to evacuate thousands of individuals from tsunami-prone areas (e.g., Seaside, Oregon; Ocean Shores, Washington). In areas where ground-based evacuations may not be feasible (due to short times before inundation and substantial distances to higher ground) and where there are no existing structures or features capable of serving as a vertical refuge, workshops are being held in coastal communities (e.g., Cannon Beach, Oregon, in September, 2009; Long Beach, Washington, in January 2010) to further discuss the opportunities and constraints of vertical evacuation structures for tsunamis (e.g., buildings, engineered berms). Although engineering guidelines have been published and officials in some coastal communities are expressing interest in new structures (Federal Emergency Management Agency, 2008), the committee found no case studies that delve into the social and economic aspects of vertical evacuation structures in at-risk communities. In addition, because there is very limited information about current velocities as the tsunami interacts with the built environment (and currently no measurements are being taken), it is difficult to estimate the forces involved with a tsunami flow field and to assess what structures might remain intact during a tsunami. Before communities commit significant time and funds to these structures, communities need a careful determination of the feasibility and issues related to near-field tsunami evacuations.

To date, no structures have been built in the United States specifically to serve as vertical evacuation sites for tsunamis. In communities with high-rise buildings in low-lying areas (e.g., Honolulu, Los Angeles), vertical evacuation into existing buildings to avoid inundation from far-field tsunamis is promoted by local emergency managers. In communities threatened by near-field tsunamis, officials should examine the structural integrity of buildings that may be used for vertical evacuation. Earthquakes that precede the tsunami may make certain building types (e.g., unreinforced masonry) unsafe for entry or for tsunami refuge and the subsequent tsunami waves may overtop or destroy wood-based buildings that survive the initial earthquake.

Strong ground motions, ground failure, and land subsidence from earthquakes that precede near-field tsunamis may also damage key egress routes, bridges, and critical facilities in coastal communities, thereby putting additional constraints on an individual's ability to evacuate a tsunami-prone area. Communities in Alaska, the Pacific Northwest, and Puerto Rico are likely to experience several minutes of strong ground motions with tsunami inundation arriving only minutes later. An initial large earthquake would likely result in damage to critical infrastructure in the evacuation zone (e.g., roads, bridges) and create barriers for individuals trying to evacuate from an imminent tsunami (e.g., toppled power lines, building debris in roads, etc.). Initial observations of the Chilean earthquake in February 2010 indicate this was the case in many coastal communities. In addition, critical facilities, such as emergency management offices, police stations, and fire stations, could be destroyed by the original earthquake or blocked by earthquake-related debris, possibly leaving emergency responders unable to manage local evacuations. Radio and television stations and the towers that transmit their signals could also be damaged, thereby limiting the dissemination of warning or all-clear messages. The number of critical facilities in tsunami-prone areas has been documented in several studies (e.g., Charland and Priest, 1995; Lewis, 2007; Wood, 2007) and hazard mitigation plans (e.g., State of Hawaii Multi-Hazard Mitigation Plan). However, these efforts are simple inventories that do not delve into evacuation and response consequences of earthquake-damaged infrastructure and facilities or whether there are redundant facilities and access routes.

Conclusion: Although many communities in the United States are threatened by a tsunami that originates from a source at close or intermediate distance, few evacuation studies have been conducted to evaluate the ability of at-risk individuals to reach higher ground before tsunami waves arrive. A related problem is that there have been no studies to assess the potential impact of local earthquakes that generate near-field tsunamis on egress routes, supporting infrastructure (e.g., bridges), or facilities considered critical in response efforts. Without such information, emergency managers are not able to identify where targeted outreach is needed and where potential vertical evacuation structures (e.g., buildings, engineered berms) may be warranted.

Recommendation: For all communities with close or intermediate proximity (i.e., arrival times ranging from minutes to about an hour) to a potential tsunami source, the NTHMP should conduct evacuation modeling studies to assess the likelihood of successful horizontal evacuations. These studies should include the potential impacts of preceding earthquakes on key egress routes and consideration of any special-needs populations in tsunami-prone areas. In communities where the time required for at-risk individuals to reach higher ground is likely greater than predicted tsunami wave arrival times, the NTHMP should conduct feasibility and effectiveness studies of various vertical evacuation strategies (e.g., buildings, engineered berms) that include engineering considerations and social and economic constraints of at-risk communities.

CHAPTER THREE

Education and Preparedness of Individuals, Communities, and Decision Makers

SUMMARY

This chapter reviews progress in education and emergency management in preparation for future tsunamis. Effective education and emergency management have been credited with saving thousands of lives in recent tsunamis elsewhere and can also save lives in future tsunamis that strike U.S. communities. Ultimately, the ability to survive a tsunami hinges on at-risk individuals having the knowledge and ability to make correct decisions and act quickly. For local tsunamis, waves will arrive within minutes after generation, and at-risk individuals need to understand that natural cues (prolonged ground shaking and shoreline draw down) may be their only warning. Local officials will not be capable of assisting them in the initial moments or even potentially for days, so individuals need to know how to respond with no official guidance. The knowledge and readiness they acquire through pre-event education could save their lives. For distant tsunamis, waves will arrive several hours after generation and individuals need to understand where official warnings may come from, how they may receive the warnings, what those warnings might say, and what they need to do in response to those warnings.

Although much has been done to educate at-risk individuals, prepare communities, develop and deliver warning messages, and coordinate agency procedures, the committee concludes that these efforts could be more effective with improved coordination, baseline assessments of the target audience, evaluations of effectiveness, transfer of best practices among the National Tsunami Hazard Mitigation Program (NTHMP) members, and use of evidence-based[1] approaches in the social and behavioral sciences of education, warning messaging, and emergency management. The committee commends the intent of the federally administered TsunamiReady program to coordinate community preparedness efforts but finds major gaps between stated program goals and current accomplishments. The recommendations listed here in summary form include:

---

[1] A program is judged to be evidence-based if (a) evaluation research shows that the program produces the expected positive results; (b) the results can be attributed to the program itself, rather than to other extraneous factors or events; (c) the evaluation is peer-reviewed by experts in the field; and (d) the program is "endorsed" by a federal agency or respected research organization and included in its list of effective programs (Cooney et al., 2007).

- Systematic and coordinated perception and preparedness studies of communities with near-field tsunami sources.
- Consistent education among NTHMP members using evidence-based approaches in the social and behavioral sciences that is evaluated and archived.
- A TsunamiReady Program that is based on professional and modern emergency management standards.
- A review of the format, content, and style of tsunami warning center (TWC) warning messages, and how dispatchers and emergency personnel understand the messages.
- The consolidation of the two TWC messages.
- Formal attention and planning given to outreach efforts at the TWCs.
- Strong local/state working groups that share best practices and lessons learned.
- Guidelines on the design and an inventory of tsunami-related exercises.

INTRODUCTION

Tsunamis are natural events that threaten coastal communities. Effective public education and emergency management can prepare individuals and reduce the likelihood of fatalities when tsunamis occur. Education is credited for saving thousands of lives during the 2004 Indian Ocean tsunami, the 2009 Samoan tsunami, and the 2010 Chilean tsunami (Box 3.1), and education will save lives in future tsunamis that strike U.S. communities. Ultimately, the ability to survive a tsunami hinges on at-risk individuals having the knowledge and ability to make correct decisions and act quickly. For local tsunamis, waves will arrive in minutes after generation and at-risk individuals need to understand that natural cues (e.g., prolonged ground shaking, shoreline draw down) may be their only warning, that local officials will not be capable of assisting them, and that the knowledge and readiness they acquire through pre-event education could save their lives. For tsunamis generated at greater distance from coastal communities, the ground shaking might be too weak to alert residents of the imminent danger, but waves may arrive anywhere from an hour to many hours after generation. In these instances, individuals need to understand where official warnings may come from, how they may receive the warnings, what those warnings might say, and what they need to do in response to those warnings.

Regardless of tsunami sources, integrated public education and preparedness planning provide the context in which individuals will perceive, process, and react to future warnings. Education and planning are long-term, ongoing efforts that strive to make tsunami knowledge and preparedness commonplace and ingrained into local culture and folk wisdom. Enculturation requires a major commitment and diverse efforts to achieve this goal; however, once accomplished, it can perpetuate itself. This chapter discusses four areas in which targeted education-related efforts can increase the likelihood that people will be able to evacuate before tsunamis arrive and that agencies will be able to execute effective evacuations, such as:

- Educating at-risk individuals in advance about what they need to know to prepare for and respond to tsunamis;

- Preparing communities for future tsunamis;
- Developing and delivering effective warning messages; and
- Improving interagency coordination, as well as coordination among all segments of the community (public and private), in preparing for and responding to tsunamis.

EDUCATION OF AT-RISK INDIVIDUALS

Tsunami education in U.S. coastal communities is a major challenge because it requires reaching hundreds of coastal communities that contain hundreds of thousands of residents, employees, and tourists. There are 29 NTHMP partner states, territories, and commonwealths, each a sovereign entity with sub-jurisdictions (e.g., counties, cities) that have individual needs, priorities, and resources for tsunami education. Tsunami education needs to also adequately convey the different tsunami threats and proper responses to each—local tsunamis that require instantaneous, self-protective action to reach higher ground based on the recognition of natural cues and distant tsunamis that involve orderly evacuations over several hours that are managed by officials and informed by the tsunami warning centers.

The NTHMP Mitigation and Education Subcommittee (M&ES) is tasked with assessing tsunami mitigation and education needs for the nation, addressing these needs through targeted products and activities, and then sharing these products with other at-risk coastal states, territories, and commonwealths. An NTHMP-approved strategic implementation plan for tsunami mitigation projects (Dengler, 1998, 2005) identifies education as a critical element in mitigation and states that effective education projects define the audience and their needs, assess existing materials, and define a strategy for sustained support. This plan also discusses the need for a resource center to provide information exchange and coordination. With guidance from the M&ES, NTHMP members develop their individual education projects to support the goals and objectives of the subcommittee and often collaborate on regional products that address common issues between members. This section provides an overview of the factors that influence the effectiveness of education and reviews progress in NTHMP education efforts. Conclusions and recommendations in this section center on the need to assess the needs and knowledge of the at-risk audience and on making NTHMP education efforts more coordinated, consistent, and subsequently, more effective.

Factors That Increase the Effectiveness of Education

A rich research base has been developed to address the question of how to enhance what the public knows and to motivate them to take actions to prepare for future hazards (Mileti and Fitzpatrick, 1992; Mileti et al., 1992; National Research Council, 2006). Based on the current literature, the committee highlights 10 practical steps to increase public knowledge of and readiness for tsunamis (Box 3.2). Effective public education on hazards has been found to correlate with many factors: dissemination content and channels, social and physical cues, the status and role of the recipient, past experience with hazard(s), beliefs about the informa-

BOX 3.1
Cautionary Tales and Education Saves Lives from Tsunamis

Traditional knowledge saves lives in Aceh, Indonesia, during the 2004 Indian Ocean tsunami:

Some 78,000 people were living on Simeulue Island, off the west coast of Aceh, Indonesia, at the time of the 2004 Indian Ocean tsunami. Most lived along the coast in villages the tsunami would strike. The tsunami began coming ashore as soon as eight minutes after the shaking stopped and too soon for official warnings. Although hundreds of thousands of lives were lost elsewhere, only seven people on Simeulue died. What saved thousands of lives was knowledge of when to run to higher ground. This knowledge had been passed down within families over the years by repeating tales of smong—a local term that entails earthquake shaking, the withdrawal of the sea beyond the usual low tide, and rising water that runs inland. Smong can be traced to a tsunami in 1907 said to have taken thousands of Simeulue lives and reminders of that event reinforced the story, such as victims' graves, a religious leader's grave untouched by the tsunami, and coral boulders in rice paddies. After any felt earthquake, a family member would mention the smong of 1907 and often concluded with this kind of lesson: "If the ground rumbles and if the sea withdraws soon after, run to the hills before the sea rushes ashore." By contrast on mainland Aceh, where education had suffered from years of military conflict, only a tiny fraction of the population used the giant 2004 earthquake as a tsunami warning. After the initial earthquake, many people gathered outdoors, fearing further damage from aftershocks. Most missed their opportunity to evacuate—a time window of 20 minutes on western mainland shores and 45 minutes in downtown Banda Aceh.[1]

Elementary education from afar saves lives in Phuket, Thailand, during the 2004 Indian Ocean tsunami:

More than 100 tourists and locals on Maikhao Beach in Phuket, Thailand, were saved when a 10-year-old girl from England persuaded them to evacuate to higher ground after the initial earthquake. While other tourists watched as the tide rushed out and boats in the distance bobbed up and down, Tilly Smith, who was in Phuket on holiday with her parents and younger sister, recognized these as natural cues of an imminent tsunami. Just two weeks earlier, Tilly had studied tsunamis in her prep-

---

1 Adapted from McAdoo et al., 2006. Mainland tsunami arrival times from Lavigne et al., 2009.

tion, perceived risk, perceived effectiveness of actions, and warning confirmation (Mileti and Sorenson, 1990). Recent work suggests that education effectiveness primarily depends on the quality and quantity of educational materials received by the public and the physical and social cues observed. The other factors (e.g., status, roles, experience) play a role when information is of low quality and of insufficient quantity (Linda Bourque, UCLA, personal communication). Each of the factors is briefly described below.

Information dissemination. The effectiveness of education is increased when verbal and written information is frequently disseminated from multiple sources over multiple communication

school geography class in Surrey, England, and quickly realized everyone was in danger. She convinced her parents that everyone needed to evacuate, who then alerted other tourists and hotel staff, and people quickly evacuated. The waves started to flood the area a few minutes later, but no one on the beach was killed or seriously injured (The Daily Telegraph, 2005).

School and community education saves lives in American Samoa during the September 2009 tsunami:
    The tsunami of September 29, 2009, took 34 lives in American Samoa but could have taken far more in the absence of tsunami education. September had been emergency preparedness month and tsunami education efforts, supported by the TsunamiReady program, included videos of the 2004 Indian Ocean tsunami and school tsunami evacuation practices. Long-term education efforts of the American Samoa Department of Homeland Security, in collaboration with Department of Public Works and National Weather Service Pago Pago, included school evacuation plans and awareness campaigns for agencies, schools, and businesses (Laura Kong, International Tsunami Information Center, written communication). After the initial earthquake ended, schools and community members knew to evacuate, and many did (Laura Kong, International Tsunami Information Center, written communication). In the community of Amenave, the mayor credited an earlier workshop for village mayors on tsunami hazards for his ability to recognize and then personally warn with a bullhorn his constituents of the potential for a tsunami after the earthquake (Earthquake Engineering Research Institute, 2010).

Signage and other education products save lives in Chile during the February 2010 tsunami:
    Initial observations of post-tsunami survey teams suggest tsunami-savvy residents knew to use the parent earthquake as a natural warning to run to high ground. Several towns had posted tsunami hazard and/or evacuation-zone signage, some communities had practiced drills, and others had held preparedness workshops. Some survivors cite their memory of the Valdivia earthquake in 1960, while others cited various books, television, documentaries, and other media information as the source of their awareness (Lori Dengler, Humboldt State University, written communication).

channels with consistent information regarding what recipients need to know and about actions that they should take (Mileti and Fitzpatrick, 1992; Linda Bourque, UCLA, personal communication).

Physical and social cues. Observing cues—when consistent with the verbal and written information that is being disseminated—can reinforce learning. Physical cues that reinforce knowledge include tsunami evacuation route signage and NTHMP-related household products (e.g., coffee mugs, refrigerator magnets); and social cues include preparedness drills and community workshops (Wood et al., 2002; Connor, 2005; Alexandra et al., 2009).

BOX 3.2
Practical Steps to Grow Public Knowledge and Readiness

The following are recommendations for maximizing the effectiveness of tsunami public education, based on social science evidence (Mileti and Sorenson, 1990; Linda Bourque, UCLA, personal communication) and lessons learned from tsunami education efforts in Hawaii (Alexandra et al., 2009) and Oregon (Connor, 2005):

(1) Use evidence-based approaches.
(2) Brand the message and work with other information providers to eliminate inconsistent messages.
(3) Use multiple sources, forms, dissemination channels, and settings because the public will be more likely to prepare if they receive the same information multiple ways and times.
(4) Focus the messages on what the public should do, how their actions can reduce their risk, and where to seek additional information instead of only focusing on convincing people that they are at risk.
(5) Customize education by identifying levels of knowledge of and preparedness for the hazard, and the special needs of the intended audience (e.g., language translation), and by incorporating personal stories of tsunami survivors to provide context.
(6) Encourage people to talk about readiness with each other and to practice protective actions, because this dialog results in people owning ideas about what to do to get ready and builds community capacity, which greatly facilitates taking action.
(7) Sustain education efforts because effective education is an ongoing process.
(8) Position physical and social cues around the community because people copy each other's behavior.
(9) Designate a lead entity for the public education program, as multiple parties with different priorities will have difficulty providing standardized, consistent messages delivered through multiple channels.
(10) Evaluate efforts by measuring the baseline of public awareness and preparedness and subsequent changes to determine program effectiveness and to revise efforts.

Statuses and roles. Factors that correlate with public hazard education effectiveness relate to status (e.g., having higher income, education, and occupational prestige, not being either young or old, being white, being female, and being native born) and roles (e.g., being in a partnership relationship, belonging to a larger family, and being responsible for children). A demographic analysis of at-risk population composition and distribution is a first step in developing targeted education for demographic sub-groups where education is not as effective (e.g., the very young, low-income families, foreign-born).

Experience. People are more inclined to be educated about and/or prepare for hazards that they have experienced. In communities where there haven't been recent tsunamis to give individuals any personal experiences with tsunamis, community memory of past events can be sustained through oral histories of tsunami preparedness passed down through the generations (McMillan and Hutchinson, 2002; Box 3.1), disaster memorials (Iemura et al., 2008; Nakaseko et al., 2008), and survivor stories from recent tsunamis, such as the growing archive of survivor stories at the Pacific Tsunami Museum in Hilo, Hawaii (Dudley, 1999). Tsunami survivor stories and oral histories not only build hazard awareness but also increase the perception that tsunamis are survivable if certain actions are taken (Paton et al., 2008). Although experience can increase the likelihood that people prepare, personal experience also biases people to interpret educational information in the context of their own experience, which can either support or contradict their notion of the risk's reality and severity. Prevalent myths and misunderstandings need to be addressed in education efforts because existing misperceptions may serve as obstacles and prevent people from hearing and correctly interpreting information (Connor, 2005; Alexandra et al., 2009).

Perceived risk and action effectiveness. At-risk populations have their own perceptions of risk which rarely match the calculations described by experts. Perceiving increased probabilities for events did not increase public readiness action-taking (Kano et al., 2008). Instead, an intentions-to-prepare model suggests people are more inclined to act on hazard education information when they believe their present actions can mitigate their future losses (Paton et al., 2008). Education efforts that dwell only on the uncontrollable aspects of tsunami hazards, specifically event probabilities, do not influence public action. Instead, risk awareness should be framed to include information on uncontrollable tsunami hazards and controllable individual consequences if a tsunami occurs, where individual actions can reduce these consequences. An example of this is information included on tsunami evacuation maps (e.g., maps in Oregon, Washington, and California) on how to prepare for tsunamis, develop emergency kits, and evacuate to safe areas if individuals recognize natural cues or receive an official warning.

Warning confirmation process. This process refers to individuals talking about educational topics with others, seeking more information from other sources and places on their own, and then making their own decisions about what they will think, do, and not do prior to taking any action (Quarantelli, 1984; Mileti, 1995). It is part of understanding how individuals convert information into actions (Quarantelli, 1984). Effective education incorporates activities that encourage people to talk about getting ready with each other, such as discussion groups during workshops (e.g., Wood et al., 2002; Connor, 2005; Alexandra et al., 2009).

Understanding the Local Risk Conditions and the Target Audience

Effective public education for tsunamis begins with an understanding of the risks that tsunamis pose to coastal communities (see Chapter 2) and of the existing knowledge and beliefs of the target audience. For example, all at-risk communities would benefit from evacuation signage and educational programs regardless of tsunami source. Education to prepare individuals for far-field tsunamis would emphasize official warnings disseminated by tsunami warning centers and organized evacuations managed by local officials, whereas those for near-field tsunamis would instead emphasize the public's ability to recognize natural cues and take timely protective actions for their own survival. Distinctions between warnings for near- and far-field tsunamis are important to convey to at-risk populations, because the public is often confused by differences between the two and this confusion can create false expectations (Connor, 2005).

The format and dissemination of education products also vary based on the intended audience. As discussed in Chapter 2, the demographics of the audience, such as age, income, or educational background, influence the ability of an individual to anticipate and react to a natural hazard (Wisner et al., 2004) and therefore are important considerations when designing evacuation signs and public education efforts. An education campaign designed for residents capitalizes on their familiarity with their surroundings, emphasizes household preparedness strategies, and could be delivered through existing social networks. An education campaign designed for tourists focuses on easily identifiable landmarks, assumes individuals would have no local friends or relatives to assist them in an evacuation, and would be delivered by employees in the tourist industry and through posted information on road-side signage, along coastlines, and in commercial establishments. The challenge of having employees serve as tsunami educators was made clear in a recent survey of hotel employees along the southwest Washington coast that indicated only 22 percent of interviewees said they had been trained about how to respond to tsunamis and had tsunami-related information available for guests (Johnston et al., 2007). However challenging, educating tourists and the businesses that serve them is critical—initial observations from the February 2010 Chilean tsunami suggest that tourists, specifically campers on an island campground, represented a significant percentage of the fatalities (Lori Dengler, Humboldt State University, written communication).

In addition to taking the local risk conditions into account, effective tsunami education is built upon an understanding of what the target audience already knows and believes. Building this knowledge requires conducting routine assessments (such as Dengler et al., 2008) of the at-risk population's perception, knowledge, and capacity to respond, which provides officials with a baseline for measuring progress in awareness and preparedness. It is also useful in evaluating an educational program's effectiveness, highlighting areas for improvement, and guiding officials in their evacuation planning. Case studies suggest that segments of coastal communities are aware of tsunami hazards, but may have difficulty evacuating if an event were to occur (Gregg et al., 2004, 2007; Johnston et al., 2005, 2007). A survey in Oregon and Washington revealed that although public officials and coastal business owners consider near-field tsunamis related to Cascadia subduction zone earthquakes to be significant threats, they had done little to make their own organization or office less vulnerable to these hazards (Wood and Good, 2005). Other studies confirm that current dissemination activities increase awareness but are inadequate to translate into increased preparedness or appropriate evacuation actions (Johnston et al., 2005, 2007; Gregg et al., 2007). Baseline measurements and post-outreach assessments documented positive changes in tsunami knowledge and preparedness of at-risk populations after a series of recent tsunami outreach efforts in Seaside, Oregon (Box 3.3; Connor, 2005).

Knowledge assessments of the at-risk population can also be used for determining the effectiveness of warning systems. For example, a survey of 956 individuals from across Hawaii found that 59 percent of respondents did not understand the meaning of the tsunami-alert sirens, even though 69 percent of respondents also said that some sort of official warning would be their signal to evacuate from a tsunami (Gregg et al., 2007). Similar confusion of what sirens signify has been expressed during educational workshops in Hawaii (Alexandra et al., 2009). Surveys of Hilo, Hawaii, residents who survived the 1960 tsunami indicate that only 40 percent of people who heard warning sirens evacuated, whereas many people waited for additional information from other information sources (e.g., television, relatives) before evacuating (Bonk et al., 1960; Lachman et al., 1961). A survey of 2004 Indian Ocean tsunami survivors in Padang, West Sumatra, indicates that the majority of people received information through social networks and not through official channels (Birkman et al., 2008).

These isolated case studies highlight the need for additional perception, knowledge, and preparedness surveys of at-risk populations to assist in developing and implementing effective education efforts, particularly in communities that are threatened by near-field tsunamis because of the lack of adequate warning time. The committee commends the NTHMP for citing the need for evaluations and surveys to determine the effectiveness of tsunami education products and the level of preparedness of at-risk populations in its draft 2009-2013 strategic plan. The committee encourages the NTHMP to focus future preparedness assessments on communities threatened by near-field tsunamis, where successful evacuations will be more the result of a well-informed population taking self-protective actions and less from official response procedures.

Conclusion: For far-field tsunamis, successful evacuations will depend on at-risk individuals understanding official warnings and following instructions given by local agencies. For near-field tsunamis, successful evacuations will depend on the ability of at-risk individuals to recognize natural cues and to take self-protective action. The committee concludes that previous knowledge gained through sustained education efforts will likely play a larger role in saving lives from near-field tsunamis than warnings issued by the tsunami warning centers, given the current scientific and technological constraints on issuing warnings fast enough. Regardless of the kind of tsunami, understanding the needs and abilities of at-risk populations is a critical element in developing effective education. Although numerous isolated studies have been conducted in coastal communities, the NTHMP has not systematically assessed the perception, knowledge, and levels of preparedness of at-risk individuals. Lacking this information, the NTHMP has limited baseline information from which to gauge the effectiveness of education efforts, to tailor future efforts to local needs, or to prioritize limited funds.

Recommendation: Faced with limited resources, the NTHMP should give priority to systematic, coordinated perception and preparedness studies of communities with near-

BOX 3.3

Developing and Evaluating Tsunami Public Education: An Example from Seaside, Oregon

(A)   (B)

Small-group discussions with (a) Seaside Tsunami Outreach Coordinator (in black) and middle-school students and (b) adults at a public workshop on tsunami preparedness in Seaside, Oregon. SOURCE: Connor, 2005; image courtesy of DOGAMI.

Educating at-risk populations on how to prepare for future tsunamis and to react properly during an event is challenging for local officials because of the dynamic mix of residents, employees, and tourists in tsunami hazard zones. In addition, it is difficult to assess whether awareness campaigns and educational efforts have any effect on changing the tsunami knowledge and preparedness of at-risk populations.

To address these challenges, the City of Seaside, Oregon, partnered with the Oregon Department of Geology and Mineral Industries (DOGAMI) and Oregon Emergency Management (OEM) on

field tsunami sources, in order to discover whether at-risk individuals are able to recognize natural cues of tsunamis and to take self-protective actions. Consistent, evidence-based approaches from the social and behavioral sciences should be used in the various study areas to allow the NTHMP to compare communities and prioritize future education efforts and resources.

Increasing the Effectiveness of Public Education of Tsunamis

Although tasked to review the availability and adequacy of tsunami education and outreach for children, adults, and tourists, the committee discovered it could not fully comment on

a pilot tsunami awareness program in 2004 (above figure; Connor, 2005). The goal was to develop a comprehensive tsunami outreach program that reached various segments of the community through multiple channels and outreach types. Baseline measurements followed by post-outreach assessments were integral to gauging the influence of outreach efforts on public knowledge of and capacity to respond to future tsunamis. The outreach efforts were managed by a tsunami outreach coordinator, made possible with NTHMP funding, and were primarily driven by the involvement of more than 50 volunteers, including local students, retired residents, and officials.

The tsunami awareness program was based on five outreach strategies designed to reach target audiences and provide multiple channels for learning: a neighborhood educator project had volunteers going door to door to discuss tsunami issues with homeowners; a business workshop focused on improving the business community's emergency plan and preparedness planning; a school outreach program educated elementary-school children through auditorium-style presentation and activities and middle-school youth through small-group discussions; a public workshop was geared for involving the community and tourists in discussing tsunami preparedness; and a tsunami-evacuation drill was run at the end of the outreach program as a chance for individuals to practice what they had learned.

Surveys were conducted before and after the various outreach strategies to determine their influence on public understanding of tsunamis and their preparedness to future events. Post-outreach surveys indicate that 68 percent of Seaside households received information and more than 2,200 people participated in outreach events. The surveys documented measurable differences in tsunami knowledge and preparedness of Seaside community members because of the various outreach efforts. The project demonstrated that each of the five strategies served a different role to fully prepare the community and create a culture of awareness. Project organizers concluded that program success was largely due to the "people-to-people, face-to-face discussions" at each event. An important next step is to see if and how these lessons could be transferred to larger communities (e.g., Los Angeles, Honolulu) where social networks are more complicated and the magnitude of people in tsunami hazards is much greater.

this topic for several reasons. One obstacle to this task is that the true breadth of U.S. tsunami education efforts is not currently known by the NTHMP. There is no existing compilation or inventory of NTHMP-related tsunami education efforts, nor is there a physical or electronic repository for aggregating education efforts. Lacking an existing compilation or national assessment of tsunami education efforts, the committee compiled a list of efforts that demonstrates the breadth of activity across the NTHMP and outside of the program (Appendix E). Based on this incomplete list of examples, it is clear that tsunami education is being done by various organizations (e.g., county and state emergency management departments, K-12 educators, International Tsunami Information Center (ITIC), Pacific Tsunami Museum, United Nations, nonprofit organizations) in various ways (e.g., coloring books, DVDs, fairs, school curriculum, brochures, business planning guides, media kits, websites, museums, online games, newsletters, workshops) to various audiences (e.g., children, adults, households, business owners, officials, tourists) and at various scales (e.g., villages, unincorporated towns, cities, counties, states).

The committee commends all those involved in tsunami education for their individual efforts to raise tsunami awareness in coastal communities. However, the lack of NTHMP mechanisms to systematically compile, evaluate, and disseminate these efforts at a national scale breeds the potential for the duplication of efforts and for conflicting messages. For example, the committee found that several states (e.g., Hawaii, Oregon, Washington) are developing their own tsunami education guidelines (often referred to as "train the trainer" workshops) with little communication among the parties and uneven application of evidence-based approaches.

With regard to new education efforts, a web-based repository for tsunami education material is a vital first step for avoiding duplication of efforts, transferring lessons learned and best practices, and identifying gaps in education coverage. Duplication of efforts could be reduced if education efforts were conducted involving several NTHMP members to develop common tsunami education materials for certain sectors (e.g., broadcasters, hotel owners, tourists, households, and schoolchildren). Local differences can be added to education materials to reflect local needs, but a core set of materials with national relevance could be developed and maintained by the NTHMP to ensure a more consistent message, which—as previously noted—increases the effectiveness of educational efforts.

A second obstacle that prevents the committee from fully commenting on the status of tsunami education is the lack of tsunami education programs that explicitly included pre- and post-outreach evaluations of effectiveness, such as the evaluation associated with a series of tsunami outreach efforts in 2005 in Seaside, Oregon (see Box 3.3). Because there are few studies that have documented the perceptions, knowledge, and capacity to prepare at-risk populations, there is no consistent baseline from which to gauge the effectiveness of education programs. That an education effort occurred could be confirmed, but there is no information on whether the knowledge of participants increased because of an effort.

Besides lacking education evaluations, the NTHMP also lacks standards and criteria for evaluating the multitude of tsunami education efforts occurring in member states, territories, and commonwealths with regard to their information content, dissemination process, presentation style, and enculturation of new information. For example, tsunami education products have not been evaluated for the level at which they discuss tsunami hazards, the vulnerability of individuals and communities to tsunamis, how at-risk populations can reduce their vulnerability, and how people should act if a tsunami occurs. A national tsunami research plan (Bernard et al., 2007) also notes that there has been little analysis on the effectiveness of education programs or coordination among states to define messages with desired outcomes. Having consistent evaluation criteria is critical if the NTHMP is to evaluate isolated education efforts and to determine funding priorities for future education programs, in terms of location (e.g., Alaska versus Puerto Rico) and focus (e.g., tourists versus residents).

The need for measurable outcomes and standards for educational programs was also noted in the NTHMP five-year review (National Tsunami Hazard Mitigation Program, 2007) and the 2007 national tsunami research plan (Bernard et al., 2007). Currently, school curricula are evaluated for their compliance with education standards (e.g., National Research Council, 1996b), but other tsunami outreach efforts (e.g., workshops, media kits, hotel training guides) do not undergo any formal evaluative review by the NTHMP. This committee commends the NTHMP for recognizing these and other deficiencies in tsunami education in its 2009-2013 draft strategic plan. The plan cites several performance measures related to tsunami education. Such performance measures include an inventory of current efforts, an education implementation plan, and electronically available curricula by 2009 (which has not been met). Goals for 2010 include guidelines for tsunami education and a national tsunami media toolkit. Further goals include outreach materials for coastal businesses and tourists, integration of tsunami information into K-12 education through at least one state pilot project by 2011, and a web-based repository for NTHMP-related products by 2012. The draft strategic plan also recognizes the need for evaluation, and it plans to conduct evaluations that determine the effectiveness of tsunami education products and programs in 10 selected communities by 2010.

The National Oceanic and Atmospheric Administration (NOAA) Tsunami Program and the NTHMP are not alone in their mission to safeguard coastal communities from hazards. Other organizations, including other NOAA entities, also share the mission of educating communities about coastal hazards. For example, the NOAA Sea Grant College Program strategic plan (2009-2013) includes a focus area on hazard resilience in coastal communities that includes research aimed at increasing the availability and utility of hazard-related information and the development of comprehensive education programs of coastal hazards and how to prepare for them. The NOAA Office of Education's strategic plan (2009-2029) cites a need for a NOAA education community that functions in a unified manner and coordinates with agency extension, training, outreach, and communications programs. The NOAA Coastal Services Center works with private and public sector partners to address coastal issues, such as resilience to natural hazards, and offers training and information on stakeholder involvement in local management, needs assessments, project/program evaluation, and resilience assessments. In the course of this review, the committee heard of little, if any, interaction between the NOAA Tsunami Program and other NOAA efforts devoted to coastal hazards education (e.g., NOAA Office of Education, NOAA Sea Grant College Program, NOAA Coastal Services Center). The committee found no evidence that TWC staff interact with staff at other NOAA warning centers (e.g., National Severe Storms Laboratory [NSSL], National Hurricane Center [NHC], Aviation Weather Center [AWC], Storm Prediction Center [SPC]), representing missed opportunities to learn best practices in educating the public about extreme events and warning messaging.

> **Conclusion:** Current tsunami education efforts are not sufficiently coordinated and run the danger of communicating inconsistent and potentially confusing messages. The committee was not able to fully evaluate the effectiveness of current educational efforts because the NTHMP lacks an inventory of education efforts or evaluative metrics. The committee concludes that current tsunami education efforts of each NTHMP member are conducted in an ad-hoc, isolated, and often redundant nature and without regard to evidence-based approaches in the social and behavioral sciences on what constitutes effective public risk education and preparedness training. The lack of NTHMP mechanisms

to systematically compile, evaluate, and disseminate these efforts at a national scale breeds the potential for the duplication of efforts and for conflicting messages. There is little evidence that the NOAA Tsunami Program or the NTHMP are leveraging the education efforts and expertise of other NOAA entities that also focus on coastal hazards, extreme events, and warning messaging.

Recommendation: To increase the effectiveness of tsunami education, the NTHMP should do the following:

- Develop consistent education efforts among its members using evidence-based approaches in the social and behavioral sciences. The goal of this education should be to teach at-risk people to correctly interpret natural cues and understand official warnings, to motivate them to appropriately prepare for tsunamis evacuations, and to make that knowledge and motivation a permanent part of the local culture.
- Tailor tsunami education to local circumstances and the specific needs and abilities of at-risk individuals in a community, including tourists.
- Create and maintain an online repository of education efforts.
- Develop and implement a program to evaluate the effectiveness of education efforts and use conclusions from evaluations to make education programs even more effective.
- Leverage the hazard education efforts and expertise of other agencies and NOAA offices.

COMMUNITY PREPAREDNESS EFFORTS

While tsunami detection and warning messages are the responsibility of the federal tsunami program, local and state officials are responsible for preparing communities for future tsunamis, issuing evacuation orders, and managing evacuations and response efforts. Community preparedness and emergency management of tsunamis are largely the responsibility of county emergency managers. Other agencies and organizations play roles as well, including K-12 educators, land-use and building regulatory agencies (e.g., location and type of construction, inspection of construction, and mitigation/abatement of existing hazards), health and social services agencies (e.g., post-disaster care and sheltering, care for special-needs populations), economic-development agencies (e.g., business continuity planning, post-disaster loans), and multi-department working groups (e.g., post-disaster community recovery planning). Because of the significant breadth and depth of actions that can be implemented by various actors to prepare communities for tsunamis, the committee restricted its review of community tsunami preparedness to the NOAA TsunamiReady program. This program is considered by NOAA to be the primary vehicle for preparing communities for future tsunamis.

The NOAA TsunamiReady program is modeled after the National Weather Service (NWS), StormReady program and its objective is to help communities reduce the potential for tsunami-related disasters through redundant and reliable warning communications, public readiness through community education, and official readiness through formal planning and exercises. Jurisdictions (e.g., cities, counties, or states) are recognized after they submit an application to the NOAA NWS and a local TsunamiReady Advisory Board verifies information in the application. Since the NTHMP approved the implementation of the TsunamiReady program in 2001, it has been constantly under review to increase its ability to effectively measure and take into account the goals and objectives of the NTHMP. Draft TsunamiReady criteria currently being discussed within the NTHMP broaden the emergency management scope of the original criteria and include aspects of mitigation through land-use planning and regulation; promulgation of inundation maps and their use in public education; alert and warning systems to notify the public of potential dangers; training emergency response and management staff in the nature of tsunami impacts and their roles and responsibilities in public notification, response, and recovery; and sustained education on evacuation procedures, routes, and refuge areas.

Based on the committee's discussion with representatives from the NOAA Tsunami Program, the TsunamiReady Program, the emergency management community, and a review of the original and new draft standards of the TsunamiReady Program (draft November 2008), the committee observed the following:

Standards. The TsunamiReady program lacks a professional standard (such as NFPA 1600 or the Emergency Management Accreditation Program) regarding what actually constitutes tsunami readiness and how success would be evaluated. The new, draft criteria for TsunamiReady include an extensive list of possible actions, but local officials have no guidance on which actions are most effective and how to prioritize the various actions. The actions taken during a tsunami on June 13, 2005, in Crescent City (see Box 3.4 and Table 3.1) illustrate how current standards might not result in sufficient community readiness. There is no current national database of actions taken by each TsunamiReady community; therefore, it is not possible to identify best practices, lessons learned, or additional needs in community resilience. Maintaining an electronic database of TsunamiReady applications, for communities new to the program and for those seeking re-certification, would provide the NTHMP with the ability to conduct annual needs assessments across the nation and to identify where additional efforts may be warranted.

Accountability for standards. The practice of local committees (led by a local Warning Coordination Meteorologist) to verify that communities have met program standards allows communities to implement a flexible program but requires a level of accountability to be maintained by the national program. In the course of our review, the committee observed situations in which communities are not satisfying mandatory criteria (e.g., hazard or evacuation signage) but are still recognized as TsunamiReady communities. The current metric for the TsunamiReady program is the number of communities that are annually recognized, yet if mandatory criteria are being ignored in the recognition process, the effectiveness of the program and its success criteria are questionable. Some states (e.g., Washington) have recognized the issue of accountability and now require state emergency management agency approval of TsunamiReady applications before a community is recognized.

BOX 3.4
Evaluation of a TsunamiReady Community's Response to a Tsunami

The primary missions of the TsunamiReady program are to educate at-risk individuals on what to do when a tsunami warning is issued and to mitigate, if possible, potential losses. Evaluating the effectiveness of the program is difficult, given the infrequency of tsunami warnings. Therefore, as a case study to provide insight on the program (as opposed to a full evaluation), the committee reviewed the actions taken in the community of Crescent City, California, a recognized TsunamiReady community during two tsunami events—the June 14, 2005, tsunami (which originated offshore Crescent City within the Gorda Plate) and the November 15, 2006, tsunami (which originated in Russia's Kuril Islands) (based on a California Emergency Management Agency [CalEMA] internal action report). Both events illustrate the limited effectiveness and challenges of the current TsunamiReady program.

Crescent City was no stranger to tsunamis when it received warnings in 2005 and 2006. The 1964 Good Friday earthquake and tsunami in Alaska inundated Crescent City harbor and parts of its business district, resulting in extensive damage and the loss of lives. After the 1991 Petrolia, California, earthquake, a focused study of tsunami potential and preparedness centered in Humboldt and Del Norte Counties, where Crescent City is located. Extensive media coverage of the earthquake and tsunami threat resulted in funding a public education and preparedness campaign and developing an earthquake/tsunami scenario to support state and local emergency planning efforts. The Federal Emergency Management Agency (FEMA) provided funding to prepare public information brochures and to make improvements to the local siren warning systems. The designation of Crescent City as TsunamiReady was a direct result of the heightened awareness and commitment of the community in preparing for a tsunami.

Observations of the 2005 and 2006 tsunami events suggest the community has more work to do with regard to community education and mitigation (Table 3.1). During the 2005 event, there were spontaneous car-based evacuations by the public *into* areas of potential inundation, and some overwhelmed dispatch offices failed to follow notification procedures. In 2006, alert and warning procedures were followed, but extensive damage was incurred because of deferred maintenance of port facilities. Despite its TsunamiReady recognition, the community observed significant weaknesses in its ability to effectively respond to tsunamis during the 2005 and 2006 tsunamis. As is likely the case in all communities, sustaining public awareness and maintaining this knowledge of evacuation procedures is a significant challenge for local officials.

Effectiveness. There are no assessments of the effectiveness of the prescriptive readiness actions, the sustainability of the readiness capabilities at the local level, or whether mitigation actions reduce exposure to losses. The program does not identify baselines against which to assess progress other than a count of the number of recognized communities. Additionally, there has not been an assessment of how TsunamiReady communities have performed in actual events or if the criteria used in the program relate to performance.

TABLE 3.1 Actions During Tsunami Events on June 14, 2005, and November 15, 2006

	Event	Prescribed Action	Actual Response
June 14, 2005	Felt earthquake	Evacuation on foot to high ground along prescribed evacuation routes	Spontaneous evacuation of population, in vehicles, on routes into areas of potential inundation
June 14, 2005	Notification of tsunami warning	Public safety dispatch to activate sirens and provide notification	Dispatch was overwhelmed by 911 calls and failed to follow activation/notification procedures
June 14, 2005	Verification of notification by state	Public safety dispatch to receive call and verify receipt of warning	Dispatch could not answer calls from state authorities
November 15, 2006	Distant earthquake and tsunami warning issued by west Coast/Alaska Tsunami Warning Center (WC/ATWC)	Notification and cancellation of warning by WC/ATWC, based on updated information from Deep-ocean Assessment and Reporting of Tsunamis (DART) buoy data and projections from Pacific Marine Environmental Laboratory (PMEL) model	Alert and warning procedures followed. Extensive damage occurred in the Crescent City marina where deferred maintenance resulted in structural deterioration of dock facilities

SOURCE: Committee member, based on CalEMA internal action report.

Entities. TsunamiReady Program recognition currently can only be given to legally recognized jurisdictions (e.g., states, counties, tribes), excluding other entities such as unincorporated communities. Previous studies have shown that significant percentages of individuals living in tsunami-prone areas are in unincorporated villages (Wood, 2007; Wood et al., 2008).

Training. NWS Warning Coordination Meteorologists (WCMs) are the designated local points of contact and advocates for the TsunamiReady Program. Proposed TsunamiReady standards seek to frame the program as an effort to increase community resilience, yet WCMs are trained meteorologists and receive little technical training in preparedness, emergency management, education, planning, and risk. Although NOAA is currently working on a training course with the WC/ATWC to inform WCMs on center operations, it is not apparent that WCMs are receiving training in mitigation, education, or other approaches to build resilience that are now being touted as the new direction of TsunamiReady.

Resources. The TsunamiReady Program objective to encourage tsunami resilience of all coastal communities in our nation is commendable, but the program has access to few resources and little staff time to accomplish this objective. As a result, the TsunamiReady Program at the national and local level becomes an additional duty of a WCM instead of the primary duty of a scientist trained in emergency management or community resilience. The situation has been exacerbated with the addition of Atlantic and Caribbean communities without an increase in overall program budgets.

Public education. The committee found that the current education criteria of inventorying the number of annual tsunami awareness programs is vague and lacks specific guidance on the target audience (e.g., residents, employees, tourists) and on how to accomplish goals. Proposed standards for outreach include a commendable list of potential activities, such as incorporating materials into public utility bills or providing tsunami safety training to local hotel staff. However, jurisdictions can meet the mandatory criteria for outreach by implementing just "one or more" of the several potential activities. Therefore, a jurisdiction could technically meet the outreach criteria simply by passively posting tsunami information on an agency website and ignoring more active efforts, such as training local hotel staff and working with faith-based and civic organizations.

Incorporation of Social Science. The TsunamiReady Program currently lacks involvement from researchers trained in the social sciences, land-use planning, and emergency management. The NOAA Tsunami Program needs to be more proactive in incorporating social science findings in program deliberations, evaluation, and criteria development.

During deliberations about the TsunamiReady Program, the committee compared TsunamiReady to other approaches to improving community preparedness to natural hazards. The Emergency Management Accreditation Program (EMAP)[2] is one such program that also aims to mitigate the risk from natural hazards, but unlike TsunamiReady, EMAP is more broadly geared toward all-hazards mitigation. EMAP is the nationally recognized standard for emergency management and establishes "a common set of criteria for disaster management, emergency management, and business continuity programs … (and) provides criteria to assess current programs or to develop, implement and maintain a program to mitigate, prepare for, respond to and recovery from disasters and emergencies."[3] Standards are established by the EMAP Standards Committee in a process complying with procedures and processes as prescribed by the American National Standards Institute (ANSI). Compliance with the standard is voluntary, and assessments and accreditation are applied through a review process conducted by peers from the emergency management profession.

Although the most current TsunamiReady draft (November 6, 2008) includes many elements of the above EMAP standards (http://www.emaponline.org/), it is currently a mix of requirements that are not well structured. At the time this report was written, the draft

---

[2] Supported by FEMA, National Emergency Management Association (NEMA), International Association of Emergency Managers (IAEM), Council of State Governments (CSG), National Association of Governors (NCG), National League of Cities (NLC), National Conference of State Legislatures (NCSL), U.S. Department of Justice Office of Justice Programs (OJP), and U.S. Department of Transportation. EMAP is managed by the CSG and is overseen by an appointed commission. http://www.emaponline.org/.

[3] EMAP Standard, April 2006,

TsunamiReady community requirements do not fit within the concept, terminology, and format of a standard, but they would more appropriately be described as a detailed assessment scorecard for rating a local government. EMAP provides a concept and structure that could serve as a model for TsunamiReady. The documentation and costs of EMAP assessments and accreditation may preclude its application below the level of state emergency management agencies, but the process of developing standards, use of baseline assessments, peer review, periodic assessment of program sustainability, and continuous improvement of both programs and standards could be a model for the TsunamiReady Program. Alternatively, the NOAA Tsunami Program could simply encourage larger jurisdictions to become accredited through EMAP, with the requirement that tsunami events be part of the multi-hazard plan. The NTHMP, through EMAP, could initiate an ANSI-compliant standard development process, within which the existing scorecard would be restructured and simplified. At a minimum, the NTHMP could consider the five elements of EMAP as mandatory requirements, in addition to a public education requirement based on the latest social science on hazard education: (1) hazard identification, risk assessment, and consequence analysis; (2) a mitigation program to reduce structural vulnerability in areas subject to tsunami inundation; (3) emergency operations, recovery, and continuity of operations plans and procedures; (4) redundant communication systems capable of communicating alerts; and (5) training of officials and responders, exercises, program evaluations, and development of corrective action plans.

Conclusion: The primary mechanism for increasing community preparedness for tsunamis in the United States is the NOAA TsunamiReady Program. The current success metric for the program is the number of communities that are annually recognized. The committee questions the effectiveness of the program and its success criteria because the program lacks the following elements: (1) a professional standard to guide its development, (2) metrics to assess baseline readiness and community needs, (3) evaluative criteria to assess community performance during a tsunami, (4) accountability measures to ensure recognized communities meet mandatory requirements, (5) local points of contact with training in community preparedness, and (6) criteria and guidance on what constitutes effective public outreach and preparedness efforts. Although signage is considered mandatory under TsunamiReady, the use of tsunami signs is inconsistent among NTHMP members, suggesting the need for greater TsunamiReady program accountability and likely creating the incorrect impression that areas without signage are not tsunami hazard zones.

Recommendation: The NOAA Tsunami Program should strengthen the TsunamiReady Program by making the following changes:

- Implement professional and modern emergency management standards following the example of EMAP.
- Develop evaluative criteria for the assessment of community performance during actual tsunamis.
- Increase the level of accountability required of communities in order to maintain their TsunamiReady status.

- Increase program transparency by creating a publicly accessible, digital database of actions taken by TsunamiReady communities.
- Conduct baseline assessments of readiness for all at-risk communities.
- Rate communities on several levels of readiness rather than a simple ready-or-not status.
- Have local points of contact who are trained in risk communication regarding public warning messaging and public warning dissemination, emergency management, and community preparedness.
- Develop guidelines as to what constitutes effective public outreach, make these guidelines publicly available, and regularly evaluate public outreach efforts.
- Ensure that program criteria are evidence-based by including social scientists in the development of the criteria.
- Provide guidance to states and local communities on state-of-the-art preparedness plans that elicit appropriate public protective actions.

The committee believes that many of these recommendations can be accomplished if the NOAA Tsunami Program folded the TsunamiReady Program into the all-hazard Emergency Management Accreditation Program, instead of continuing to maintain its own, separate recognition program.

DEVELOPING AND DELIVERING EFFECTIVE WARNING MESSAGES

Long-term education and community preparedness efforts set the stage for changing the behavior of at-risk individuals and, in the case of near-field tsunamis, may be the only guidance people receive to help them evacuate. However, the likelihood of individuals evacuating tsunami-prone areas is also influenced by the official warning message they receive from the tsunami warning centers and from local and state emergency management agencies. To be effective, these messages need to (1) contain the necessary information that leads individuals to take appropriate protective action, and (2) reach at-risk people in a timely fashion. This section reviews progress in developing and delivering effective warning messages that motivate individuals to take protective action. Topics discussed below represent active fields of research across all hazards, and the tsunami community can draw from lessons learned particularly in the severe weather (e.g., Weather and Society*Integrated Studies (WAS*IS); http://www.sip.ucar.edu/wasis/) or the earthquake community.

Developing Effective Messages

Because some people believe disasters will not happen in the near future and will happen to someone else, warning messages need to first overcome people's natural belief in their own safety and then guide them to take protective actions that are inconsistent with their perception of safety (Slovic et al., 1980; Burningham et al., 2008). An effective warning message provides content about what to do using language that allows a person to visualize the response (e.g., "Climb the nearest slopes until you are higher than the tallest buildings" instead of "Evacuate to higher ground"). It also informs people when they should start and complete the recommended protective action, which groups are in harm's way and must take action, and how protective actions will reduce pending consequences of inaction. Effective warning message style is simply worded, precise, authoritative, and non-ambiguous, even when discussing uncertainty in forecasts (e.g., "We cannot know exactly how high the tsunami will be when it reaches our shores, but all experts agree that it is likely high enough that everyone should evacuate now"). Accurate messages are critical because information errors confuse people and affect their response to a pending disaster. Consistent messages (both consistent internally and across messages from different sources) are needed to reduce the public's choices regarding risk. The protective action to be taken and changes from previous messages need to be clearly explained (Mileti and Sorensen, 1990).

To date, the committee is not aware of any previous efforts to formally review the messages from the TWCs relative to evidenced-based approaches from the social and behavioral sciences. The current and future warning messages of both the TWCs and the NWS would both benefit from review and improvements based on the latest social and behavioral sciences about how and why such messages influence the behavior of people at risk and from being rendered consistent with this knowledge base.

The committee reviewed the standard messages that are composed and delivered by the TWCs and observed that several of the principles for effective warning-message content and style have not been followed. For example, the information statement issued at 12:57AM on February 27, 2010 (see Box 3.5), does not clearly identify who needs to take action because it describes the affected area twice using different descriptions (listing individual West Coast states by name once and then referring to the U.S. West Coast once). Although most informational statements of the WC/ATWC were issued when the earthquake was too small to generate a tsunami, this statement was more like a preliminary statement that at the end said to stay tuned for more. The message was also ambiguous and internally inconsistent—it stated "A tsunami is not expected" at one point and then later stated "A tsunami has been generated that could potentially impact the U.S. West Coast/British Columbia and Alaska." Recommended actions for states to take were largely absent in this message, except for the last sentence, which encouraged people to see the warning center website. A true warning message includes required next steps; therefore, this message was more of an alert of a physical process.

With regard to providing recommendations for required actions, the committee recognizes the limitations placed on the TWCs. Because of existing laws, the TWCs, as part of the federal government, cannot order evacuations. Therefore, unless new national policies are implemented, the TWCs are limited to what they can say for recommended next steps and official warning messages will continue to lack specificity with regard to what protective actions need to be taken. Agreements have been made in certain states (Hawaii and Washington) where warning messages issued by the TWCs can automatically trigger sirens, but only in cases where tsunamis are expected to arrive within minutes of generation. In all other cases, however, it is the responsibility of county emergency managers, in close consultation with state emergency managers,

---

BOX 3.5
Example of a Message from a Tsunami Warning Center

WC/ATWC Information Statement – TW
WEAK53 PAAQ 270857
TIBAK1

PUBLIC TSUNAMI INFORMATION STATEMENT NUMBER 3 NWS WEST COAST/ALASKA TSUNAMI WARNING CENTER PALMER AK
1257 AM PST SAT FEB 27 2010

...A STRONG EARTHQUAKE HAS OCCURRED BUT A TSUNAMI IS NOT
   EXPECTED ALONG THE CALIFORNIA/ OREGON/ WASHINGTON/
   BRITISH COLUMBIA OR ALASKA COASTS...

NO WARNING... NO WATCH AND NO ADVISORY IS IN EFFECT FOR THESE AREAS.

A TSUNAMI HAS BEEN GENERATED THAT COULD POTENTIALLY IMPACT THE U.S. WEST COAST/ BRITISH COLUMBIA AND ALASKA.
   THE WEST COAST/ALASKA TSUNAMI WARNING CENTER IS INVESTIGATING THE EVENT
TO DETERMINE THE LEVEL OF DANGER. MORE INFORMATION WILL BE ISSUED AS IT BECOMES AVAILABLE.

A TSUNAMI HAS BEEN OBSERVED AT THE FOLLOWING SITES

---

to issue evacuation orders based on the messages released by the TWCs. Given the separation in responsibilities and authorities, close coordination between the TWCs, states, and local jurisdictions is needed to ensure that the public receives information about the threat and proper protective action (see the section on interagency coordination for more discussion). Observations during the 2010 Chilean event suggest that confusion still exists among the public about actions to take in response to the TWC messages (Wilson et al., 2010). Initial observations of the warning messages issued related to the 2010 Chilean earthquake and tsunami suggest that California jurisdictions did not have a consistent understanding of lines of communication, what an "advisory" alert level means in terms of recommended next steps, or who should be involved in taking next steps (Wilson et al., 2010).

The committee also found inconsistencies between the warning products of the TWCs and those of the NWS. For example, the NWS issues a "watch" for an event that has an 80 percent chance of becoming a warning, but this is not the case with the TWCs whose watches rarely

LOCATION		LAT	LON	TIME	AMPL
30MIN	TALCAHUANO C	36.7S	73.1W	0659UTC	1.8M/6.0FT
22MIN	VALPARAISO C	33.0S	71.4W	0659UTC	1.69M/5.7FT
20MIN	CORRAL CHILE	39.9S	73.4W	0729UTC	1.45M/4.8FT
14MIN	SAN FELIX C	26.3S	80.1W	0809UTC	0.69M/2.3FT

TIME - TIME OF MEASUREMENT
AMPL - TSUNAMI AMPLITUDES ARE MEASURED RELATIVE TO NORMAL SEA LEVEL.
    IT IS ...NOT... CREST-TO-TROUGH WAVE HEIGHT.
    VALUES ARE GIVEN IN BOTH METERS (M) AND FEET (FT).

AT 1034 PM PACIFIC STANDARD TIME ON FEBRUARY 26 AN EARTHQUAKE WITH PRELIMINARY MAGNITUDE 8.8 (REVISED FROM 8.6) OCCURRED NEAR THE COAST OF CENTRAL CHILE.

THE PACIFIC TSUNAMI WARNING CENTER IN EWA BEACH HAWAII WILL ISSUE A TSUNAMI WARNING FOR REGIONS NEAR THE EPICENTER.

MESSAGES WILL CONTINUE TO BE ISSUED HOURLY BY THE WC/ATWC UNTIL THE EVENT STATUS IS UPGRADED TO A WARNING OR ADVISORY OR UNTIL THE CENTER HAS DETERMINED THAT THE EVENT POSES NO THREAT TO THE CALIFORNIA/ OREGON/ WASHINGTON/ BRITISH COLUMBIA AND ALASKA COASTS. SEE THE WEB SITE WCATWC.ARH.NOAA.GOV FOR BASIC TSUNAMI INFORMATION - SAFETY RULES AND TSUNAMI TRAVEL TIMES.

become warnings. Another current inconsistency is how the TWCs and the NWS deal with "all-clears." The TWCs cancel a bulletin, which could be read by the public as a signal that it is safe to return, which is not the same as an "all-clear" issued by the NWS. The NWS will soon move from using the "alert bulletin system" to an "impact-based system," which will introduce another inconsistency with the TWCs. An additional potential source of confusion stems from the fact that "information statements" are used for conveying two different types of messages: first to convey some initial information and the notion to stay tuned for more, and second to convey that a certain earthquake has not generated a tsunami.

Recommendation: The NOAA/NWS should remedy current differences between the TWC's and the NWS's warning products and ensure consistency in the future. A mechanism should be put in place so that pending and future inconsistencies are quickly identified and acted upon so that products from the TWCs and the NWS match.

Once the TWCs issue warnings, watches, or advisories, it is the responsibility of county emergency management and/or public-safety agencies to issue their own messages to individuals in tsunami-prone areas (except in the case of near-field tsunamis where evacuations will be spontaneous and local agencies will have to react to, and instead of lead, the evacuations). In many cases, after the TWCs issue their messages, county and state agencies discuss the situation and potential strategies with the TWCs through teleconferences. Following these conversations, individual jurisdictions will then make their own decisions regarding whether to evacuate tsunami-prone areas or not. Pre-event tsunami education of local public-safety officers is important because their knowledge of tsunami threats, the vulnerability of local populations, and the time and logistics required to evacuate these populations all play a factor in the evacuation decision making process.

Disparities in knowledge and risk tolerance at the local level can lead to different decisions. Although the TWCs are releasing information to an entire region (e.g., the Cascadia subduction zone from northern California to Washington), individual county jurisdictions will decide how to use this information in their own warning messages sent to people along their coasts. In the response to the warning messages released during the June 14, 2005, event, there were cases where adjacent counties in Oregon received the same information from the TWCs and yet made different decisions. One county called for an evacuation, while the adjacent county did not. To the public, these disparities in response are inconsistent and create confusion.

Effective Delivery of Warning Messages

Actions taken by the public are influenced by the warning delivery method because of the time it takes people to convert pre-warning perceptions of safety into current perceptions of risk. The frequency of warning-message communications and the increasing number and types of communication channels are shown to positively impact people's warning response. There is no one single credible source of information, because various groups attach credibility to different spokespeople, perceptions of credibility change with time, and credibility and warning message belief are not identical. Consequently, it is vital to create diverse sources of public warnings, an effort that requires pre-event emergency planning and agreement from many partners to disseminate the same warning message long before an event occurs.

Effective dissemination of warnings involves multiple organizations using multiple channels to frequently deliver the same message. Both TWCs disseminate their messages over multiple channels, such as the National Warning System (NAWAS), State Warning Systems (e.g., Hawaii Warning System (HAWAS), California Warning System (CALWAS)), the Global Telecommunications Service (GTS), the NOAA Weather Wire, the Aeronautical Fixed Telecommunications Network, the military-related Gateguard circuit, emails, faxes and telex, and phone calls (among others). With two warning centers sending messages over multiple channels, it is most important that messages be consistent in order to minimize confusion by the public. However, because the TWCs have different areas of responsibility (AORs), warning messages from the two centers are designed for different audiences and contain different information (see more discussion on this AOR issue in Chapter 5). Although these messages are not intended for public consumption (Paul Whitmore, NOAA, personal communication), products from both TWCs are distributed to members of the public and the media who have signed up on the TWCs' websites to receive these alerts. Consequently, the messages are immediately disseminated to the public via television, radio, and the Internet—as was the case during the 2009 Samoan tsunami (Appendix I) and the 2010 Chilean tsunami (Appendix J).

The generation of two sets of warnings, although technically appropriate due to different areas of responsibilities, can be a major source of confusion among the emergency management community and the public, as illustrated by the June 14, 2005, event (Appendix F). During the 2005 event, media outlets in the Pacific Northwest received messages from both TWCs that appeared to contradict each other because the distinction between areas of responsibility was not well understood. It is likely that media outlets and the public will continue to misunderstand this jurisdictional distinction in future warnings. Therefore, it is central to the success of the TWCs that they further improve the consistency and clarity between their messages to prevent any confusion resulting from the distinct AOR. Alternatively, the issuance of a single message after internal consultation between the TWCs could definitively eliminate the potential for confusion from differences in message content.

Local officials will receive warning messages from the TWCs typically within five minutes of an event. Information will come via the NAWAS to the state warning centers and via state versions of the warning system (e.g., HAWAS) to county Public Safety Access Points (PSAPs or 911 dispatch) if a warning has been issued. If the jurisdiction is an incorporated municipality, warning messages are sent from the PSAP and law enforcement teletype to the local dispatch. If local officials are directly linked to seismic network displays (e.g., the California Integrated Seismic Network (CISN)), then they will receive confirmation of the earthquake via both ShakeMap[4] and TWC documentation. Once local officials decide to issue an evacuation order, they will issue it via pre-determined channels (e.g., sirens, AHAB (All-Hazard Alert Broadcast), reverse-911 calls, etc.).

It is important to note that current dissemination routes and plans described by the TWCs and local emergency management resemble the old paradigm of a linear message pathway from the warning center to the local emergency officials, who then notify the public and order an evacuation. Such a linear information transfer can no longer be assumed with the rise of the Internet and other telecommunications technology. Instead, communication networks resemble a web of sources with information coming from multiple systems, both official (e.g., local sheriff, NOAA Weather Radio (NWR)) and unofficial (e.g., TV, Internet, friends). As mentioned earlier, the media and many members of the general public now receive alerts directly from the TWCs, thereby removing local emergency management from the communication path.

Another factor likely to change warning messaging is mobile social networking technology (e.g., Twitter, Facebook, and Google Maps). These technologies have all been harnessed

---

[4] ShakeMap is a product of the USGS Earthquake Hazards Program in conjunction with regional seismic network operators. ShakeMap sites provide near-real-time maps of ground motion and shaking intensity following significant earthquakes.

through grassroots efforts to disseminate frequent updates on personal safety and relief support after a disaster, such as the 2007 San Diego wildfires and the 2008 Virginia Tech University shootings (Hughes et al., 2008; Winerman, 2009) and hold great promise in complementing current warning dissemination methods for communities threatened by both near- and far-field tsunamis. For at-risk individuals who may only have minutes to escape tsunami-prone areas, being warned by social networking technology used by other people in tsunami hazard zones may be a more realistic and timely way to quickly disseminate information than traditional message-dissemination paths. The number of people using these technologies will surely grow in the future, and their applications to disaster warnings and response efforts will be more prevalent.

The use and role of social networking and mobile technologies in emergency, crisis, and disaster management is an active research area (International Community in Information Systems for Crisis Response and Management, 2008, 2009). A persistent concern about their use is the potential for inconsistent information that promotes confusion, and additional research is needed to contend with this problem. The future of tsunami warning likely involves a concerted effort by local, state, and federal agencies to integrate and leverage social networking technologies with the current message dissemination methods. Public agencies and officials with disaster warning and response duties could also monitor the spread of social networking technologies in coastal communities threatened by near-field tsunamis. Unofficial messages from these social networks could confirm official warnings, minimizing the amount of time people typically take for the warning confirmation process and before they evacuate (Mileti and Sorenson, 1990; International Community in Information Systems for Crisis Response and Management, 2008, 2009). Collaborative web-based tools (e.g., chat rooms, blogs, wikis, instant messaging) could assist in maintaining situational awareness and clarify concerns at the state or local level.

Although social networking technology holds great promise in supporting near-field tsunami evacuations, the technology is not currently embraced by many local or federal officials. The incorporation of social networking technologies into official emergency response efforts may be difficult as federal and local disaster response agencies operate under the Incident Command System—a standardized protocol that includes a top-down chain of command for information flow (Winerman, 2009). The committee reviewed a draft white paper from the Pacific Tsunami Warning Center (PTWC) on the topic of the use of social networking and is encouraged that the TWCs are at least considering such new technologies. Although there is evidence of the TWCs investigating the potential use of collaborative information technologies with emergency managers, the committee saw little evidence that they were also embracing mobile social networking technologies that empower the general public to warn each other.

Conclusion: Messages from the two tsunami warning centers do not completely follow evidence-based approaches in format, content, and style of effective messages. The generation of two different TWC warning messages to accommodate different areas of responsibility has created confusion among the media and the general public and will likely continue to do so. Little formal attention has been paid to the use of traditional, nontraditional, or next generation technologies (e.g., mobile and social networking) in support of community outreach and dissemination.

Recommendation: The NWS should establish a committee of experts in the social science of warning messaging to review the format, content, delivery channels, and style of TWC messages. If distinct messages are to be produced by the two TWCs, then the messages should be consistent. Ideally, the committee recommends that one message be released by the two TWCs that internally covers information for all areas of responsibilities.

IMPROVING COORDINATION OF PREPAREDNESS NEEDS AND EVACUATION PROCEDURES

Because tsunami evacuations involve multiple actors (e.g., the at-risk individual, TWCs, media outlets, critical facilities, schools, and local, state, and federal officials), significant pre-event planning, coordination, and testing of procedures are necessary to increase the likelihood that evacuations are successful. As the June 14, 2005, tsunami warning case study demonstrates (Appendix F), warning dissemination and coordination of responses is not trivial. The next section discusses efforts to ensure effective communication within the NTHMP and to test interagency coordination in the event of a tsunami.

Improving Communication Among TWCs and NTHMP Members

Because the TWCs can only provide the public with alerts about the hazard and local officials are responsible for the public response (e.g., issue evacuation orders and facilitate the evacuation), the TWCs need to establish and maintain partnerships with agencies responsible for managing evacuations. Because information flow is no longer linear or hierarchical (i.e., TWC to emergency manager to public), the TWCs need to consider not only emergency managers, but also the media and the general public as an audience when refining the warning and dissemination plans. To date, the TWCs and the NTHMP have done a great deal to engage with the customers and establish community connections, including the following actions.

- The creation of the NTHMP Tsunami Warning Coordination Subcommittee (WCS), which enables members to give input on TWC warning products and dissemination, coordinates major tsunami exercises and tsunami end-to-end tests, exchanges experiences of past events, and discussses improvements related to operational products and dissemination.
- NAWAS is routinely tested, including communication between the TWCs, states, and local jurisdictions. The test results and issues resolved are published by the TWCs and disseminated to all stakeholders.
- The TWCs and the NTHMP support the development of "State Alert and Warning

Center Standard Operating Procedures" by Washington State to ensure that all stake-holders' procedures are well coordinated and tsunami bulletins are efficiently and effectively disseminated.

- The TWCs and the NTHMP also give briefings at the quarterly Washington State/Local Tsunami Workgroup on current operations and issues.
- The TWCs and the NTHMP are involved in the dissemination of messages in support of tsunami exercise scenarios for regional, state, and local governments, including table tops, drills, and functional and full-scale exercises, and they collaborate with emergency management staff to develop such exercises.
- There are yearly end-to-end tsunami communication tests by TWC staff, regional NWS personnel, and state and local officials to coordinate testing procedures and reporting requirements.
- The TWCs and the NTHMP are involved in public fora to educate people on tsunami messages and their dissemination.

The TWC staff is to be commended for their efforts to establish connections with external groups, especially considering their multiple responsibilities, and to revise their procedures and products based on customer feedback (e.g., after the June 14, 2005, event; Johnson, 2005). However, counter to recommendations in a Tsunami Warning Center Reference Guide issued in 2007 by the U.S. Indian Ocean Tsunami Warning System Program, the U.S. TWCs do not have dedicated public affairs officers, and as a result, the important duties of establishing and maintaining relationships with customers are performed as collateral duties by natural scientists. There are no formal outreach plans for media training or working with county officials, no formal training interactions for TWC watchstanders and state civil defense officers, and no formal standard operating procedures for evaluating the effectiveness of warning-message content or channels. The committee finds that relatively few resources have been allocated to maintaining partnerships with customers and that these efforts are ad hoc and secondary to the technical aspects of the warning centers. The committee agrees with the June 14 assessment team (Johnson, 2005) and endorses its call for (1) a review of warning message format, content, and update cycle; (2) formalized and routinely practiced procedures at NWS offices; (3) public tests of the tsunami warning system paired with increased outreach through TsunamiReady and other awareness programs; and (4) enhanced coordination of the NWS and with its partners.

Conclusion: TWC staff is committed to establishing connections with external groups and is to be commended for their efforts considering their multiple responsibilities. However, relatively few resources have been dedicated to maintaining partnerships with customers, and efforts to do so are performed as collateral duty by natural scientists on an ad hoc basis and are secondary to the technical aspects of the warning centers. There are no formal outreach plans for media training or working with emergency management and response personnel, no formal training and interactions for TWC watchstanders and state civil defense officers, and no formal standard operating procedures for evaluating the effectiveness of warning message content or channels.

Recommendation: The NOAA/NWS should better integrate TWC warning functions with those of the states, counties, and cities with regard to the content and dissemination of public warning messages. The NOAA/NWS should develop formal TWC outreach plans, approaches to assess needs and priorities of TWC customers, and evaluative criteria for examining the effectiveness of warning-message content and delivery after significant events.

The NOAA/NWS should formally study and develop a plan to incorporate the use of traditional, nontraditional and next generation technologies (e.g., mobile and social networking) in support of community outreach and dissemination related to tsunamis. This plan should be one component of the enterprise-wide technology and organizational planning effort recommended by the committee and described more fully in Chapter 5. It should reflect an understanding of the rich literature evaluating the use and effectiveness of various technologies, including emerging social networking technology.

A better integration with the states, counties, and cities could be achieved by modeling the outreach approach used by the NWS to reach warning partners, which uses NWS field offices to facilitate such interactions, training sessions, and more with state, county, and city warning partners. To do this, communication, education, and outreach require more attention and resources to accomplish the TWCs' goal of warning people and protecting lives. The NOAA/NWS needs to establish dedicated TWC positions for public affairs officers who have expertise in the social science of risk communication regarding warning-message creation and dissemination, needs assessments, program evaluation, and emergency management.

After significant tsunami warnings are issued to U.S. communities (e.g., the 2010 Chilean event), the NOAA/NWS should initiate an independent review of TWC actions and the TWC's integration with its partners and customers through an external science review board.

Communicating Local Community Needs to NTHMP and the TWCs

Just as the NTHMP provides a forum for state and federal agencies to discuss issues and needs related to tsunami education and warnings, several NTHMP members (as part of the NTHMP M&ES) have created working groups to facilitate communication, coordination, and planning among local and state agencies. Some regional groups are making important contributions to the coordination of educational efforts. Some NTHMP members (e.g., Washington) have used these groups to develop risk reduction priorities for future NTHMP funding. The following is a brief summary of tsunami working groups within various NTHMP member states.

- **Washington State/Local Tsunami Working Group:** Established in 1996, the Washington State/Local Tsunami Working Group is a forum that meets quarterly to identify tsunami preparedness, response and recovery, and education and outreach needs and to develop the direction of the state tsunami program. The state of Washington has a tsunami advisor to the Washington congressional delegation who also advises at meetings of the governor's office support workgroup. Working group organizers credit the group for enabling rapid buy-in and implementation of assessment tools, planning documents, education efforts, preparedness outreach, warning systems and dissemination protocols, and neighborhood mapping efforts. The workgroup also adopted NWR "All-Hazards" Warning System, and it was instrumental in adding a repeater to the NWR system. It also developed a new notification system to target the public on beaches and in high traffic areas (Crawford, 2005).

- **Tsunami Advisory Council for Oregon:** Established in 2008 by DOGAMI, the Tsunami Advisory Council for Oregon (TACO) is a mechanism for DOGAMI to receive advice from a broad spectrum of coastal users (e.g., planners, elected officials, emergency responders) on hazard and risk assessment products and risk reduction strategies for tsunami hazard mitigation. Current TACO efforts include the development of a statewide outreach strategy, community support for achieving TsunamiReady recognition, improved evacuation route signage, presentations to public officials and the general public, web-posting of evacuation maps and tsunami hazard zone data, an online mapping application to display evacuation zones and routes, and an online tsunami information clearinghouse.

- **Hawaii Tsunami Technical Review Committee:** Established in 1998 with funding from Hawaii State Civil Defense and the NTHMP program, the Hawaii Tsunami Technical Review Committee (TTRC) provides a forum for reducing tsunami risk in Hawaii and for improving coordination and information exchange among members. TTRC subcommittees include public awareness, warning systems, technical oversight, and zoning, codes, and guidelines. The TTRC originally met twice a year but has been less active in recent years.

- **California State Tsunami Steering Committee:** Established originally using National Earthquake Hazards Reduction Program (NEHRP) funds, the California State Tsunami Steering Committee (STSC) exists to increase tsunami awareness, build a constituency for tsunami risk reduction efforts in California, provide a platform for communication and ongoing collaborations, and create state strategic plans for tsunami preparedness. The STSC also serves as the mechanism for "after action" assessments of event responses and has produced guidance for local government planning with templates, scripts, county-level training, and table-top exercises to address integration of inundation maps and response planning.

- **Puerto Rico Tsunami Technical Review Committee:** Established in 2004, the Puerto Rico Tsunami Technical Review Committee (PRTTRC) focuses on tsunami hazard and risk identification, tsunami warning protocols, emergency management and mitigation, and public awareness. The PRTTRC is coordinated by the University of Puerto Rico at Mayaguez as part of the Puerto Rico Tsunami Warning and Mitigation program, which is supported by the Puerto Rico State Emergency Management Agency (International Tsunami Information Center, 2004).

Conclusions: The committee cannot fully evaluate the effectiveness of the existing working groups because of the ad hoc nature of most groups and the lack of documentation. However, based on interviews and discussions with various local and state officials, the committee concludes that tsunami working groups are useful mechanisms for coordinating and communicating the needs and abilities of at-risk communities to state emergency management agencies and federal tsunami programs. These working groups also provide fora for improving the dissemination of tsunami warning messages and for reviewing new products. Most coastal states have tsunami working groups, but the level of activity varies significantly among the groups. The committee observed that there is little to no interaction between state working groups, thereby limiting the sharing of lessons learned and likely creating redundant efforts and discussions.

Recommendation: The NTHMP should actively encourage all members to develop and maintain strong tsunami working groups to help facilitate and coordinate tsunami education, preparedness, and warning dissemination. The NTHMP should work to communicate efforts of various working groups across the NTHMP and help disseminate best practices. To ensure local efforts are evidence-based, state working groups should actively encourage the involvement of social scientists trained in risk communication regarding public education to increase knowledge about hazards and motivate preparedness, tsunami risk, and emergency management.

Practicing Evacuation Procedures and Protocols

Tsunami evacuations will involve multiple actors making decisions in limited time that will affect hundreds to thousands of individuals. Reviews of past tsunamis, such as the 1960 tsunami in Hilo, Hawaii (Johnston, 2003) and the June 14, 2005, event (Appendix F) indicate that the lack of coordination among government agencies led to confusion among response agencies and affected parties. To ensure that evacuations minimize unnecessary social and economic interruptions, it is important for public safety and emergency management agencies to practice and coordinate response procedures and protocols. Exercises present opportunities to foster communication and seamless operations, as formal response plans mean little if agencies and affected parties fail to train and improve upon them (Sutton and Tierney, 2006). This section discusses the two primary approaches to improve response procedures and protocols—table-top exercises to discuss evacuation and response coordination among agencies and functional exercises to test agency procedures with a live simulation. In limited situations, it may be useful to having the public practicing actual evacuation behavior as discussed below.

Recent table-top and functional exercises, such as Pacific Peril 2006, Exercise Pacific Wave 08 (Intergovernmental Oceanographic Commission, 2008), Exercise Lantex 2009 (National Tsunami Hazard Mitigation Program, 2009b), and functional exercises in 2009 in northern California, Hawaii, and Washington have been conducted to test interagency communication and to provide a forum for discussing potential issues if a real event were to occur. Reviews of the Pacific Peril 2006 exercise indicate the need for a regional campaign for citizen preparedness programs (as opposed to fragmented state approaches) and a need for the federal government to take a stronger role in facilitating partnerships across political boundaries (U.S. Department of Transportation, 2006). These conclusions highlight the additional benefits from regional exercises and drills in identifying gaps in coordination and improving coordination for events that transcend state boundaries.

The impact of and learning from this and other exercises would be greatly improved if independent social scientists were formally evaluating interagency communication patterns during the exercises, as opposed to the current system of organizers and participants writing up their own lessons learned. In addition, the committee cannot fully comment on the effectiveness of these efforts because it could not find any information on the costs and benefits of conducting tsunami table-top exercises or post-exercise evaluations relative to pre-event conditions. Also, learning from these exercises is likely temporary and limited to the participants involved because the NTHMP currently lacks an inventory of past exercises or a repository of lessons learned from each exercise.

In addition to exercises conducted by agencies responsible for managing evacuations, some communities (e.g., Seaside, Yachats) have conducted voluntary evacuation drills where the public practices actual evacuation behavior. A 2004 tsunami evacuation drill held in Seaside, Oregon, was considered a success by organizers because of broad participation by the community (e.g., residents, employees, and tourists), a successful evacuation (set at 30 minutes by organizers) for all but 2 of the 436 participants, and more than 90 percent of participants were comfortable with the procedures (Connor, 2005). Aside from Connor's 2005 summary of the Seaside experience and media accounts, the committee found no NTHMP documentation or inventory of past tsunami evacuation drills. The committee was also unable to find any NTHMP guidelines for how to design or evaluate a drill that involves the public, and it did not find any documentation on the costs and benefits of these drills. Without evaluation or documentation of past drills, the committee cannot comment on whether there has been any long-term impact on tsunami public education.

Based on evidence from drills run for different hazards (e.g., hurricanes, nuclear-power plants), the committee does not endorse drills involving the public in most situations. For far-field tsunamis, evacuations will likely take place over several hours, and the public will be given explicit instructions and guidance on where and how to evacuate. The media will provide continuous updates on conditions, and public safety officers will be in the streets managing evacuations. Therefore, the significant financial resources it takes to stage these events and the social and economic costs of disrupting a community (especially large communities like Honolulu, Los Angeles, and Seattle) outweigh the benefits of having people practice orderly evacuations, given that they will have hours to do so if an event occurs. The certainty of significant business disruption and the potential for injuries or possibly fatalities in a large-scale evacuation exercise preclude the need to stage public exercises.

For near-field tsunamis, input from a few emergency managers suggests that voluntary drills that involve the public practicing evacuation behavior are useful in promoting tsunami awareness and in training certain populations, such as schoolchildren, to evacuate when time is of the essence. However, as previously discussed, the committee knows of no cost-benefit studies to evaluate evacuation drills that involve the public regarding whether they increase public readiness and knowledge and, if they do, whether these benefits outweigh the social and economic costs of staging them. Nevertheless, the committee identifies some benefits of having individuals, households and neighborhoods practice evacuation behavior. In very small communities that may have less than 30 minutes to evacuate and limited vertical evacuation options, these voluntary drills may be useful in promoting tsunami awareness, providing social cues, and building social networks. However, the utility of these voluntary exercises is likely limited to very small communities. Also the transferability and applicability is minimal, and perhaps dangerous, to larger communities with far-field threats and greater options for vertical evacuation.

California and Alaska use "live code" tsunami tests, in which the communication system is tested from the initiation through to issuing a public test message. Such end-to-end tests, if done well, could provide great benefits. For example, such live code tests would help to "integrate" the different components of the system, contribute to improved coordination, "enculturate" the hazards with the local population, and potentially reduce confusion during a real event. However, there are also major risks associated with such tests. If the tests were designed poorly, e.g., distribute a "bad" message to the public that could confuse people, the tests would lead to potentially negative societal impacts. Therefore, care would need to be given to properly design these tests (good emergency planning) and to appropriately implement them (adequate training provided to those who would conduct the test). Important consideration in designing the test include using the correct message content distributed to the public as well as the appropriate dissemination mechanisms (e.g., reverse 911, text messages to mobile devices, traditional media, Internet, etc.). Thus, pre-test planning will need the involvement of not only emergency managers but also social scientists with knowledge of how to design such tests.

Because actual tsunami events are opportunities to test and exercise all components of the tsunami program (including the technology, organizations, and people), evaluating these events presents a good opportunity for learning. The benefits and requirements for such a research and evaluation effort is further discussed in Chapter 4.

> **Conclusion:** Practicing evacuation procedures and protocols is important in order to minimize confusion in future evacuations. Current efforts to practice evacuation procedures and protocols include community-led evacuation drills, table-top exercises among emergency management agencies to discuss response coordination, and functional exercises to field test interagency communication and coordination. The committee concludes that the importance of these approaches varies based on local conditions and the tsunami threat that communities face.
>
> - **Far-field tsunami threats:** Evacuations will be managed by multiple agencies over several hours; therefore, exercises involving agencies to discuss and test coordination and communication are important to reduce the potential for confusion during a tsunami. The NTHMP currently lacks guidelines on how to evaluate exercises, and there

is no repository for past exercises to increase the effectiveness and transferability of experiences. There is limited literature on the cost-effectiveness of table-top and functional exercises. The committee concludes that evacuation drills that have the public practice actual evacuation behavior are not advisable to prepare the public for far-field tsunamis. The financial costs, significant planning requirements, unnecessary individual and business interruptions, and the number of hours to respond during an actual far-field evacuation far outweigh the benefits of having individuals practice walking to higher ground or vertically evacuating.

- **Near-field tsunami threats:** Initial evacuations will be self-directing after at-risk individuals recognize natural cues. The committee concludes, however, that table-top and functional exercises are still important because of the significant response and relief operations after the initial tsunami wave arrives. For small communities, the committee concludes that these community-led, voluntary drills may be useful in promoting tsunami awareness, providing social cues, and building social networks, but only in very small communities or villages that may have less than 30 minutes to evacuate and limited vertical evacuation options.

Recommendation: To ensure that managed evacuations for far-field tsunamis are effective and that they minimize societal and economic interruptions, the NTHMP should develop guidelines on the design of effective exercises for use by emergency management agencies. The NTHMP should also evaluate these exercises from an economic and social cost-benefit perspective and should provide a repository for exercise evaluations in order to increase the transferability of observations. Public tests of the tsunami warning system paired with increased outreach through TsunamiReady and other awareness programs should be undertaken regularly and reported to the repository. For small communities in Alaska, the Pacific Northwest, Puerto Rico, and U.S. territories in the Pacific Ocean that have near-field tsunami sources, short tsunami arrival times, and limited vertical evacuation options, voluntary evacuation practices (e.g., households, schoolrooms, neighborhood gatherings) may be useful elements in larger tsunami-education efforts. To the extent possible, the NTHMP should provide evidence-based guidance on how to include them in a community-based education program.

POST-EVENT RECONNAISSANCE

Post-event field surveys are crucial for gaining understanding of tsunami characteristics, behaviors, impacts, and people's behavior that reduced the impact of the tsunami. Measured run-up height distributions and flow patterns are critical data that can be used for the validation of both hydrodynamic and tsunami source models. Because it is difficult and likely too costly to prepare, install, and maintain adequate instruments to measure the effects at enough onshore locations and times, field surveys are likely the best alternative to collecting this important data.

Solving the inverse problem, the tsunami source can be estimated from the measured run-up data. The estimated source condition is analyzed to determine whether or not the earthquake mechanism inferred from fault dislocation models is consistent. This type of analysis for the 1992 Nicaragua tsunami led Kanamori and Kikuchi (1993) to propose the mechanism of "slow-slip tsunami earthquakes"—deceptively mild quakes that generate anomalously large tsunamis. Understanding such a phenomenon is critical for adequate tsunami risk assessment. The measured run-up data can also be used as a benchmark to validate the hydrodynamic models. For example, the measured run-up data for the 1993 Okushiri tsunami were used for the model validation exercise at the community workshop (Yeh et al., 1996). This benchmark problem is adopted in the recent model validation guideline by the NTHMP (OAR PMEL-135, Synolakis et al., 2007).

Tsunami surveys in the past have revealed many tsunami characteristics. For example, locally high anomalous run-up resulting from the 1992 Flores and the 1998 Papua New Guinea tsunamis indicated the possible occurrence of earthquake-induced submarine landslides (Yeh et al., 1993; Synolakis et al., 2002). The field survey in Babi Island—a small cone-shaped island where 263 people were killed in the normally safe lee side of the island by the 1992 Flores tsunami—led the subsequent numerical simulations (Liu et al., 1995) and large-scale laboratory experiments (Briggs et al., 1995). The comprehensive study revealed the unexpected tsunami behavior (Yeh et al., 1994). When it hit the island, the tsunami split in two. The split tsunami wrapped around the island and joined to create a new, larger wave that crashed into the lee side of the island. This phenomenon that is unique to tsunami is also adopted as one of the benchmark problems in OAR PMEL-135.

Tsunami surveys are also needed for other important observations: flow effects on man-made structures and natural geomorphologic features, social impacts, and identifications of all salient features for the use of future tsunami loss reduction. Tsunami field surveys also provide us with evidence that tsunamis are capable of transporting sediments, rocks, and boulders (Bourgeois et al., 1999). Such information and data are important not only for future prevention of scouring and structure damages, but also for the assessment of geological evidence of prehistoric tsunami events.

Systematic and organized field surveys specifically aimed at the social impacts were initiated for the first time in response to the 2004 Indian Ocean tsunami event (Suter et al., 2009). Appropriate social science post-event research audits hold the potential to document important lessons to be learned. Such social science research of this type would cover a range of topics; it would be conducted at different points in time after an event; and it would be performed by researchers with varied and specific training, expertise, and experience. The range of topics benefiting from this post-event investigation includes but not limited to how well the warning system functioned as a system across the varied players involved in the system, e.g. the TWCs, state and local government, and the public; the adequacy of TWC and state and local government messages to each other and the public in terms of how those messages influenced protective action-taking; and much more. An adequate social science research agenda would include both quick-response reconnaissance research to capture perishable data and longer-term research conducted months or longer after an event.

The committee views it as essential that this social science research be fully informed by and performed by trained and experienced social science warning researchers so that it would be of the highest quality and hold the potential to produce the most useful results. There is long-standing and now mature precedent in the nation for organizing an appropriate mix of appropriately trained interdisciplinary teams to conduct post-event research audits. For example, the Learning from Earthquakes Program in the Earthquake Engineering Research Institute represents an excellent current approach to conducting this type of research that uses a mix of researchers from varied disciplines in the physical and social sciences and engineering.

Recommendation: Social science post-event research audits should be performed after all tsunami "warning events" that hold the potential to document important lessons to be learned.

Tsunami survey teams have been organized on an ad hoc basis primarily on the initiative of the individual scientists. There exists no systematic funding mechanism to support the survey efforts, although NSF, the USGS, NOAA, the Earthquake Engineering Research Institute (EERI) (through NSF), and the American Society of Civil Engineers (ASCE) have been the sponsors on an event-by-event basis. The USGS and NOAA have supported primarily their own personnel. The lack of coordination often resulted in collecting duplicate data and information by multiple survey teams, and at the same time, failing to collect important data and information. In addition, the lack of coordination does not lead to a good balance of skills, experience, and disciplines for the survey teams.

Conclusion: Tsunami reconnaissance field surveys are crucial to gain understanding of tsunami effects, and the findings directly improve tsunami risk assessment. This knowledge in turn helps reduce the impacts of future tsunamis. To make the future field surveys more effective and efficient, coordination by a lead agency is needed.

Recommendation: Tsunami field surveys should be conducted by multi-disciplinary personnel including physical and social scientists, engineers, disaster mitigation planners, and sociologists. A quick dispatching capability is crucial for tsunami surveys, in order to capture as much information as possible. Tsunami run-up marks, destruction patterns, and other detailed tsunami-affected features can disappear within a few weeks. NOAA should take a more proactive role in the coordination for tsunami surveys with other agencies, in particular the USGS and NSF.

CHAPTER FOUR

Tsunami Detection and Forecasting

SUMMARY

An incoming tsunami may be anticipated in many ways, from direct human recognition of cues such as earthquake shaking or an initial recession of the sea, to technological warnings based on environmental sensors and data processing. This chapter reviews and evaluates the technological detection and forecasting capabilities of the U.S. tsunami warning centers (TWCs) paying specific attention to the infrastructure of the earth and ocean observation networks and to the data processing and tsunami modeling that occur at the TWCs. The next chapter discusses the centers' operations, their human resources, and the infrastructure for their warning functions.

The initial decisions by the TWCs to issue an initial tsunami advisory, watch, or warning after an earthquake are based on analyses of data from a global seismic detection network, in conjunction with the historical record of tsunami production, if any, at the different seismic zones (see Weinstein, 2008; Whitmore et al., 2008 for greater detail on the steps taken). Although adequate for most medium-sized earthquakes, in the case of very large earthquakes or tsunami earthquakes[1] the initial seismological assessment can underestimate the earthquake magnitude and lead to errors in assessing the tsunami potential (Appendix G). Far from the tsunami source, data from sea level networks provide the only rapid means to verify the existence of a tsunami and to calibrate numerical models that forecast the subsequent evolution of the tsunami. Near the source, a tsunami can come ashore before its existence is detected by the sparse sea level observation network.

Two separate U.S. TWCs monitor seismic activity and sea levels in order to detect tsunamis and warn of their presence. Based on their own data analysis, the TWCs independently decide whether to issue alerts to the emergency managers in their respective and complementary areas of responsibility (AORs). The TWCs must not only provide timely warnings of destructive tsunamis, but also must obviate needless evacuations that can cost money and even lives. An ideal warning would provide emergency managers with the necessary information to call for an evacuation in a timely fashion at any particular location in the projected tsunami path. The ideal product would also be clearly worded so that the general public easily understands the threat and who is affected by the threat. This information includes predictions of the time of arrival of the ocean waves, the duration of the occurrence of damaging waves, when the larg-

---

[1] An earthquake that produces an unusually large tsunami relative to the earthquake's magnitude (Kanamori, 1972).

est wave is expected to arrive, the extent of the inundation and run-up, and the appropriate time to cancel the warning. Whether a call for evacuation is practicable, and how soon the "all clear" can be sounded, will depend on many factors, but especially on how soon the tsunami is expected to arrive and how long the damaging waves will continue to come ashore. Therefore, the warning system needs to be prepared to respond to a range of scenarios. They range from a near-field tsunami that arrives minutes after an earthquake to a far-field tsunami that arrives many hours after a triggering, distant earthquake yet lasts for many more hours due to the waves' scattering and reverberation along their long path to the shore. In the case of the near-field tsunami, major challenges remain to provide warnings on such short timescales.

The committee concludes that the global networks that monitor seismic activity and sea level variations remain essential to the tsunami warning process. The current global seismic network is adequate and sufficiently reliable for the purposes of detecting likely tsunami-producing earthquakes. However, because the majority of the seismic stations are not operated by the TWCs, the availability of this critical data stream is vulnerable to changes outside of the National Oceanic and Atmospheric Administration's (NOAA's) control. The complex seismic processing algorithms used by the TWCs, given the available seismic data, quickly yield adequate estimates of earthquake location, depth, and magnitude for the purpose of tsunami warning, but the methodologies are inexact. Recommendations to address these two concerns fall under the following categories: (1) prioritization and advocacy for seismic stations; (2) investigation and testing of additional seismic processing algorithms; and (3) adoption of new technologies.

The tsunami detection and forecasting process requires near-real-time[2] observations of tsunamis from *both* coastal sea level gauges *and* open-ocean sensors (such as provided by the Deep-ocean Assessment and Reporting of Tsunamis (DART) network). The committee finds that the upgrades enabled by the enactment of the Tsunami Warning and Education Act (P.L. 109-424) to both coastal sea level gauges and the DART network have significantly improved the capacity of the TWCs to issue timely and accurate tsunami advisories, watches, and warnings. Furthermore, these sensors provide researchers with the essential data to test and improve tsunami generation, propagation, and inundation models after the fact.

The new and upgraded DART and coastal sea level stations have closed significant gaps in the sea level observation network that had left many U.S. coastal communities subject to uncertain tsunami warnings. Although both sea level gauge networks have already proven their value for tsunami detection, forecasting, and model development, fundamental issues remain concerning gaps in coverage, the value of individual components of the network, and the risk to the warning capability due to coverage gaps, individual component failures, or failures of groups of components. Of special concern is the relatively poor survivability of the DART sta-

---

[2] The report generally uses the term near-real-time rather than real-time. Near-real-time data are returned by geophysical instruments after a variety of intermediary processes including filling a data buffer (e.g., with a length of a second or more) and transferring data through various switches and routers in the Internet. Normally the resulting latency can be as little as a second, several seconds, or minutes associated with the Internet connection modality (e.g., satellite, fiber optics, or network switches). Real-time data can generally be achieved only with very special sampling and transmission protocols.

tions that currently average a little over one year before failure, compared to a four-year design lifetime. Additional open questions include dependence of U.S. tsunami warning activities on sea level data supplied by foreign agencies and on sea level data derived from U.S. and foreign gauges that do not meet NOAA's standards for establishment, operation, and maintenance.

Looking to the future, the committee concludes that the numbers, locations, and prioritizations of the DART stations and coastal sea level gauges should not be considered static, in light of constantly changing fiscal realities, survivability experience, maintenance cost experience, model improvements, new technology developments, and increasing or decreasing international contributions. The committee finds of great value NOAA's continual encouragement and facilitation of researchers, other federal and state agencies, and nongovernmental organizations (NGOs) who utilize their sea level observations for novel purposes. The committee believes that stations with a broad user base have enhanced sustainability.

The committee is optimistic that continued enhancements to the sea level monitoring component of the U.S. Tsunami Program can measurably mitigate the tsunami hazard and protect human lives and property for far-field events. The committee's recommendations for the DART and coastal sea level gauge networks fall under the following categories: (1) assessment of network coverage; (2) station prioritization; (3) data stream risk assessment and data availability; (4) cost mitigation and cost prioritization; and (5) sea level network oversight.

Similar to open-ocean tsunami detection, tsunami forecast modeling has only recently become operational at the TWCs, as described below. The committee anticipates that further development and implementation of numerical forecast modeling methodologies at the TWCs will continue to help improve the tsunami warning enterprise.

As described below, the rapid detection of a tsunami striking within minutes to an hour, either for the purpose of providing an initial warning or for confirming any natural warnings that near-field communities have already received, will likely require consideration of alternative detection technologies, such as sensors deployed along undersea cabled observatories and coastal radars that can detect a tsunami's surface currents tens of kilometers from the shore. Finally, examples of other new technologies and methodologies that have the potential to improve both estimation of earthquake parameters and tsunami detection are discussed at the end of this chapter.

DETECTION OF EARTHQUAKES

All initial tsunami warnings are based on rapid detection and characterization of seismic activity. Because of the fundamental differences in nature between the solid earth in which an earthquake takes place and the fluid ocean where tsunami gravity waves propagate, the vast majority of earthquakes occurring on a daily basis do not trigger appreciable or even measurable tsunamis. Nevertheless, some smaller earthquakes could trigger submarine landslides that can result in local tsunamis. It takes a large event (magnitude >7.0) to generate a damaging tsunami in the near-field and a great earthquake (magnitude >8.0) to generate a tsunami in the far-field. However, the generation of a tsunami is affected not only by the magnitude of an earthquake, but also by material conditions at the source, such as source focal geometry, earthquake source depth, and water depth above the fault-rupture area.

Although estimating the size of a tsunami based on the magnitude of an earthquake has severe limitations (see Appendix G), the initial warning from a seismically generated tsunami is still based on the interpretation of the parent earthquake for several reasons:

- most tsunamis are excited (or initiated) by earthquakes;
- earthquake waves are easy to detect, and seismic instrumentation is available, plentiful, and accessible in near-real time (latencies of seconds to a few minutes);
- most importantly, seismic waves travel faster than tsunamis by a factor of 10 to 50, thereby allowing an earthquake to provide an immediate natural warning for people who feel it while leaving time for instrumental seismology to trigger official warnings for coasts near and far from the tsunami source; and
- earthquakes have been studied, and their sources are reasonably well understood.

Although most tsunamis result from earthquakes, some are triggered by landslides or volcanic eruptions. Technological warning of a tsunami that has been generated without a detectable earthquake will likely require detection of the tsunami waves themselves by water-level gauges.

Seismic Networks Used by the Tsunami Warning Centers

Both TWCs access the same extensive seismic networks that provide near-real-time information on earthquakes from around the world. Currently, about 350 independent channels of seismic data are monitored and recorded by the TWCs (National Oceanic and Atmospheric Administration, 2008a; Figure 4.1). Seismic networks that provide these data are operated and funded by many different agencies and organizations, including the U.S. Geological Survey (USGS), the National Science Foundation (NSF), the National Tsunami Hazard Mitigation Program (NTHMP), the UN Comprehensive Nuclear Test-Ban Treaty Organization (CTBTO), various universities in the United States, non-U.S. networks, and stations run by the Pacific Tsunami Warning Center (PTWC) and the West Coast/Alaska Tsunami Warning Center (WC/ATWC) themselves. Many of the networks used by the TWCs are part of the USGS/NSF Global Seismographic Network (GSN), which currently comprises more than 150 globally distributed, digital seismic stations and provides near-real-time, open access data through the Data Management System (DMS) of the Incorporated Research Institutions for Seismology (IRIS). The IRIS DMS also serves as the primary archive for global seismic data. GSN is a partnership between the NSF/IRIS and the USGS. The TWCs access seismic network data through dedicated circuits, private satellites, and the Internet.

The GSN is widely recognized as a high-quality network, having achieved global coverage adequate for most purposes, with near-real-time data access as well as data quality control and archiving (National Science Foundation, 2003; Park et al., 2005). GSN stations have proven

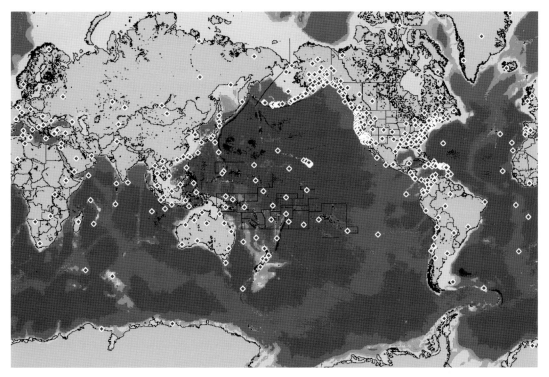

FIGURE 4.1 Data from approximately 350 seismic stations are accessed by the TWCs. SOURCE: West Coast/Alaska Tsunami Warning Center, NOAA.

to be reliable, with current (2009-2010) data return rates of 89 percent. The GSN is sufficiently robust to support warnings for events far from the recording devices and provides good global coverage (U.S. Indian Ocean Tsunami Warning System Program, 2007). The USGS was provided funding through the Emergency Supplemental Appropriations Act for Defense, the Global War on Terror, and Tsunami Relief, 2005 (P.L. 109-13) to expand and upgrade the GSN for tsunami warning. For redundancy, the TWCs also receive seismic data from many other vendors on multiple communication paths. Given the wide array of uses of the existing seismic networks, GSN can generally be viewed as a data network that is likely to be continued, well-maintained, and improved over the long-term. A future broad upgrade of seismometers in the GSN may be important for tsunami warning.

Nevertheless, the TWCs' heavy reliance on data networks from partnering agencies exposes them to some degree of vulnerability to potential losses of data availability in the future. For example, much of the seismic data crucial to the operation of the TWCs comes from GSN stations whose deployment and maintenance have been and are currently funded primarily from NSF cooperative agreements with IRIS, renewable every five years. The Scripps Institution of Oceanography's (SIO's) International Deployment of Accelerometers (IDA) project with

NSF/IRIS funding operates 41 of the total 150 GSN stations through this mechanism. There can be no assurance that this funding will be sustained at current levels in the future. GSN stations have been operating since the mid-1980s (see Appendix G); much of their hardware is out of date and increasingly difficult to maintain. Operations and maintenance budgets regularly decrease and, except for events like the 2004 tsunami, modernization funds are generally not available to boost the data return rates including the necessary hardware. The more modern NSF EarthScope Transportable Array (with more than 400 telemetered broadband stations), for example, boasts data return rates in excess of 99 percent. Unfortunately, the TWCs could be among the most vulnerable of the IRIS clients in a constrained budget environment, because the TWCs are among the users needing some of the most remote seismic stations, which are difficult, hence expensive, to maintain.

To meet the requirements for detection of near-field tsunami events, the TWCs have supplemented existing seismic networks with their own local stations. The WC/ATWC maintains a network of 15 sites throughout Alaska, and most stations were upgraded to satellite communications and broadband seismometers after 2005 (National Oceanic and Atmospheric Administration, 2008a). The PTWC, in collaboration with other partners, is also working to enhance an existing seismic network in Hawaii to improve tsunami and other hazard detection capabilities through a Hawaii Integrated Seismic Network (Shiro et al., 2006).

NOAA's Tsunami Program Strategic Plan (2009-2017; National Oceanic and Atmospheric Administration, 2008b) recommends that the TWCs "monitor critical observing networks, establish performance standards, and develop a reporting protocol with data providers" (e.g., the USGS and the NTHMP) and effect "complete upgrades of Alaska and Hawaii seismic ... networks." The committee agrees with these recommendations; however, to be strategic with limited resources, it is essential to determine and prioritize seismic stations that are critical to tsunami warning (e.g., oceanic stations in known tsunamigenic source regions or within 30°-50° from potential tsunami source areas to allow the more rapid determination of the tsunami potential).

Algorithms for Estimating an Earthquake's Tsunami Potential

Once data from the seismic networks have been received, the data are analyzed by the TWCs to determine three key parameters for evaluating tsunamigenic potential: location, depth, and magnitude of an earthquake. Algorithms for determining the geographical location and depth of an earthquake source from seismic arrival times are based upon the concept of triangulation (U.S. Indian Ocean Tsunami Warning System Program, 2007). With the network of stations available to the TWCs, automatic horizontal locations are routinely obtained within a few minutes of origin time with accuracy on the order of 30 km. This is more than satisfactory to determine tsunami source locations, given the fact that earthquakes of such high magnitudes have much larger source areas. The three seismic parameters are used for issuing the initial bulletin. The focal mechanism characteristics are later obtained through moment tensor inversion of broadband seismic data if the data quality is adequate (see below). In the present configuration of worldwide networks, the large number of available stations provides robust location determination, although losing a significant number of seismic stations could affect the accuracy of earthquake location and depth.

A great earthquake on a subduction thrust tends to nucleate beneath shallow water, or even beneath land in the case of the giant 1960 Chile and 1964 Alaska earthquakes. The source of such an earthquake, and of the ensuing tsunami, extends far beyond the earthquake's point of nucleation (the hypocenter, on the fault plane; the epicenter, if projected to the earth's surface). What matters for earthquake size, and for tsunami size as well, is the fault-rupture area, which extends seaward into deep water as well as coastwise. The hypocenter is much like the match that initiates a forest fire in which the damage depends on the total area burned. The tendency to instead equate an earthquake with its hypocenter contributed to confusion during the near-field tsunami from the February 27, 2010, Chilean earthquake of magnitude 8.8. Partly because this earthquake's hypocenter was located near the coast, the Chilean government retracted a tsunami warning before the largest waves came ashore.

Depth determination is crucial to assessing an earthquake's tsunamigenic potential because sources deeper than about 60 km generally pose no tsunami threat and are well resolved by location algorithms. Finer resolution of depth for shallower earthquakes remains a general seismological challenge, particularly in near-real time. This parameter can have some influence on the generation of tsunamis in the near-field; however, for far-field tsunamis generated by megathrust earthquakes, theoretical studies (Ward, 1980; Okal, 1988) have shown that the probability of tsunami excitation is moderate for depths less than 60 km. This somewhat paradoxical result reflects the fact that a shallower source may create a locally larger deformation of the ocean floor, but over a smaller area. This acts to compensate for the effect on the generation of the tsunami, which is controlled by the integral of the deformation over the whole ocean floor. Given the techniques and data available, the committee found that the location techniques used at the TWCs (Weinstein, 2008; Whitmore et al., 2008) were adequate in the context of tsunami warning.

Determining an earthquake's magnitude is a more problematic aspect of the initial earthquake parameterization. The concept of magnitude is probably the most popular, yet most confusing, parameter in seismology. In simple terms, it seeks to describe the size of an earthquake with a single number. Reliable and well-accepted determinations of earthquake size (the "moment tensor solution"—or the product of fault area with the amount of slip) are possible, but these estimates are necessarily based on long-period surface waves arriving too late to be useful for tsunami warning, which strives for initial estimates within five minutes of the first measurements having been received. Most seismologists agree that it is not currently possible to predict how much of a fault will ultimately break based on the seismic waves propagating away from the point of nucleation (the epicenter), and that only when the slip ends can the true size or moment be inferred. For an event such as the Sumatra earthquake, the propagation of breakage along the fault surface alone takes nearly eight minutes (e.g., de Groot-Hedlin, 2005; Ishii et al., 2005; Lay et al., 2005; Tolstoy and Bohnenstiehl, 2005; Shearer and Bürgmann, 2010). Magnitudes determined at shorter times will necessarily underestimate the true size of the earthquake.

In this regard, the major challenge for tsunami warning is that tsunamis are controlled by the lowest frequency part of a seismic source, with periods of 500 to 2,000 seconds, whereas routinely recorded seismic waves have energy in the treble domain, with periods ranging from 0.1 to 200 seconds, exceptionally 500 seconds. In addition, seismic waves fall into several categories. Body waves travel through the interior of the earth at average velocities of 10 km/sec, take seconds to minutes to reach recording stations, and their high-frequency components are a good source of information. By contrast, surface waves travel around the surface at considerably slower speeds (3-4 km/sec) and take as much as 90 minutes to reach the most distant stations. The surface waves carry low-frequency signals; that is, the part of the spectrum most relevant to tsunami warning, although high-frequency body wave methods can also resolve event duration and rupture length (e.g., Ishii et al., 2005; Ni et al., 2005). For this latter case, the high-frequency body waves have not yet been exploited by the USGS's National Earthquake Information Center (NEIC) or the TWCs. In short, the evaluation of earthquake size for tsunami warning faces a double challenge: extrapolating the trebles in the earthquake source to infer the bass, and doing this as quickly as possible to give the warning enough lead time to be useful.

Magnitudes can be obtained from various parts of the seismic spectrum, and expectedly such different scales have been "locked" to each other to quantify an earthquake with a single number. This is achieved through the use of "scaling laws," which assert that the spectrum of a seismic source (the partitioning of its energy between bass and treble) is understood theoretically and can be estimated as a function of earthquake size. However, this universal character of scaling laws is far from proven, especially in its application to mega-earthquakes, which trigger the far-field tsunamis of major concern. In addition, scientists have identified a special class of generally smaller events, dubbed "tsunami earthquakes" by Kanamori (1972), whose source spectra systematically violate scaling laws (see Appendix G). Therefore, characterizing an earthquake source with a single number representing magnitude cannot describe all its properties, especially in the context of tsunami warning.

A detailed technical review of these topics is given in Appendix G, and the special case of tsunami earthquakes is reviewed in Appendix H. A summary of the conclusions of Appendix G are:

- Classical magnitudes routinely determined by conventional seismological methods are inadequate for tsunami warning of great and mega-earthquakes.
- The authoritative measurement of earthquake size, the moment tensor solution, is based on normal modes and long-period surface waves arriving too late to be used for tsunami warning.
- The TWCs currently use an algorithm named M_{wp} which integrates the long-period components of the first arriving P-waves to infer the low-frequency behavior of the seismic source.
- PTWC has recently implemented the use of the "W-phase" algorithm as well as the M_{wp} algorithm.

Although the use of M_{wp} is satisfactory for the majority of the (small, non-tsunamigenic, and medium) events processed, M_{wp} has very serious shortcomings in its application to great earthquakes (magnitude greater than 8.0), to mega-earthquakes (magnitude greater than 8.5; Appendix G), and to the anomalous tsunami earthquakes (Whitmore et al., 2002; Appendix H).

Thus, the committee is concerned that the TWCs have relied on a single technique applied without sufficient attention to its limitations discussed above. Other approaches are presently being studied including the "W-phase" algorithm, which could eventually be implemented after both the theoretical and operational bases of the approach are established and the limitations of current technologies are understood (Appendix G). Improvements are urgently needed for the determination of the tsunami potential of mega- and tsunami earthquakes.

Potential Use of Earthquake Alerts from the NEIC

While NOAA and the NTHMP lead the efforts relevant to tsunamis, the USGS and the National Earthquake Hazards Reduction Program (NEHRP) lead the efforts in research and re-ducing impacts from earthquakes. The USGS' Earthquake Hazard Program provides and applies earthquake science information to mitigate potential losses from earthquakes. This separa-tion in mission runs the risk of developing tsunami efforts that neglect the earthquake hazard within NOAA and vice versa within the USGS.

One service the USGS provides through its NEIC is to rapidly determine the location and size of earthquakes around the world. The NEIC in Golden, Colorado, derives initial solutions, not made public, within seconds after arrival of the seismic data. The NEIC monitors the GSN and other stations and produces accurate seismic analysis within minutes of an event, which it disseminates to a broad range of customers (national and international agencies, academia, and the public). In a development that may influence the methods and roles of the TWCs, U.S. seismology is on the verge of being able to warn of earthquakes while they are still under way. The drive toward such earthquake early warning includes the NEIC. USGS sources say that the NEIC, which began operating 24/7 in January 2006, plans to support this warning function by developing a back-up center at a site other than Golden. At present, the two TWCs do not use the epicentral, hypocentral, or magnitude estimate provided by the NEIC. Instead, each TWC uses its own mix of seismic processing algorithms and as described above develops its own seismic solutions. The TWCs may correct their initial estimates, which are often made public faster than the NEIC's solutions, to be more consistent with the NEIC's solutions and at times confer with NEIC staff during an event to ensure consistency. With the availability of the new tsunami forecasting methods and sea level observations (as described below), the TWCs rely more on sea level data and numerical models than on details of earthquake parameters after the issuance of the initial warning product. Therefore, the committee discussed whether it remains necessary for the TWCs to run their own independent seismic analysis. For the forecast models, the TWCs require little more than location, rough magnitude, and time of the event, which could come directly from the NEIC.

The TWCs in-house analysis offers the benefit of obtaining solutions much faster than the NEIC's publicly available solution, which might take tens of minutes longer. In addition, the TWCs' assessment of the tsunami potential of any given earthquake depends on knowing the depth of the earthquake and the earthquake's geometry, neither of which are as high of a priority for the NEIC.

Regardless, there are many benefits to leveraging research and development at the TWCs and the NEIC and to more broadly find synergies in the tsunami and earthquake hazard reduction programs.

Conclusion: The current global seismic network is adequate and sufficiently reliable for the purposes of detecting likely tsunami-producing earthquakes. Because the majority of the seismic stations are not operated by the TWCs, availability of this critical data stream is vulnerable to changes outside of NOAA's control. Furthermore, as discussed in Appendix G, many of the STS-1 seismographs in the GSN are now more than two decades old, and because the STS-1 is no longer manufactured, spares are not available.

Recommendation: NOAA and the USGS could jointly prioritize the seismic stations needed for tsunami warnings. These needs could be communicated with partner agencies and organizations to advocate for upgrading and maintenance of these critical stations over the long-term.

Conclusion: The complex seismic processing algorithms used by the TWCs, given the available seismic data, quickly produce adequate estimates of earthquake location, depth, and magnitude for the purpose of tsunami warning. The methodologies are inexact, partly because of the physically variable nature of tsunami-generating earthquakes (one model does not fit all), and partly because of the need for rapid determination of earthquake parameters that may not be certain until the entire rupture process is complete (potentially minutes). For example, the methodologies applied by the TWCs do not properly reflect the tsunami-generating potential of mega-earthquakes or tsunami earthquakes.

Conclusion: In parallel with their own analyses, staff at the TWCs and at the Tsunami Program could avail themselves of earthquake locations and magnitudes that are estimated within minutes of an event from the USGS's NEIC. An interagency agreement could be established to make these initial estimates available on secure lines between the USGS and NOAA.

Recommendation: Among the methodologies employed by the NEIC is the W-phase algorithm for estimating earthquake magnitude. The committee recommends that the TWCs work jointly with the NEIC to test the potential utility of the W-phase algorithm in the tsunami warning process, using both a sufficient dataset of synthetic seismograms and a set of waveforms from past great earthquakes, paying particular attention to the algorithm's performance during tsunami earthquakes and to the assessment of a lower-magnitude bound for its domain of applicability.

DETECTION OF TSUNAMIS WITH SEA LEVEL SENSORS

Because the seismic signal is the first observation available to the TWCs, seismic detection provides the basis for the initial evaluation of the potential for a tsunami. The decision about the content of the first message from the TWCs is based solely on seismic parameters and the historical record, if any, of tsunamis emanating from the neighborhood of the earthquake. However, as previously noted, this indirect seismic method is limited in the accuracy of its estimates of the strength of the tsunami, usually underestimating the tsunami potential of large earthquakes and tsunami earthquakes. In acknowledgment of this bias, and because forecasters must err on the side of caution when human lives may be at stake, the TWCs use conservative criteria to trigger advisories, watches, or warnings based on this initial seismic assessment (e.g., Weinstein, 2008), as seen in the PTWC's far-field forecast of the tsunami from the Chilean earthquake of February 27, 2010 (Appendix J). However, these conservative assessments might cause unwarranted evacuations, which can cost millions of dollars and might threaten lives. A TWC must, therefore, not only provide timely warning of a destructive tsunami, but also must avoid causing unnecessary evacuations with their attendant negative impacts.

The detection and forecasting process requires real-time observations of tsunamis from both coastal sea level gauges and open-ocean sensors (such as provided by the DART stations). The combination of the open-ocean and coastal sea level stations, which provide direct observations of tsunami waves, are important for adjusting and canceling warnings as well as for post-tsunami validation of models of the tsunami propagation and inundation (U.S. Indian Ocean Tsunami Warning System Program, 2007). These sea level networks can also detect tsunamis from sources that fail to generate seismic waves or are generated by an earthquake on land that generates a sub-aerial and/or a seafloor landslide. Progress to expand the ocean observing network and advances in oceanographic observing technologies allow the TWCs to incorporate the direct oceanographic detection of tsunamis into their decision processes.

Conclusion: An array of coastal and open-ocean sea level sensors is necessary until such time, in some distant future, when the capability exists of observing the entire tsunami wave-front in real-time and with high horizontal resolution (e.g., perhaps with satellites) as it expands outward from its source and comes ashore.

The Tsunami Warning Decision Process Before and After Enactment of Public Law 109-424

A majority of the funds authorized by the Tsunami Warning and Education Act (P.L. 109-424) have been used to manufacture, deploy, and maintain an array of 39 DART stations (not counting the 9 purchased and deployed by foreign agencies; http://www.ndbc.noaa.gov/dart.shtml), establish 16 new coastal sea level gauges, and upgrade 33 existing water level stations (National Tsunami Hazard Mitigation Program, 2008; http://tidesandcurrents.noaa.gov/1mindata.shtml). All these new and upgraded sea level stations, especially the DART sites, have closed large gaps in the sea level observation network that had left many U.S. coastal communities subject to uncertain tsunami warnings. Among TWC personnel and tsunami warning researchers, it is common to find sentiments echoing the following statement in Whitmore et al. (2008): "Since 2005, the amount and quality of both tide gage data and DART data [have] greatly improved. These data are critical to verify the existence of tsunamis and to calibrate models used to forecast amplitudes throughout the basin. Depending on the source location, it can take anywhere from 30 minutes to 3 hours to obtain sufficient sea level data to provide forecasts for wave heights outside the source zone, or to verify that no wave has occurred and cancel the alert. Within the AOR, upgraded sea level networks have dropped the verification time to 30 minutes in some regions."

The implementation of the EarthVu tsunami forecast system and the Short-term Inundation Forecasting for Tsunamis (SIFT) system into the TWCs (e.g., Weinstein, 2008; see Section *Forecasting of a Tsunami Under Way*) places additional emphasis on the importance of the proper operation of the sea level stations, especially the open-ocean DART stations whose sea level observations of the tsunami waves are not distorted by bathymetric irregularities and local harbor resonances that affect the coastal sea level observations. With these models and data from the sea level networks, it has become possible to make reasonably accurate predictions of the amplitude of the first tsunami wave that arrives at a given shoreline, enabling the issuance of more timely and more spatially refined watches and warnings (e.g., Titov et al., 2005; Geist et al., 2007; Whitmore et al., 2008).

Furthermore, the array of DART stations, when properly functioning, enables unique and important capabilities for both tsunami detection and forecasting as described below. Whether the current DART and coastal sea level networks are sufficient for both rapid detection of tsunamis and accurate tsunami forecasting with respect to all U.S. coastal territories is addressed below.

> **Conclusion:** The expansion and upgrades to the DART and coastal sea level network have closed large gaps in the sea level observation network that had left many U.S. coastal communities subject to uncertain tsunami warnings. These enhancements to the detection system have significantly improved the TWCs ability to detect and forecast tsunamis in a timely and more accurate fashion.

> **Conclusion:** Based on the analysis described below, the coastal and DART sea level gauge networks have proven their value for the forecasting and warning of far-field tsunamis, especially when coupled with numerical propagation and inundation models.

> **Conclusion:** Despite the improvements in detection and forecasting, some fundamental issues remain concerning gaps in coverage, the value of individual components of the network, and the risk to the warning capability due to the gaps and from individual component failures, or failures of groups of components.

The Economic Value of the DART Network

Although the foremost concern for emergency responders is the protection of human lives in the event of large tsunamis, another significant value of the DART stations is to provide assurance that a large wave has not been generated by a seismic event, permitting an initial watch or warning to be canceled expeditiously. Thus, the DART stations help to prevent unnecessary public concern and economic disruption.

Two estimates of economic benefits have been derived for Hawaii. In one, the cost of a needless evacuation in the state of Hawaii was put at $58.2 million in 1996 dollars (Hawaii Research and Economic Analysis Division, 1996, cited in Bernard, 2005). A second estimate is based on nearly identical earthquakes off the Aleutian Islands before and after the existence of the DART network. On May 7, 1986 (pre-DART), a magnitude 8.0 earthquake near the Aleutian Islands precipitated a full coastal evacuation in Hawaii at an estimated cost of $30-$40 million in lost productivity, emergency provider expenses, and other costs (Hawaii Research and Economic Analysis Division, 1996; National Science and Technology Council, 2005), yet tsunami amplitudes did not exceed 0.6 m. On November 17, 2003, a DART station offshore of the Aleutian Islands clearly showed that a sizable tsunami was not generated by a magnitude 7.8 earthquake in a similar location near the Aleutian Islands, and the watch was canceled (the subsequent maximum tsunami height reached only 0.33 m in Hawaii). Adjusting the 1986 figure for inflation, the cost to Hawaii's government and businesses in 2003 could have been $70 million had an evacuation been ordered. These findings are consistent with cost estimates associated for unnecessary hurricane evacuations along the U.S. coastline between Maine and Texas (Centrec Consulting Group, LLC, 2007).

Clearly, unwarranted evacuations can cost millions of dollars; and, although the costs associated with the loss in public confidence are less easy to quantify, the effectiveness of a warning system is ultimately grounded in credibility. Therefore, a tsunami warning system should not only provide timely warning of a destructive tsunami, but also should avoid issuing "false alarms."

Although the DART stations have their greatest value in discerning tsunami propagation characteristics in the open ocean, the inundation problem requires, ideally, sea level sensors along tsunami-prone coastlines because of the spatial variations in tsunami height that are produced by local bathymetry, coastal geometry, and the resultant system responses (e.g., coastal and harbor resonances).

Description of the Coastal Sea Level Gauge Network

Although coastal sea level stations were originally installed for monitoring tides for navigational purposes, most now serve a broad range of uses (including tsunami detection) that have contributed to their continued support and upgrades. Stations are commonly located deep within harbors or bays, where nonlinear hydrodynamic effects and local geographic complexity strongly alter the structure and amplitude of any impinging tsunami waveform. These nonlinear effects hamper the determination of open-ocean tsunami wave parameters (e.g., Titov et al., 2005) without eliminating the stations' utility for the TWCs (e.g., Whitmore, 2003).

Tide stations were typically configured to measure sea level height in a stilling well, a vertical pipe that is secured to a piling, pier, wharf, or other shore-side structure. These pipes have a small orifice(s) to allow water to enter relatively slowly thus filtering out the short period (3-30 seconds) wind waves, and even tsunamis, so that the hourly recorded sea level values from within the pipe are not aliased by the short period variability. This technology works well for measuring tides and other long period phenomena, but even if the sampling rate is increased from hourly to minutes the true tsunami signal may not be well observed given these filtering effects. Furthermore, a large tsunami can overtop a well and render it useless in extreme events. Consequently, sea level observations intended for tsunami detection are now often accomplished inside a tsunami-hardened station equipped with a rapid-sampling pressure, acoustic, or microwave sensor with an orifice set apart from the structure (National Tsunami Hazard Mitigation Program, 2008).

The most important roles for coastal sea level data in the tsunami forecasting and warning process are currently the initial detection of a tsunami, scaling the tsunami forecast models in near-real time, and post-tsunami validation of tsunami models (see Weinstein, 2008; Whitemore et al., 2008). These roles require accurate, rapidly sampled sea level observations delivered in near-real time via an appropriate telemetry system. In practice, these requirements translate into a need for sea level averages at least as often as every minute that are made available in near-real time (U.S. Indian Ocean Tsunami Warning System Program, 2007), and a need for assiduous maintenance of the sea level gauges so that near-real-time data can be trusted and will be available most of the time. Furthermore, subsequent to collection, the data need to be carefully processed through a set of rigorous quality control procedures to maximize the value for model validation after the fact (U.S. Indian Ocean Tsunami Warning System Program, 2007). As an example of the importance of high temporal data resolution, Figure 4.2 shows how sea level data sampled every six minutes completely missed the largest component (the third crest and trough) of the Kuril Islands tsunami of November 15, 2006 (the modeled wave heights of which are shown in Figure 4.3).

Coastal sea level data used by the TWCs originate from a number of different networks (PTWC, WC/ATWC, National Ocean Service (NOS), and University of Hawaii Sea Level Center (UHSLC)), which are maintained by various national and international organizations (Figure 4.4).

Ideally, these stations are maintained to the standards listed in the Tsunami Warning Center Reference Guide (U.S. Indian Ocean Tsunami Warning System Program, 2007) for sea level stations that are intended to provide data for tsunami warning. For coastal tide gauge stations, the requirements are:

- independent power and communications, for example, solar and satellite;
- fault-tolerant redundant sensors (multiple sensors for tsunami, tides, and climate);
- local logging and readout of data (local back-up of data);
- warning center event trigger (ramping up of sampling rate and transmission upon detection of a tsunami);

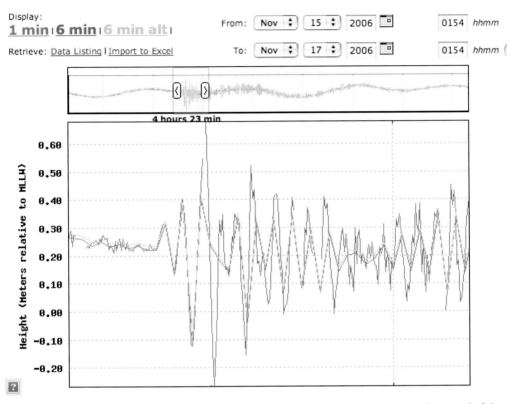

FIGURE 4.2 Sea level data from Midway Island for a short time period encompassing the arrival of the November 15, 2006, Kuril Island tsunami. One-minute samples are shown in red; two different gauges providing 6-minute samples are shown in green and orange. Note that the 6-minute samples completely miss the highest amplitude component (the third crest and trough) of the tsunami. SOURCE: http://co-ops.nos.noaa.gov/tsunami/; NOAA.

- establishment of a system of surveying benchmarks;
- locating gauges in protected areas that are responsive to tsunamis, such as wide-mouthed harbors (sustainability and filtering); and
- standard sampling of 1-minute averages and a continuous 15-minute transmission cycle via the World Meteorological Organization's (WMO) Global Telecommunications System (GTS) to the Japan Meteorological Agency (JMA), PTWC, and other appropriate warning centers/watch providers.

NOS Sea Level Stations for Tsunami Detection

In the several decades leading up to 2004, NOAA's NOS Center for Operational Oceanographic Products and Services (CO-OPS; http://tidesandcurrents.noaa.gov/) operated long-

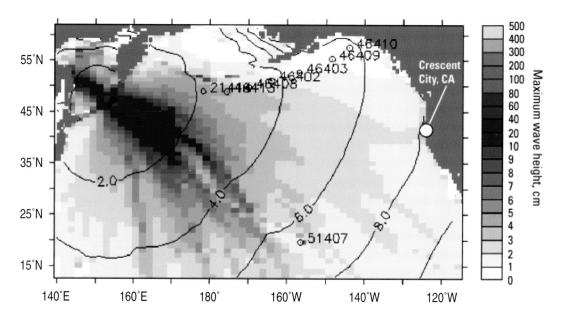

FIGURE 4.3 North Pacific Ocean, showing predicted maximum wave heights (indicated by color) and arrival times (contour lines labeled with numbers representing hours after the triggering earthquake) of tsunami waves generated by a magnitude 8.3 earthquake near the Kuril Islands on November 15, 2006. The predicted wave heights illustrate the phenomenon of "tsunami beaming"—the tendency of tsunami waves in the open ocean to be highest along azimuths approximately perpendicular to the subduction zone where the triggering earthquake occurred. Note the minor beam aimed at Crescent City, California, where the boat harbor was damaged, largely by secondary tsunami waves. SOURCE: Geist et al., 2007; with permission from Vasily Titov, NOAA/PMEL.

term tide stations, and the National Weather Service (NWS) utilized the data to support the national tsunami warning system. However, following the devastating 2004 Indian Ocean tsunami, and with the support authorized in P.L. 109-424, CO-OPS began a system-wide upgrade of its instrumentation. This upgrade increased the rate of data collection to 15-second and 1-minute sampling (National Tsunami Hazard Mitigation Program, 2008) and increased the rate of transmission (to every 6 minutes) at its coastal National Water Level Observation Network (NWLON; http://tidesandcurrents.noaa.gov/nwlon.html) stations. The increased data sampling and transmission rates advance the objectives of tsunami detection and warning, as well as to provide critical inundation model input. In addition to upgrading equipment at 33 existing long-term NWLON stations, CO-OPS collaborated with the TWCs and the Pacific Marine Environmental Laboratory (PMEL) to establish 16 new tide stations at high-priority locations in Alaska, the Pacific Islands, the U.S. West Coast, and the Caribbean, increasing the geographic coverage of water level observations in tsunami-vulnerable locations. This initiative was completed in 2007 (National Tsunami Hazard Mitigation Program, 2008; http://tidesandcurrents. noaa.gov/1mindata.shtml).

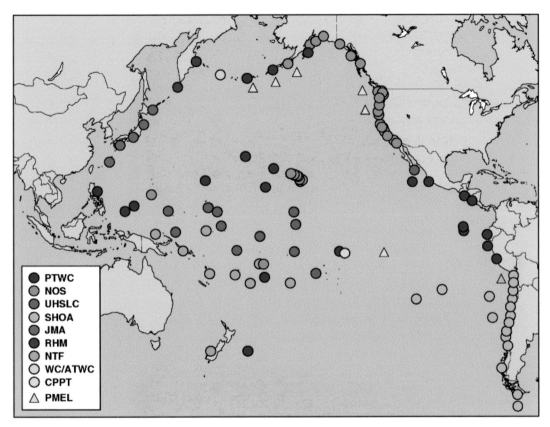

FIGURE 4.4 Map of the coastal sea level stations in the Pacific basin that provided sea level data at sufficient temporal resolution and quality for use in the PTWC's tsunami detection activities in 2008. Color codes indicate the authorities responsible for gauge maintenance. U.S. authorities include PTWC, WC/ATWC, NOS, and UHSLC. Non-U.S. authorities include the following: Centre Polynésien de Prévention des Tsunamis (CPPT; France); Servicio Hidrográfico y Oceanográfico de la Armada de Chile (SHOA); Japan Meteorological Agency (JMA); ROSHYDROMET (RHM; Russia); and National Tidal Facility (NTF; Australia). The positions of the original six DART buoys (yellow triangles) existing in 2005 before the enactment of P.L. 109-424 are also displayed. SOURCE: Weinstein, 2008; Pacific Tsunami Warning Center, NOAA.

At the current time, CO-OPS operates tide stations on all U.S. coasts in support of tsunami warning. Upgraded tide stations are equipped with new hardware and software to enable the collection and dissemination of 1-minute water level sample data. The TWCs can receive this data in near-real time either via Geostationary Operational Environmental Sattelites (GOES) over the National Weather Service Telecommunication Gateway (NWSTG) or via the Tsunamis Stations' website (http://tidesandcurrents.noaa.gov/tsunami/). Although near-real-time data are not subjected to the National Ocean Service's quality control or quality assurance procedures and do not meet the criteria and standards of official National Ocean Service data, the stringent maintenance procedures for the NWLON stations maximize the probability of a reliable data stream in near-real time. In addition to having access to raw water level data via satellite transmission, CO-OPS collaborated with the TWCs to develop a webpage (http://co-ops.nos.noaa.gov/1mindata.shtml) to disseminate 1-minute water level data. This webpage allows users to view both 6- and 1-minute data numerically or graphically for all tsunami-capable tide stations in increments of up to 4 days (Figure 4.2 is one example). Like the near-real-time data, all water level data displayed through the CO-OPS tsunami webpage are raw and unverified at this time. However, verified 6-minute sea level data are available through another website (http://tidesandcurrents.noaa.gov/station_retrieve. shtml?type=Historic+Tide+Data), usually within 2 months of collection, which enables the user to easily evaluate the quality of the 1-minute data, although well after the occurrence of the tsunami. The 15-second data, potentially more useful for model validation, are not telemetered on a regular basis, but are available to the TWCs via remote phone dial-in.

The NOAA/NOS has developed and rigorously follows a set of standards for the establishment, operation, and maintenance of its critical NWLON coastal sea level stations. As well, NOAA describes in its Tsunami Warning Center Reference Guide (U.S. Indian Ocean Tsunami Warning System Program, 2007) the performance and maintenance standards it recommends for sea level stations that are intended to aid tsunami detection, forecasting, and warning activities. Unfortunately, the high-quality NOS NWLON stations make up only a small portion of all the sea level observation stations needed for tsunami detection (Figure 4.4). Whether sea level gauges operated and maintained by other U.S. agencies satisfy, or can be upgraded to, the standards of the NWLON stations, or whether these other U.S. stations should be operated and maintained under the NWLON program, are questions that remain unanswered. In addition, the committee is not aware of any process by which the non-NOS sea level stations (U.S. or international) are evaluated or certified relative to these standards. How much of a risk occurs as a result of the TWC's reliance on un-certified sea level gauges is not known.

The University of Hawaii Sea Level Center (UHSLC) Stations

The UHSLC (http://ilikai.soest.hawaii.edu/) maintains and/or operates a worldwide array of sea level observing stations, some of which are employed in the tsunami detection and warning process (for the Pacific Ocean, see Figure 4.4). The UHSLC is a research facility of the University of Hawaii/NOAA Joint Institute for Marine and Atmospheric Research (JIMAR) within the School of Ocean and Earth Science and Technology (SOEST). The mission of the UHSLC is to collect, process, distribute, and analyze in-situ sea level gauge data from around the world in support of climate research. Primary funding for the UHSLC comes from NOAA's Office of Global Programs (OGP). In recent years, the UHSLC, recognizing the potential importance of its stations to tsunami hazard mitigation, has upgraded many of its stations to short period sampling and reporting (http://ilikai.soest.hawaii.edu/RSL1/index.html).

Because of the UHSLC's climate research mission, which includes ascertaining the small (typically, 1-3 mm) annual sea level rise associated with global warming, the UHSLC strives for high operational standards and data quality. It is not known whether the UHSLC's operational standards meet or exceed the NOS NWLON maintenance standards.

TWC Sea Level Stations

The TWCs operate a small subset of coastal tide stations (Figure 4.4). The WC/ATWC operates seven stations along southern Alaska and the Aleutian Islands with data being archived for public use at National Geophysical Data Center (NGDC) (http://wcatwc.arh.noaa.gov/ WCATWCtide.php). The PTWC stations are distributed throughout the Pacific and Hawaii. In Hawaii, PTWC maintains 14 sea level gauges solely for local predictive and diagnostic value; the data from these gauges are archived under separate NOAA support (http://ilikai.soest.hawaii. edu/arshsl/techrept/arshsl.html). In general, the TWC stations are not maintained to the specifications of the NWLON but have historical precedence and fill gaps in the observing array or fill specific local needs. For example, the PTWC gauges on the Big Island of Hawaii will provide about 20 minutes of warning for Honolulu should a large amplitude tsunami be generated by an earthquake or landslide on the Big Island. The TWCs have indicated they do not have the resources to properly maintain these gauges or to process, distribute, and archive the data.

International Sea Level Stations

The Intergovernmental Oceanographic Commission (IOC) Global Sea Level Observing System (GLOSS) has about 290 stations worldwide, and many are configured for near-real-time reporting of rapidly sampled data relevant to tsunami applications. After the Indian Ocean tsunami of December 26, 2004, the IOC established a centralized Sea Level Station Monitoring Facility (http://www.vliz.be/gauges/), where most of the needed, rapidly sampled coastal sea level observations are now available and reported in near-real time over the World Meteorological Office's (WMO's) Global Telecommunications System (GTS). The website serves as a central clearinghouse of data from a range of international providers, including the data sources mentioned above. The objectives of this service are to provide information about the operational status of global and regional networks of near-real-time sea level stations and to provide a display service for quick inspection of the raw data stream from individual stations.

Since 2007, the Sea Level Station Monitoring Facility (SLSMF) also has the information necessary to determine data stream reliability. SLSMF is an appropriate place to obtain such reliability information because it lists only data that were initially made available in near-real time over the GTS, not what was eventually available after internal memory was finally accessed during a maintenance operation. The sea level data that the TWCs employ in their tsunami detection activities and which are acquired via the GTS are essentially the same data now disseminated and archived at SLSMF, excluding the TWCs' own stations discussed above. As with the data received by the TWCs via the GTS after a tsunami-producing earthquake, the data flowing through SLSMF are not quality controlled, but the website provides additional metadata for most of the non-U.S. stations. To the committee's knowledge, the level of adherence of international stations used by the TWCs to either NWLON or Tsunami Warning Center Reference Guide (U.S. Indian Ocean Tsunami Warning System Program, 2007) performance and maintenance standards has not been determined.

Adequacy of the Geographical Coverage of the Coastal Sea Level Gauge Network

Following the disastrous 2004 Indian Ocean tsunami, many additional global sea level observing stations have become available for the purpose of tsunami detection and warning, including those enabled in the United States by P.L. 109-424. Despite this increase in the number of near-real-time-reporting, rapid-sampling coastal sea level gauges, a map of the sea level station coverage (e.g., Figure 4.4 for the Pacific Ocean) reveals that large regions with no coverage remain, such as Central America and southern Mexico, the Kuril Islands north of Japan, and most of the Caribbean Islands, as pointed out previously (Bernard et al., 2007). In addition, this dependence on data supplied by foreign agencies, although mitigated somewhat by the redundancies and overlaps in coverage, exposes a vulnerability of the tsunami detection and warning activities to potential losses in data availability.

A recent earthquake in the Caribbean illustrates the issue of coverage. On May 27, 2009, a magnitude 7.3 earthquake occurred off the coast of northern Honduras. Eight minutes after the earthquake, the PTWC issued a Tsunami Watch for Honduras, Belize, and Guatemala. Worst-case-scenario tsunami forecast models suggested tsunami amplitudes up to nearly 1 m given initial earthquake source parameters. No rapidly sampled, near-real-time sea level gauges exist in the western Caribbean, so the PTWC could only wait for visual reports. After 74 minutes, the PTWC canceled the watch based on the following, in the PTWC's own words: " … This center does not have access to any real-time sea level gauges in the region that would be used to quickly detect and evaluate the tsunami if one were present. However, enough time has passed that any nearby areas should already have been impacted. Therefore, this center is canceling the tsunami watch it issued earlier" (Pacific Tsunami Warning Center Message, May 27, 2009).

Gaps in the coastal sea level network exist, such as revealed by the Honduran earthquake in May 2009. No analysis has been undertaken to evaluate critical coverage gaps with regards to the tsunami warning decision process. Furthermore, no analysis has been undertaken to determine the relative importance of each existing coastal sea level gauge to the tsunami warning decision and evacuation decision processes. Although there is some degree of redundancy in coverage in the current sea level gauge network for some purposes, there has been no evaluation of the associated risk and the vulnerability of the warning process to failures of single or multiple stations.

The spacing of sea level gauges for the purpose of tsunami detection is sparse, because it is now known that tsunamis can be quite directional, focusing the majority of their energy within a narrow sector, perpendicular to the seafloor rupture direction. For instance, Figure 4.3 displays the modeled beam pattern of a small tsunami generated by a large (magnitude 8.3) Kuril Islands earthquake on November 15, 2006. Given the array of sea level gauges in Figure 4.4, it is obvious that the maximum amplitudes of this tsunami were not observed in near-real time. Because DART stations were not yet in place off the Kuril Islands, only the Midway Island (28.2° N, 177.4° W) station at the far northwest end of the Hawaiian archipelago provided significant advance notice to forecasters of the possible size of the tsunami at the main Hawaiian Island to the southeast. Had the Midway Island station been temporarily inoperative, forecasters would have been forced to issue a warning at Hawaii, given the magnitude of the earthquake, with a subsequent costly and time-consuming evacuation of coastal zones. As it was, the Midway Island record confirmed that the tsunami was not going to significantly threaten lives or property in the main Hawaiian island, and no evacuation order was issued.

After a similar Kuril Island earthquake on October 4, 1994, the lack of direct confirmation of the existence of a tsunami (including lack of high-resolution sea level data from the temporarily inoperative Midway Island station) resulted in the issuance of a warning that precipitated an unnecessary evacuation of Hawaii's coastal zones.

Although many gaps exist in the sea level network for rapid tsunami detection, limitations in U.S. and international resources preclude immediate closure of all gaps, and some of these gaps are more important than others. A sophisticated analysis is needed to evaluate critical coverage gaps for coastal sea level gauges to inform the warning decision process. Ideally, such a study would include an evaluation of a region's tsunami-producing potential, sensitivity analysis of source location, tsunami travel time, local population density, timing for initial warning versus evacuation decision process for communities at risk, and warning/evacuation time gained for additional station coverage. Such an analysis could also determine the relative importance of each existing coastal sea level gauge to the tsunami warning decision and evacuation decision processes. Although there is some degree of redundancy in coverage in the current sea level gauge network, there has been no evaluation of the associated risk and the vulnerability of the system to failures of single or multiple stations. It is possible that isolated gauges near historically tsunami-producing seismic zones would be considered highly important, while individual gauges among a relatively compact group of gauges might be considered less important (although the need for at least one gauge within the group might be considered highly important). Such an assessment of the relative importance of existing gauges could then be the basis of prioritization for maintenance schedules and enhancement opportunities, and for the identification of critical stations that are not under U.S. control and may require augmentation with new U.S. gauges as well as operations and maintenance support.

In order to mitigate the cost of enhancing and maintaining tsunami-useful sea level monitoring stations, the U.S. Tsunami Program could continue coordinating with other programs interested in monitoring sea level variability for other purposes, such as climate variability. Sea level stations maintained by the NOS, UHSLC, etc., have evolved from their primary missions to include higher sampling and reporting rates to serve the tsunami community. Coastal stations with a broad user base have enhanced sustainability.

Reliability of the Coastal Sea Level Gauge Network

International coastal sea level networks vary greatly in station density, transmission rates, and data quality. Improved near-real-time international sea level data observations are crucial to proper TWC response for events distant to U.S. territories, and are necessary for the TWCs to provide advice to their international customers.

Recommendation: Two important concerns regarding the entire coastal sea level network employed by the TWCs in their warning activities need to be addressed soon, as follows:

- A priority list of the coastal sea level stations should be constructed, based at first on the experience of the TWC forecasters, and later updated from the results of the more objective coverage analysis described in the previous section.
- A risk assessment of the data flow from the highest priority stations should be performed.

U.S. or international stations deemed high priority with a high risk that the data flow could be interrupted for more than very short periods of time should thereafter be carefully monitored and, if possible, upgraded by the appropriate authority (national or international) to meet all requirements for a tsunami monitoring sea level station that are listed in the Tsunami Warning Center Reference Guide (U.S. Indian Ocean Tsunami Warning System Program, 2007). As an example of prioritization, note that as of June 26, 2009, all five DART stations covering the Aleutian Islands west of the Dateline, and the Kuril Islands to Hokkaido, had been inoperative for nearly all of 2009. Such failures meant that the Midway Island coastal station at 28.2° N, 177.4° W, was the only sea level station that forecasters had available during the first six months of 2009 to evaluate whether a tsunami created in the Kuril Island, and directed toward the southeast (e.g., Figure 4.3), was bearing down on Hawaii. Therefore, the Midway Island station is a strong candidate for high-priority status.

Compliance with the Reference Guide's recommendations would be a good starting point for assessing the risk in the data flow from each high-priority sea level station. Much of the needed information is now available at the IOC's SLSMF (http://www.vliz.be/gauges/) discussed previously. SLSMF also has the information needed to determine data stream reliability, at least since 2007. SLSMF is actually a very appropriate place to obtain such reliability information because it lists only data that was initially made available in near-real time over the Global Telecommunications System, not what was eventually available after internal memory was finally accessed during a maintenance operation.

Coastal Sea Level Data Processing

In January 2008, NTHMP issued a report (National Tsunami Hazard Mitigation Program, 2008) intended to identify vulnerabilities in the U.S. environmental data streams needed by the TWCs to effectively detect tsunamis and make accurate tsunami forecasts. The data streams under consideration included, among others, sea level data from DART buoys and from U.S. coastal gauges. The committee identified findings in NTHMP (2008) with respect to processing, distribution, archiving, and long-term access to tsunami-relevant sea level data that remain highly relevant today including the following issues:

- There is currently no routine acquisition of the 15-second CO-OPS data, which are most relevant for model validation, and there is no routine retention of these data.
- Fifteen-second data are only collected on request and have no quality control or archive.

- One-minute data are not currently quality controlled to the same level as the six-minute data.
- No formal long-term archive for the TWC coastal water level data is in place, although a minimal-service archive of the PTWC Hawaiian sea level data is being maintained and some of the TWC data reach the IOC's Sea Level Station Monitoring Facility.
- Retrospective data from the TWCs cannot be easily accessed.

The absolute time accuracy of 15-second data (30s Nyquist period) should be 0.035 seconds if archival or even near-real-time data are to be processed between stations using correlation or coherence methods. Time accuracy at this level is required in order to preserve phase relationships at the highest observed frequencies (i.e., 1/(2*15) Hz). Such absolute accuracy is not difficult to achieve.

Recommendation: The committee endorses the following recommendations of the NTHMP report (National Tsunami Hazard Mitigation Program, 2007) for the TWCs to:

- Create a formal data archive for both CO-OPS and TWC data and metadata, including 15-second data.
- Address 1-minute and 15-second quality control issues in unison with the archive issue to ensure quality of archive.
- Enact Federal Geographic Data Committee (FGDC)-compliant station metadata.
- Create an operational website providing a portal for 15-second tsunami station water level data.

This committee did not undertake an assessment of the processing, distribution, archiving, and long-term access to tsunami-relevant sea level data originating from international sea level stations. As previously stated, the near-real-time, tsunami-relevant sea level data available to the TWCs via the GTS (and archived at the IOC's SLSMF; http://www.vliz.be/gauges/) is not quality controlled.

Conclusion: Despite the excellent accomplishments by NOAA with respect to improving the processing, distribution, archiving, and long-term access to the tsunami-relevant sea level data that it collects, there remain several inadequacies. There is currently no routine acquisition, quality control, or archiving of the 15-second NOS/CO-OPS data, which are most relevant for model validation. In addition, NOS/CO-OPS 1-minute data are not currently quality controlled to the same level as their 6-minute data; and no formal long-term archive for TWC coastal water level data exists.

Description of the Deep-ocean Assessment and Reporting of Tsunamis (DART) Network

To ensure early detection of tsunamis, especially where the coastal sea level network is sparse or nonexistent, and to acquire data critical to near-real-time forecasts, NOAA has placed

DART stations in regions with a history of generating destructive tsunamis. The DART technology was developed at NOAA's PMEL under the U.S. National Tsunami Hazard Mitigation Program (González et al., 1998; http://nthmp-history.pmel.noaa.gov/index.html) to provide early detection of tsunamis regardless of the source (http://www.ndbc.noaa.gov/dart/dart.shtm). A DART station comprises an autonomous, battery powered, bottom pressure recorder (BPR) on the seafloor and a companion moored surface buoy that forwards the data it receives acoustically from the BPR to an onshore receiver via satellite links (Figure 4.5; see González et al., 1998). The BPR collects and internally stores pressure and temperature data at 15-second intervals. The stored pressure values are corrected for small temperature-related offsets and converted to an estimated sea-surface height (the height of the ocean surface above the seafloor). The BPR water height resolution is 1 mm in water depths to 6,000 m, and the maximum timing error is 15 seconds per year.

The station has two data reporting modes: standard and event. In standard mode, data are transmitted less frequently to conserve battery power. Event mode is triggered when internal detection software in the BPR identifies anomalous pressure fluctuations associated with the passage of a tsunami. During event mode, all 15-second data are transmitted for the first few minutes, followed by 1-minute averages. If no further events are detected, the system returns to standard mode after 4 hours.

There have been two types of operational DART stations: the first generation DART stations (DART I) became operational in 2003, but all six were replaced with the second generation DART stations (DART II) by early 2008. The DART II station has two-way communications between the BPR and the TWCs/National Data Buoy Center (NDBC) using the Iridium commercial satellite communications system (Meinig et al., 2005). The two-way communication allows the TWCs to set stations into event mode in anticipation of possible tsunamis or to retrieve high-resolution (15-second interval) data in 1-hour blocks for detailed analysis, and allows near-real-time troubleshooting and diagnostics. NDBC receives the data from the DART stations and distributes the data in near-real time to the TWCs via NWS secure communications and to other national and international users via the GTS. The data is also available on the NDBC website, and event data is highlighted when a system has been triggered.

Adequacy of the Geographical Coverage of the DART Network

The NDBC completed, in a little more than two years, an upgrade and expansion of the DART array from 6 DART I stations to the present 39 DART II stations, as shown in Figure 4.6. The expansion was supported with funding from the Tsunami Warning and Education Act (P.L. 109-424). In addition, Figure 4.6 shows the locations of 9 DART stations purchased, deployed, maintained, and operated by Chile, Australia, Indonesia, and Thailand.

Planning for the deployment and siting of the expanded DART network was initiated at a workshop attended by representatives of NOAA, the USGS, and academia on July 6-7, 2005, in Seattle (Geist et al., 2005). The central goal of the workshop was to determine an optimal network configuration that would meet multiple mitigation objectives, while addressing scientific,

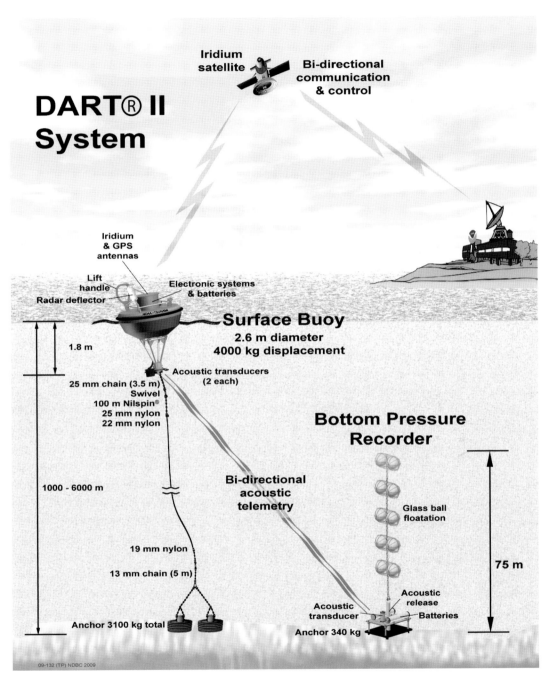

FIGURE 4.5 Schematic depicting a DART station's components: the surface buoy with acoustic transducers communicates with the BPR acoustic transducer and then transmits data via the Iridium antenna to satellites; the BPR detects changes in bottom pressure and temperature. SOURCE: http://www.ndbc.noaa.gov/dart/dart.shtml; National Data Buoy Center, NOAA.

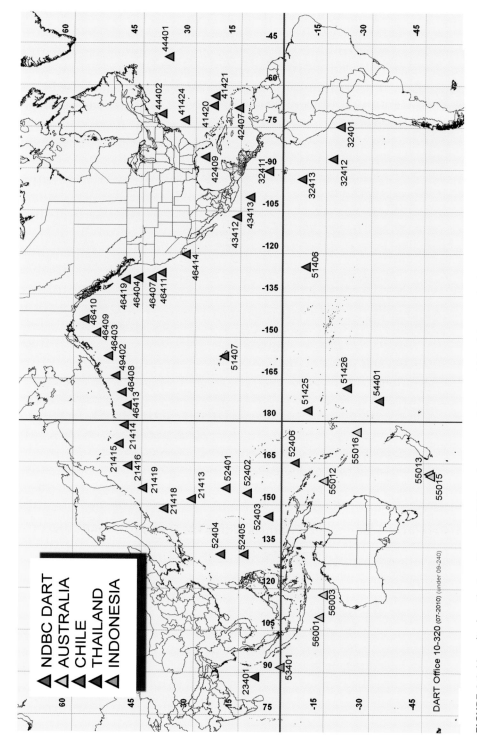

FIGURE 4.6 Map displays the locations of DART stations around the world. Red diamonds depict the 39 DART stations maintained and operated by NOAA's National Data Buoy Center (NDBC). Nine other DART stations are maintained and operated by non-U.S. agencies, as indicated in the legend. SOURCE: http://www.ndbc.noaa.gov/dart.shtml; National Data Buoy Center, NOAA.

engineering, operational, logistical, and political constraints. The process that began at this workshop was augmented by an optimization analysis, which was subsequently completed at the NOAA Center for Tsunami Research (NCTR) at PMEL. To the extent that the constraints on siting can be quantified and the benefits expressed in functional form, array design can be approached as a problem in optimization. This avenue was explored using a tool called NOMAD (Nonlinear Optimization for Mixed vAriables and Derivatives; Audet and Dennis, 2006). Although the scheme was tested for only relatively simple cases, the methodology shows promise as an example of a scientifically robust process for siting and prioritizing stations in an operational sensor network.

The methodology and final rationale for the siting of DART stations are the subjects of a NOAA technical memorandum (Spillane et al., 2008). The final siting decisions were based on the workshop recommendations, as well as site recommendation reports produced at NCTR in consultation with the TWCs, with input from the USGS, NDBC, and other interested parties. The technical memorandum provides a starting point for continued refinement of the siting decisions and extension of the DART array, if necessary, while also providing information to aid efforts by the international community to extend the network coverage.

The net result of the deliberations on the siting of the DART stations is the current array displayed in Figure 4.6. The prioritization of groups of these sites is presented in Table 4.1 (Spillane et al., 2008). Some of the more important issues involved in site selection are described in Box 4.1.

The committee does not find any serious gaps in the geographic coverage of the DART network as designed, with regard to providing timely and accurate warnings and forecasts of far-field tsunamis on U.S. coasts. It can certainly be argued that denser coverage of open-ocean sensors would provide important redundancy capacity (in light of current reliability problems discussed below) and would provide more opportunities to improve the accuracy of model-generated wave forecasts. From a more global perspective, gaps in coastal sea level station coverage (as revealed in the Caribbean region, for instance; see previous section), which expose

TABLE 4.1 Sub-Region Allocations and Priorities Within the Overall U.S. DART Array

Array Sub-Group	Instruments Assigned	Pre-Existing Sites	Priority
Alaska/Aleutians	6	3	1
Western Pacific	6	0	2
Puerto Rico/Caribbean	3	0	3
West Coast	5	2	4
Southwest Pacific	4	0	5
Central/South America	4	0	6
Atlantic	3	0	7
Gulf of Mexico	1	0	8
Northwest Pacific	5	0	9
Hawaii/Mid-Pacific	2	2	N/A

SOURCE: Spillane et al., 2008; NOAA.

BOX 4.1
Siting Considerations for DART Stations[a]

Tsunami Signal-Timeliness, Signal-to-Noise, and Signal Complexity Issues:

- Tsunamigenic zones. The likelihood that a particular fault zone will produce a tsunami is considered along with the coverage of the existing sea level network.

- Seismic wave noise. If a DART is located too close to the seismic event that generates a tsunami, the shaking of the seafloor can cause spurious BPR fluctuations (e.g., from seafloor interfacial Rayleigh waves) unrelated to the passage of tsunami water waves. This seismic noise can be reduced significantly by locating the instruments no closer than 30 minutes of tsunami travel time from the closest possible source, after which time the seismic body and surface waves will have passed.

- Timely signal. If the DART is sited too far from the tsunami source, too much time is lost between the seismic event, which is detected within a few minutes, and the arrival of an unambiguous sea surface disturbance at a DART site. In some locations, this consideration is more important than the seismic wave noise issue; DARTs have been placed as close as 15 minutes of tsunami travel time from the closest source.

- Tsunami scattering. The presence of seamounts or other major seafloor features between a DART and likely tsunami sources needs to be avoided. These bathymetric features cause zones of shadowing or of reinforcement in their lee due to tsunami wave diffraction. To the extent that these effects are imperfectly represented in the tsunami propagation model databases on which the SIFT and EarthVu tsunami forecast systems rely, forecast quality will be adversely affected.

Engineering and Survivability Issues:

- Water depth. The acoustics communications device currently in use is rated to water depths up to 6,000 m, but the narrow acoustic beam requires the surface buoy to be closely held above the BPR.

- Strong currents. Because of the need for a surface buoy, it is important to avoid strong current regimes, which could cause swamping or dragging of the buoy, or could make buoy maintenance difficult.

- Sub-surface landslides. Landslide-prone seabeds need to be avoided.

- Redundancy. Either the bottom unit or the surface buoy of a DART station may fail and, in remote locations, repair/replacement may not be an immediate option because of seasonal

---

vulnerabilities of non-U.S. territories in the TWCs' AORs, could be filled by DART stations if the resources of international partners are insufficient to fill the gaps with coastal sea level stations. However, the high cost of DART acquisition and maintenance may preclude any significant network growth.

NOAA is to be commended for having developed a prioritization scheme for DART stations and for having rapidly deployed the DART array. Looking to the future, the committee con-

foul weather or ship scheduling. Even though tsunamis do not occur frequently, redundancy in the array is still desirable. The surface buoy has two independent complete communication systems for full redundancy. In addition, in high-risk source regions, a certain amount of overlap in spatial coverage is desirable so that instrument failures may be partially compensated by having more than one DART in the region capable of providing a timely, high-quality signal.

Communication Issues:
- Bottom roughness. A DART BPR needs to communicate acoustically with its surface unit. For reliable communications, the BPR must be deployed on a reasonably flat, smooth seabed that will not produce scattering and interference of the acoustic signals.

Logistical Issues:
- Although DARTs are typically deployed for two years, and have a design life of four years, there is considerable expense associated with deploying and maintaining them in remote regions. For some sites, co-locating DART buoys with other buoy arrays might allow leveraging ship time and maintenance costs if there is no conflict with special DART requirements. For example, co-location might be considered for other sites maintained by the NDBC, such as in the equatorial Pacific near the Tropical Atmosphere Ocean Project (TAO) (http://www.pmel.noaa.gov/tao/) buoy array or near U.S. coastlines where meteorological buoys are maintained.

Other Issues:
- Other considerations in choosing buoy sites include the difficulty or ease of obtaining permissions to enter other national EEZs (Exclusive Economic Zones), shipping routes, seafloor infrastructure (e.g., communications cables that could be damaged by the mooring's anchor), and piracy or a history of damage to unattended buoys that make some areas less desirable for DART siting.

---

[a] Spillane et al., 2008.

cludes that the numbers, locations, and prioritizations of the DART stations should not be considered static. These parameters of the DART network clearly deserve frequent re-consideration in light of constantly changing fiscal realities, survivability experience, maintenance cost experience, model improvements, new technology developments (even new DART designs), increasing international contributions, and updated information on the entire suite of siting issues listed in Box 4.1. In addition, simulations of the effectiveness of the DART network, under numerous earthquake scenarios and under various DART failure scenarios, should continue to help improve the network design (Spillane et al., 2008). The potential contributions of optimization algorithms to the design process have not been exhausted.

A component of the periodic re-evaluations of the DART network needs to be the re-evaluation of the prioritization of each group of DART stations, not just individual stations, with detailed justifications for these determinations. In particular, the committee questions the rationale for the very low priority of the group of five DART stations deployed in the Northwest Pacific (Table 4.1) that provide coverage from the Dateline along the western Aleutian Islands, and past the Kuril Islands to Hokkaido. The Kuril Islands in particular have been the source of numerous tsunamis large enough to invoke tsunami watches and warnings. At the very least, DART stations covering the Kuril Islands would have a high value for the prevention of false alarms.

DART station prioritization could be refined by first distinguishing prioritization criteria based on the system's primary function in the detection process. A list of criteria might include:

- detection of a large tsunami,
- detection of a medium to small tsunami (to mitigate false alarms),
- providing data for scaling forecast models during the occurrence of a large tsunami, and
- providing data for forecast model validation after the fact.

Depending on the order of importance of criteria such as these, quite different prioritizations of the DART stations might result. For instance, the value of the DART stations in the Western Pacific south of 25° N (Figure 4.6) is primarily for scaling forecast models, since the numerous island stations in the region (Figure 4.4) can adequately perform the tsunami detection function. The value of the DART stations in the Northwest Pacific is primarily for the detection of medium to small tsunamis, in order to confirm that a large tsunami has not been generated and thus avoid the issuance of an unnecessary warning with its attendant costly evacuation. Depending on the relative importance of the criteria in the list above, the Northwest Pacific DART stations may be more important than the Western Pacific DART stations, contrary to the present prioritization represented in Table 4.1.

Conclusion: NOAA is to be commended for having developed a prioritization scheme for the distribution of the DART stations and for having rapidly deployed the DART array. There are no serious gaps in the geographic coverage of the DART network as designed, with regard to providing timely and accurate tsunami warnings and forecasts for at-risk U.S. coasts and territories. However, the vulnerabilities of non-U.S. territories in the TWCs' AORs were not a high priority in the network design, and the potential contributions of optimization algorithms to the design process have not been exhausted.

Recommendation: NOAA should regularly assess the numbers, locations, and prioritizations of the DART stations, in light of constantly changing fiscal realities, survivability experience, maintenance cost experience, model improvements, new technology developments (even new DART designs), increasing international contributions, and updated information on the entire suite of siting issues listed previously.

Reliability of the DART Network

Since the build-up of the DART network began in 2006, it has experienced significant outages that can have adverse impacts on the capability of the TWCs to issue efficient warnings, to use near-real-time forecasts, and to cancel warnings when a tsunami threat is over. The data loss also reduces post-tsunami model validation capability. Figure 4.7 indicates how network availability steadily declined to a low of 69 percent in February 2009. The number of DART stations deployed grew from 10 in July 2006 (7 new DART II systems, along with 3 older DART I systems) to 39 in March 2008 (including replacement of the original DART I systems with DART II systems). By March 2009, only a year after the DART array was completely deployed,

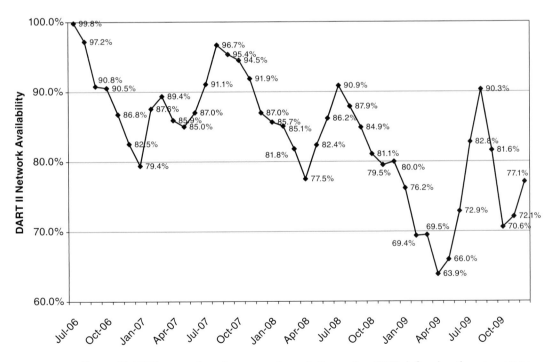

FIGURE 4.7 Chart of DART II network performance through December 2009, defined as the percentage of hourly transmissions of water column heights received vs. expected. The peaks in performance occur during Northern Hemisphere summer when maintenance is performed. Note, however, that the peak values in performance are decreasing with time as well. SOURCE: National Data Buoy Center, NOAA.

12 of the 39 DART station buoys were nonoperational despite the four-year design lifetime of the DART II systems.

Maintenance occurs in the summer months, accounting for the annual cycle in Figure 4.7. The declining trend in performance is emphasized in Figure 4.8 that depicts the median age of the deployed DART stations as increasing while the median age of failed systems has declined below the median age of deployed systems and hovers around one to two years. A system availability of 69 percent is significantly below the network performance goal of 80 percent, which perhaps is not surprising for such a large, new, and admittedly hurriedly-deployed set of complex systems that are deployed in very harsh environments. (For comparison, the effort to establish a German Indonesian Tsunami Early Warning System (GITEWS; http://www.gitews. de/index.php?id=5&L=1) has expended more than €55 million over the past five years, resulting in deployment of a single open-ocean tsunami sensor that has been operational for only six months to date, in 2007-2008.)

The issue of low network performance is exacerbated by the fact that clusters of nearby DART stations tend to be nonoperational for many months, leaving large gaps in DART coverage. For example, five stations cover the Aleutian Islands west of the Dateline, past the Kuril

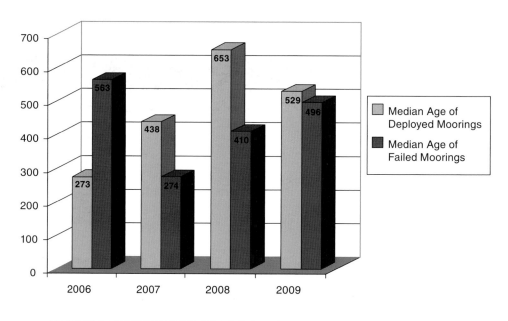

	2006	2007	2008	2009
Number of Deployed Stations:	(15)	(30)	(39)	(39)
Number of Mooring Failures:	(1)	(4)	(9)	(15)

FIGURE 4.8 Median age of deployed DART II moored systems and median age of failed DART II moored systems. SOURCE: National Data Buoy Center, NOAA.

Islands to Hokkaido. Although the Kuril Islands region produced many small basin-wide tsunamis over the past five years, all of these stations had failed by December 2008, and four had failed in October 2008, or earlier. None were repaired until late June 2009, after weather conditions had improved enough to reduce the risk of shipboard operations. As of May 2010, three of these five DART stations have been inoperative since September, 2009.

The optimization scheme used for planning the locations of the DART stations and testing their ability to detect tsunamis basin-wide is based on an assumption of nearly 100 percent performance (Spillane et al., 2008). There is a small amount of redundancy and overlap in the DART network design in case of a single DART failure, but the consequences of multiple DART failures have not been considered. Given the current geographic coverage, the DART network is only useful for tsunami detection and forecasting if it is operational nearly 100 percent of the time. In a practical sense, when one DART station is inoperative, its neighbors on either side must be operational. If two neighboring DARTs become inoperative, then there must be an immediate mitigating action. A minimum first step in rectifying this situation is to establish more explicit priorities for the DART stations in order to provide guidance for NDBC's maintenance activities. Table 4.1 from Spillane et al. (2008) provides the coarsest priorities set for the initial DART deployments, but the report does not provide justifications for the prioritizations, and they are not specific enough for the purpose of prioritization of maintenance schedules.

Figure 4.7 emphasizes that maintenance of inoperative gauges is slow and generally performed on an annual cycle irrespective of the timing of outages. Even with many DART stations inoperative in late 2008, NDBC's repair plan was to restore all nonoperational DARTs by the end of July 2009. As a consequence of the pervasive outages of the DART stations, the TWCs cannot depend on the DART network for tsunami forecasting. According to NDBC personnel, the budget only allows for annual routine maintenance and no funds are available for "discrepancy response" (that is, nonroutine maintenance for inoperative gauges) (National Data Buoy Center, personal communication, 2009). The committee has assumed that summer time maintenance cycles are, at least in large part, dictated by north Pacific weather. If this is the case, the maintenance of the high-priority DART buoys may not be practical or even possible. NDBC's budget for maintaining the DART stations decreased the past few years, despite the mandate in P.L. 109-424 for NOAA to "ensure that maintaining operational tsunami detection equipment is the highest priority." However, lack of maintenance funding explains only part of the present problem with DART station failures. The number of DART II system failures is higher than expected, with a current median time to failure of approximately one year when the design lifetime was four years (Figure 4.8).

The task of building and deploying the DART buoys in two years, by Presidential directive, has been challenging for NDBC. To meet the mid-2007 deadline, the DART II was rushed to production and deployment without the customary level of testing required for a complex system like the DART, with its relatively extreme operational environment. This rapid deployment schedule required an active reliability improvement program, concurrent with initial operations and funding, to sustain effective operations while reliability improvements were defined and implemented. However, budget cuts slowed both maintenance and reliability improvement. Furthermore, NDBC had no prior experience with seafloor instrumentation, acoustic modem communications, or taut-line surface moorings before the transfer of operations from PMEL. The committee's assessment revealed problems that reduced the effectiveness of the technology transfer from PMEL to NDBC, including a lack of training of NDBC personnel on DART deployment methods, a preference for NDBC mooring deployment procedures that conflict with PMEL's recommended deployment procedures for the DART stations, and a lack of coordination of post-transition research activities. These observations are consistent with other issues raised in a report by the Inspector General of the Department of Commerce about the need to make improvements to some of NDBC's buoy maintenance operations (U.S. Department of Commerce Office of Inspector General, 2008). The report found that technology transfers from PMEL to NDBC have not been well coordinated and planned, and it offered several recommendations to address these concerns, such as ensuring that data requirements and technical specifications are clearly defined prior to the transition and that adequate funding is available to cover the transition costs. The report also recommends better coordination on research and development projects between the two NOAA centers to avoid duplication of efforts.

DART II failure modes cut across the suite of components in the DART II stations, such as bottom pressure sensor faults; acoustic transducer failures; tilt sensor failures; CPU, acoustic modem, and interface board failures on both the BPR and buoys; and mooring hardware failures. By far the most common problem is mooring hardware failure. For example, of the 12 DART II stations listed as inoperative on May 11, 2009, 8 were listed as "adrift." In other words, the mooring line holding the surface buoy had parted so that the surface buoy had drifted away from the location of the BPR. Although NDBC has an active failure analysis program, this program needs improvement; for instance, when a buoy goes "adrift," neither it nor the mooring remnants left on site are presently recovered by NDBC, so that the cause of the mooring line failure, or other failure mode, remains undetermined. There are many possible causes for mooring line and buoy failures, such as faulty components, improperly assembled moorings, physical interference from "long-line" fishing activity, fish bite, vessel collisions resulting in buoy sinking, vandalism, extreme environmental conditions, metal fatigue, high currents towing the buoy under water, and improper mooring scope on deployment due to error in water depth determination and/or mooring line measurements (an allowable error is 1.5 percent or less). In response to these problems, NDBC held a mooring workshop in February 2010 with participants from Woods Hole Oceanographic Institution (WHOI), PMEL, SIO, Science Applications International Corporation (SAIC), and other agencies and institutions. A broad spectrum of topics was addressed, and specific issues affecting DART reliability were identified and summarized for NDBC management.

A principal objective of NDBC's effort to improve DART reliability is to reduce ship time costs. A system that requires unanticipated maintenance visits using costly ship time reduces availability of funds for other activities.

The committee analyzed the benefits and disadvantages inherent in each of these maintenance approaches. In order to maintain the current DART network configuration, adequate resources are needed for maintenance, including funding for unscheduled ship time to effect repair and replacement of inoperable DART stations. The alternative approach would be to invest the majority of resources into improving the DART station reliability to get closer to the design goal of a four-year lifetime (Figure 4.8), which would reduce the need to fund ship time for station maintenance. The second choice implies that DART stations are maintained sparingly, with only minimal attention to the integrity of the network's tsunami detection capability, until the reliability of the DART stations is improved. In this case, it must be understood and acknowledged that the DART network might be fully deployed but will not be fully functional until such time as the reliability of the DART stations gets much closer to the design goal of a four year lifetime than the present median time-to-failure of just over one year.

A partial amelioration of the draconian choices above could come from exploring new maintenance paradigms, such as (1) simplifying the DART mooring for ease of deployment from small, contracted vessels that are available, for instance, from the commercial fishing fleet and the University-National Oceanographic Laboratory System (UNOLS) fleet; and (2) maintaining a reserve of DART buoys for immediate deployment upon the occurrence of a significant gap in the network, weather permitting.

The transfer of the DART technology from research (at PMEL) to operations (at NDBC) did not include the establishment of mechanisms for scientific or TWC operational feedback into the management of the program.

> **Conclusion:** There is insufficient station redundancy in the DART network. Since the buildup of the DART network began in 2006, it has experienced significant outages that have a potentially adverse impact on the capability of the TWCs to issue efficient warnings, use near-real-time forecasts, and cancel the warnings when a tsunami threat is over. Worse, multiple, neighboring DART stations have been seen to fail in the North Pacific and North Atlantic, leaving vast stretches of tsunami-producing seismic zones un-monitored. This situation persists for long periods of time. The committee considers it unacceptable that even a neighboring pair of DART stations in high-priority regions is inoperative at the same time. Although an 80 percent performance goal may be satisfactory for the entire DART network, and for individual gauges, a much better performance is required for neighboring pairs of DART stations, especially in high-priority regions.

> **Recommendation:** In order to bring NDBC into compliance with P.L. 109-424, NDBC should engage in a vigorous effort to improve the reliability of the DART stations and minimize the gaps caused by outages.

> **Conclusion:** The transfer of the DART technology from research (PMEL) to operations (NDBC) did not include the establishment of mechanisms for scientific or operational feedback from PMEL or the TWCs into the management of the program. The DART network reliability could be enhanced by improving the technological and scientific knowledge transfer between PMEL and NDBC and the management of the continued joint development of next generation DART stations.

> **Conclusion:** Continued engineering refinements of the DART concept will allow NOAA to establish a more sustainable capability with reduced costs of construction, deployment,

and maintenance. The committee supports, and encourages the continuation of, NDBC's recent effort (February 2010 workshop) to engage industry, academia. and other NOAA agencies for help in solving its problems with DART reliability.

Recommendation: The committee encourages NDBC to establish rigorous quality control procedures, perform relentless pre-deployment tests of all equipment, and explore new maintenance paradigms, such as simplification of DART mooring deployment and maintaining a reserve of DART stations for immediate deployment.

Recommendation: NDBC should improve its efforts at failure analysis, especially through more vigorous attempts to recover both buoys that have gone "adrift" and the mooring remnants that are left on site.

Conclusion: DART presents an outstanding opportunity as a platform to acquire long time series of oceanographic and meteorological variables for use for climate research and other nationally important purposes. Potentially a DART buoy could also telemeter data acoustically from a seafloor seismograph although the demands on DART power would increase proportionally. The additional power requirements for acoustic and satellite telemetry would press the current design of the buoy thereby increasing risk to the primary goal of tsunami detection. Nevertheless, broadening the user base could enhance the sustainability of the DART program over the long term and future designs should consider additional sensors. Other programs, such as the coastal sea level network, have encouraged a broad user base to enhance sustainability of their infrastructure.

Recommendation: NOAA should encourage access to the DART platform (especially, use of the acoustic and satellite communications capabilities) by other observational programs, on a not-to-interfere basis; that is, the primary application (tsunami warning) justifies the cost, but DART presents an outstanding opportunity as a platform to acquire long time series of oceanographic and meteorological variables for use for climate research and other nationally important purposes. Broadening the user base would be expected to enhance the sustainability of the DART program in the future.

Conclusion: In a world of limited resources, a strategic decision needs to be made as to whether it is more important to maintain the current DART network at the highest level of performance or to focus on improving the DART station reliability.

A first step could be for NOAA to establish a strategic plan that determines whether (1) it is most important to maintain the DART II network at the highest level of performance right now (meaning that the first priority for resources is maintenance, including funding of costly ship time to repair and replace inoperative DART stations as soon as possible), or (2) it is most important that NDBC focus first on improving DART station reliability, at the possible expense of maintenance.

DART Data Processing

NDBC has enacted automated quality checks for DART data as it is delivered in near-real time, as well as post-processing quality analyses for archived DART data dating back to 2003 (http://www.ndbc.noaa.gov/dart.shtml). Even older (since 1986) quality-controlled BPR data can be found at NGDC (http://www.ngdc.noaa.gov/hazard/DARTData.shtml). Currently, the archived data comprise the 15-minute sea level samples from the "standard" mode of DART operation, as well as the 15-second and 1-minute samples transmitted during the "event" mode of operation. Access to the continuous 15-second sea level data that are stored internally in the DART BPRs, and are retrieved after recovery of each BPR, has not yet been automated; the data is available upon request from NGDC.

To facilitate the use of the 15-second data for studying such phenomena as atmospherically-generated "meteo-tsunamis," coastally-generated infra-gravity waves, and the earth's seismic "hum," among other phenomena, quality-controlled 15-second data could be made available from an archive center such as NGDC. The NTHMP (2008) recommendations for enhancing the quality and availability of tsunami-relevant data (see sub-section on *Coastal Sea Level Data Processing*) also apply to the DART station data.

Sea Level Data Integration into Other U.S. and Global Observation Systems

The coastal sea level data and metadata are available through the IOS Sea Level Monitoring Facility (http://www.vliz.be/gauges/index.php). However, the IOC website does refer back to Permanent Service for Mean Sea Level (PSMSL), the British Oceanographic Data Center (BODC), and the UHSLC for low-frequency and high-frequency research quality sea level data. In addition, the expanded DART array data and metadata are available globally from the NDBC website (http://www.ndbc.noaa.gov/dart.shtml), which can be reached through NGDC. However, edited bottom pressure data are not available after 2004 and are awaiting review.

However, with respect to the integration of the U.S. DART and coastal tsunami-relevant sea level stations, the committee has found no evidence that an integration with Integrated Ocean Observing System (IOOS), Global Ocean Observing System (GOOS), or Global Earth Observation System of Systems (GEOSS) is being pursued or implemented, despite a recommendation in the NTHMP (2008) report to "develop an observing system architecture to design, build, deploy and operate tsunami observation and data management systems in conjunction with IOOS and the all-hazards GEOSS. Tsunami near-real-time observation systems (including seismic, water level, and oceanographic) and data management systems (including modeling and archiving) are key elements of IOOS and GEOSS."

The DART buoy platforms present an outstanding opportunity to acquire long time-series data of oceanographic variables for nationally important research and monitoring goals, including for climate research. Giving other observational programs access to the DART platform (especially, use of the acoustic and satellite communications capabilities) provides an opportunity for leveraging resources (ship time for maintenance, long-term funding for maintenance or replacements) and is encouraged by the committee.

Similarly, there is great value in the continued coordination of U.S. tsunami-focused sea level observation efforts with other U.S. and international programs interested in monitoring sea level variability for other purposes, such as climate variability and climate change.

> **Conclusion:** Because coastal sea level stations have evolved from their primary mission to serve a broad user community, their long-term sustainability has been enhanced.

The following are conclusions and recommendations related to the detection of tsunamis with sea level sensors:

Assessment of Network Coverage for Tsunami Detection and Forecasting

> **Recommendation:** NOAA should assess on a regular basis the appropriateness of the spatial coverage of the current DART sea level network and coastal sea level network (U.S. and international), in light of constantly changing fiscal realities, survivability experience, maintenance cost experience, model improvements, new technology developments, and increasing or decreasing international contributions. Especially, NOAA should understand the vulnerabilities of the detection and forecast process to the following: (1) gaps in the distribution of existing gauges and (2) failures of single or multiple stations.

A first step in the assessment could be the establishment of explicit criteria, based on TWC forecaster experience and on the arguments outlined for the DART site selection (Spillane et al., 2008). An appropriate aid in this process would be simulations (e.g., Spillane et al., 2008) of the effectiveness of the combined sea level networks, under numerous earthquake scenarios and under various station failure scenarios. Such a study would also consider a region's tsunami-producing potential, sensitivity analysis of source location, tsunami travel time, local population density, timing for initial warning versus evacuation decision process for communities at risk, and warning/evacuation time gained for additional station coverage. The contributions of optimization algorithms to the network design process could be explored more fully as well.

Station Prioritization

> **Recommendation:** NOAA should prioritize the existing DART stations and coastal sea level gauges (both U.S. and international) according to their value to tsunami detection and forecasting for both U.S. territories and other AORs of the TWCs. Furthermore, this priority list should be merged with the results from the network coverage assessment (above) to determine the following: (1) maintenance priorities and schedules; (2) network expansion priorities; and (3) identification of critical stations that are not under U.S. control and may require either augmentation with new U.S. gauges or operations and maintenance support.

An important aspect of this activity would be to develop and publish criteria, such as the following examples: (1) value of a station for initial detection of a large tsunami near an active fault zone, to maximize warning lead time; (2) value of a station for initial detection of a medium to small tsunami, to mitigate false alarms; (3) value of a station for scaling forecast models of inundation of U.S. territories; (4) value of a station for after-the-fact model validation; and (5) density (sparsity) of the observing network in the region.

Data Stream Risk Assessment and Data Availability

Recommendation: NOAA should assess on a regular basis the vulnerabilities to, and quality of, the data streams from all elements of the sea level networks, beginning with the highest priority sites determined per the recommendations above.

Coastal station vulnerabilities can be assessed by the following: (1) whether the operating agency is committed to gauge maintenance, which can be assessed by the continuous availability (or not) of the station's data on the IOC's Sea Level Station Monitoring Facility (http://www.vliz.be/gauges/) and (2) whether the station adheres to the station requirements, processing protocols, quality control procedures, distribution, long-term archiving, and retrospective access recommendations in the Tsunami Warning Center Reference Guide (U.S. Indian Ocean Tsunami Warning System Program, 2007).

The risk assessments, along with the prioritization lists described above, could be used to determine the following: (1) whether authority for a U.S. gauge should be transferred to a different U.S. agency: for example, the TWCs have acknowledged that they do not have the resources to properly maintain the gauges under their authority; authority for maintenance of these gauges could be transferred to NOS/CO-OPS or the UHSLC, with appropriate funding; (2) whether aid should be offered to an international partner; and (3) whether a substitute gauge should be established in a nearby location.

Sea Level Network Oversight

Recommendation: In view of (1) the declining performance of the DART network, (2) the importance of both the DART and coastal sea level networks for tsunami detection and forecasting, and (3) the overlapping jurisdictions among federal as well as non-federal organizations, NOAA should establish a "Tsunami Sea Level Observation Network Coordination and Oversight Committee" to oversee and review the accomplishment of the recommendations listed above.

The committee would report to the management level within NOAA that has the responsibility and authority for ensuring the success of the U.S. Tsunami Program. The oversight committee would be most useful if its members represented a broad spectrum of the community concerned with tsunami detection and forecasting (e.g., forecasters, modelers, hardware designers, operations and maintenance personnel) from academia, industry, and relevant government agencies.

FORECASTING OF A TSUNAMI UNDER WAY

In contrast to inundation models used for evacuation planning in advance of an event (see Chapter 2), near-real-time forecast models produce predictions after a seismic event has been detected, but before tsunamis arrive at the coast, which is the ultimate goal of the monitoring and detection system. These forecast models make available to emergency managers in near-real time the time of first impact as well as the sizes and duration of the tsunami waves, and give an estimate of the area of inundation, similar to hurricane forecasting.

The entire forecasting process has to be completed very quickly. For example, Hawaii Civil Defense needs about 3 hours to safely evacuate the entire coastline. As most far-field tsunamis generated in the North Pacific take less than 7 hours to strike Hawaii, the entire forecast, including data acquisition, data assimilation, and inundation projections, must take place within 4 hours or less. Although this sounds like a comfortable margin, in fact it is quite a short time period compared to many other natural disasters, especially since it can take anywhere from 30 minutes to 3 hours to acquire sufficient sea level data (Whitmore et al., 2008). For hurricanes, forecasts are made days in advance of landfall and evolve spatially at scales over 100 times slower than a tsunami. The time window for a forecast for a near-field tsunami event is even smaller, because the first waves may arrive in less than 30 minutes (see the section on *Instrumental Detection of Near-Field Tsunamis* below).

The importance of forecasting the duration of wave arrivals, and forecasting the sizes of each arrival, is well known; for example, the largest and most destructive wave of the tsunami originating off the Kuril Islands on November 15, 2006, was the sixth wave to hit Crescent City, California. This wave hit more than two hours after the first wave arrival (Uslu et al., 2007; Barberopoulou et al., 2008; Dengler et al., 2008).

Although time-of-arrival information has been available since the 1960s (Ambraseys, 1960), only beginning in the 1990s (e.g., Kowalik and Whitmore, 1991; Whitmore and Sokolowski, 1996; Titov and González, 1997), with full development not completed until a decade later, have forecast methodologies been employed to provide estimates of inundation prior to wave arrival and of duration (see Whitmore, 2003; Mofjeld, 2009; Titov, 2009). The use of near-real-time forecasting models is only possible because of data from the coastal and open-ocean sea level networks. Modeling tsunamis based on seismic data alone is currently not very accurate, as noted in the above section on *Detection of Earthquakes*. The importance of accurate forecasts of maximum wave height was illustrated quite clearly in the wake of the recent Chilean earthquake on February 27, 2010.

In the United States, NOAA's WC/ATWC and PMEL have developed distinct tsunami forecast systems (respectively, Alaska Tsunami Forecast Model (ATFM), http://wcatwc.arh.noaa.gov/DataProcessing/earthvu.htm; and SIFT, http://nctr.pmel.noaa.gov/tsunami-forecast.html) to provide information on tsunami arrival times, wave sizes, and event durations at the shoreline. An advanced version of the ATFM is currently in development at the WC/ATWC.

These systems employ pre-computed, archived event scenarios, in conjunction with near-real-time sea level observations. The PMEL system takes the forecast a step further by providing inundation distances and run-up heights that enable even more targeted evacuations. These forecast models allow the TWCs to make more accurate tsunami wave predictions than were possible without them, enabling more timely and more spatially refined watches and warnings (e.g., Titov et al., 2005; Geist et al., 2007; Whitmore et al., 2008). The PTWC was able to forecast reasonably well the observed tsunami heights in Hawaii more than five hours in advance of the Chilean tsunami arrival (Appendix J). The models place an additional emphasis on the importance of the proper operation of the sea level stations, especially the open-ocean DART stations, whose sea level observations of tsunami waves are not distorted by bathymetric irregularities and local harbor resonances that affect the coastal sea level observations.

Japanese scientists have been leading in tsunami forecast modeling, have had forecast models in operation for a while (including for near-field events), and are able to draw from a very sophisticated, densely covered observation network. They are also very active in developing new methods for real-time forecasting (e.g., using the inversion method; Koike et al. 2003).

In brief, the SIFT model identifies an interim wave field from its database based on the seismic data (inferred source parameters and epicentral location) once an earthquake is triggered. As the tsunami arrives at sea level stations along its propagation path, tsunami amplitude data are used to improve the forecast by scaling the pre-computed free-surface distribution. Finally, the resultant scaled surface is used to initialize a boundary value problem and determine, at high resolution, the wave field, including inundation at the locations of interest. The three steps, in more detail, are as follows:

1. *A pre-computed database of wave fields from unit earthquake sources is consulted*: NOAA/ PMEL built a database of 1,299 unit earthquakes. The seafloor displacement is computed by the linear-elastic dislocation theory and is applied for each unit earthquake, each representing a magnitude 7.5 earthquake with a deformation area 100 km long by 50 km wide. Because the NOAA system was initially developed to produce forecasts for U.S. coastlines, the current database includes only events in the Pacific Ocean and the Caribbean Sea, although efforts are under way to extend the database to the Indian Ocean and the Mediterranean Sea. In addition, the database was developed for thrust events only and is now being updated for other types of earthquakes, particularly for the Caribbean region. By linearly combining the wave fields from adjacent unit sources, the most plausible and realistic tsunami scenarios are roughly inferred from the earthquake parameters. For example, a magnitude 8.7 earthquake with an approximate 400 km by 50 km deformation area requires superimposing the results from four adjacent segments. Because the unit sources are arranged in a pair of parallel rows, larger events with widths on the order of 100 km can also be represented. Each archive includes data on the spatial distribution of wave heights and fluid velocities; this information is needed to initialize the boundary conditions, which is then used to calculate in near-real time the inundation in specific locales.

2. *Data assimilation from DART station data is performed*: In this step, near-real-time measurements of the tsunami are used to scale the combined wave field constructed from the database. Once the tsunami is recorded by the DART sensor, the pre-computed wave time series (wave heights and arrival times) are compared to and scaled using the observed wave time series by minimizing a least square fit. This scaling process can achieve results as soon as the full wavelength of the leading wave is observed and is updated with observations of the full wave time series. When the wave arrives at the next buoy, the tsunami wave heights are corrected again, although the experience to

date with 16 events has shown that even a single DART buoy is sufficient to scale the pre-computed wave fields appropriately for qualitatively accurate predictions.

3. *Inundation estimates using the nonlinear model, Method of Splitting Tsunami (MOST), are developed*: Once the combinations of wave fields from the pre-computed scenarios are constrained by the DART sea level data using the least squares fit technique, the database is queried for wave height and fluid velocity time series at all sea-boundaries of the region targeted for the inundation forecast. At each boundary point, the time histories of heights and velocities are used to initialize the boundary conditions. The inundation computation proceeds using the nonlinear MOST model that includes shoaling computations of wave inundating dry topography, until inundation estimates are obtained. The process is built on the Synolakis (1987) theory of a solitary wave propagating over constant depth and then evolving over a sloping beach. The wave field of approaching waves in deep waters are assumed to be linear, so there are reasonable interim estimates for the entire flow including reflection from the beach; i.e., where the constant depth and sloping regions connect. Once there is a linear solution in the deep waters (where depths are more than 20 m), this input can be used to solve the nonlinear evolution problem on a sloping beach (Carrier and Greenspan, 1958).

Figure 4.9 displays the SIFT tsunami predictions at two stations after the February 2010, Chilean earthquake. One of these stations is at an open-ocean island (Midway Island) at the northwestern end of the Hawaiian Archipelago; the other station is at the North American coast (Santa Barbara, California). In the open ocean, SIFT-predicted amplitudes (although not the phases) agree fairly well with the observed. However, the figure also illustrates the difficulty in predicting coastal amplitudes that are very sensitive to the small-scale details of the model's bathymetry and coastal geometry. The highest observed wave at Santa Barbara, occurring about four hours after the first arrival, is missed by SIFT. For comparisons of SIFT predictions with many other observations of the Chilean tsunami, go to http://nctr.pmel.noaa. gov/chile20100227/.

For SIFT (but not for ATFM), the ability to make accurate forecasts of tsunami waves is predicated on the availability of DART sea level measurements. The method's accuracy is tied directly to receiving data from the sea floor in near-real time. ATFM can utilize sea level data from both DART and coastal stations. To date the two technologies have successfully forecast 16 tsunamis with an accuracy of about 80 percent when compared with tide gauge data. (Titov et al., 2005; Tang et al., 2008, 2009; Wei et al., 2008; Titov, 2009). Although these models forecast wave height reasonably well, forecasting the inundation remains a challenge. To date, only one of the models (ATFM) is fully operational, although the SIFT model is being transitioned. At present, based on its review the committee found no clear process by which the forecasts' skill is evaluated and improved, nor by which the differences in the forecast outputs are reconciled. As with the ensemble model approach for hurricane forecasts, the committee considers it beneficial to run and compare multiple model outputs. However, a process is needed that assists watchstanders in reconciling the differences and arriving at a single forecast output to be transmitted in the warning products. Such a process is well established in the National

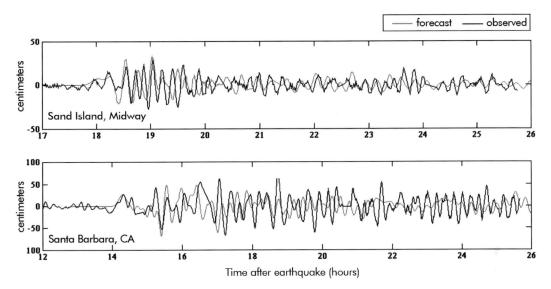

FIGURE 4.9 Comparisons of the February 27, 2010, Chilean tsunami recorded at two U.S. sea level gauges with forecasts obtained from high-resolution model runs. The forecast models were run in near-real time before the tsunami reached the locations shown. The model data for Santa Barbara exhibited a 9-minute early arrival (0.8-1 percent error accumulated during the propagation simulation) that has been removed for the purposes of this comparison. SOURCE: http://nctr.pmel.noaa.gov/chile20100227/; Center for Tsunami Research, NOAA.

Hurricane Center (NHC) or more generally the weather service, where ensemble modeling is commonplace.

> **Conclusion:** Metrics are needed to objectively measure each model performance. In addition, a process is needed by which multiple model outputs can be used to develop a single solution (e.g., ensemble model approach in the NWS and NHC).

> **Recommendation:** The TWCs and the NOAA Center for Tsunami Research at PMEL should continue to work together to bring the SIFT tsunami forecast methodologies into full operational use. The utility of the methodologies could be improved by ensuring that TWC staffs undergo a continuous education and training program as the forecast products are introduced, upgraded, and enhanced.

INSTRUMENTAL DETECTION OF NEAR-FIELD TSUNAMIS

Near-field tsunamis present a daunting challenge for emergency managers. Even if the near-shore populace is well informed about the potential for a tsunami when the ground shakes, and even if local managers receive information from forecasters of an impending tsunami within minutes after the earthquake, there will likely only be an additional few minutes before inundation, barely enough time for individuals to flee a short distance. The earthquake itself, if severe enough, may have already disrupted local communications, destroyed structures, and cut evacuation routes, as happened in Samoa during the September 29, 2009, tsunami (http://www.eqclearinghouse.org/20090929-samoa/category/emergency-management-response). Nevertheless, successful evacuations have occurred during the recent events in Samoa and Chile.

As for communities a little farther away from the tsunami source (where a tsunami might strike within an hour or so), the lack of communications could mean that tsunami forecasters will not receive data from the coastal sea level gauges that the tsunami reaches first. These communities might also be too distant from the triggering earthquake to have felt the ground shaking sufficiently to regard this as their warning. These communities depend on the detection system to very rapidly assess the threat and deliver the warning product and evacuation order.

Almost every tsunami, because their likely sources are along undersea fault zones that tend to be near the continents or islands, will have a near-field region that is affected relatively soon (within minutes) after the earthquake, as well as a whole suite of regions at varying distances that are affected from minutes to many hours after the earthquake. As an example, Figure 4.10 presents a simulation of the great 1700 tsunami that was generated by a magnitude 9.0 earthquake on the Cascadia subduction zone. After 1 hour, the leading tsunami wave crest has already inundated the local coastlines of Oregon, Washington, and Vancouver Island and has reached as far south as San Francisco. After 2 hours, the leading crest is well within the Southern California Bight.

For the benefit of the communities at intermediate and greater distances from likely tsunami source regions, and given the possibility that a near-coast earthquake will not only generate a large tsunami but also will destroy infrastructure (including sea level gauges or the telecommunication paths for their data) on the nearby coast, offshore open-ocean gauges that provide near-real-time, rapidly sampled sea level observations are needed. This need motivated the placement of five DART stations off the coasts of California, Oregon, Washington, and British Columbia (see Figure 4.6). Note that at least two of these DART stations would have observed the 1700 tsunami (Figure 4.10) well before the initial wave crest reached San Francisco.

Despite the short lead time for a near-field tsunami, there is still value in providing rapid official warning to the local populace, so long as people are not taught to wait for such a warning if they have already felt a strong earthquake. Such formal warning from every possible means (e.g., loudspeakers, TV, radio, Internet, text message, Twitter, etc.) will urge people to evacuate more quickly (the people will likely be under strained conditions instilled by the strong ground shaking). More importantly, such warning could be the only way to notify the people to evacuate in the event of a tsunami earthquake that, because of its peculiar temporal evolution, generates a tsunami of greater amplitude than would be expected from the small amount of ground shaking. The most catastrophic example is the Meiji Sanriku tsunami of 1896 in northeast Japan. The earthquake magnitude was large, $Ms = 7.2$, but the ground shaking was so weak that few people were overly concerned about the quake. More than 22,000 people

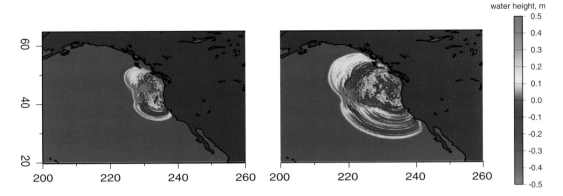

water height, m

FIGURE 4.10 Two snapshots from a simulation of the great 1700 tsunami that was generated by a magnitude 9.0 earthquake on the Cascadia subduction zone. The left panel shows the sea level displacements after one hour and the right panel after two hours. Warmer colors show wave crests; cooler colors are the troughs. After one hour, the leading crest has already inundated the local coastlines of Oregon, Washington, and Vancouver Island and has passed San Francisco Bay. On the west side of the disturbance, the initial crest is over 800 km from the coast after one hour. After two hours, the initial wave crest is well within the Southern California Bight on its way to Los Angeles. SOURCE: Satake et al., 2003; reproduced by permission of the American Geophysical Union; http://serc.carleton.edu/NAGTWorkshops/ocean/visualizations/tsunami.html.

perished in the huge tsunami that followed, which had a maximum run-up in excess of 30 m. Tsunami earthquakes are not rare. In addition to the Meiji Sanriku tsunami, Okal and Newman (2001) list the following tsunami earthquakes: the 1946 Aleutian Island tsunami; the 1963 and 1975 Kuril Island tsunamis; the 1992 Nicaragua tsunami; the 1994 and 2006 Java tsunamis; and, the 1996 Chimbote, Peru, tsunami. To detect a tsunami earthquake, direct measurements of the water-surface variations and/or water currents are required in near-real time. Such measurements are also critical for detecting tsunamis generated by submarine landslides. One way to accomplish such measurements is to utilize the data from existing and planned cabled ocean observatories.

Several cabled seafloor observatories are currently in operation or will be constructed in the near future off North America. These observatories comprise various sensors or sensor systems that are connected to each other and to the shore by a seafloor communications cable. This cable also provides power to the sensors and a pathway for high-speed data return from the sensors. The sensors gather a variety of oceanic and geophysical data that are transmitted in near-real time via the fiber optic cables from the seafloor to onshore data servers. Among the sensors are those useful for tsunami detection; for example, bottom pressure sensors, seismometers, current meters, hydrophones, gravimeters, and accelerometers. The cable can deliver relatively high amounts of electric power to support many sensors acquiring data at high sampling rates. Observatories are currently in operation off British Columbia (North-East Pacific Time-Series Underwater Networked Experiments, NEPTUNE-Canada: http://www.

neptunecanada.ca/) and in Monterey Bay, California (Monterey Accelerated Research System, MARS: http://www.mbari.org/mars/). Another large U.S. observatory has been funded by the NSF for deployment across Oregon's continental shelf, slope, and the Cascadia subduction zone, over the Juan de Fuca plate, and on to the Juan de Fuca Ridge (Ocean Observatories Initiative, OOI: http://www.interactiveoceans.washington.edu/). Both the NEPTUNE-Canada and OOI networks can be used for quantitative tsunami detection primarily via their seismometers and seafloor pressure sensors.

Off Oregon, Washington, and British Columbia, the water pressure sensors placed on the seafloor cabled observatories can readily replace or enhance the DARTs in providing warning to communities at mid- to far-ranges from the tsunami-producing Cascadia subduction zone. In addition, because the seismic data from the observatories can be used in near-real time by automatic computer algorithms in order to separate seismic and tsunami signals in the pressure data, the pressure gauges can be placed very near, and even on top of, the expected tsunami source regions. This can yield very rapid determination of the generation (or not) of a sizable tsunami, thus providing a capability for producing some modicum of warning to the near-field coasts.

From a pragmatic operational point of view, the utilization of NEPTUNE-Canada and the OOI sensors for tsunami detection could be expected to eliminate the need for the DART buoys off Washington and Oregon, thus freeing up those resources for other purposes.

In Japan, cabled observatories already exist that are focused on collecting measurements of earthquakes and tsunamis. For example, Japan Agency for Marine-Earth Science and Technology (JAMSTEC) has installed three observatories and is constructing a fourth, called Dense Ocean-floor Network System for Earthquakes and Tsunamis (DONET), that specifically aims at capturing the data from the next Tokai earthquake and tsunami. One exceptional event has already occurred on one of JAMSTEC's observatories, the Tokachi-oki site, which was located atop the source area of the 2003 Tokachi-Oki earthquake; for the first time ever, seafloor sensors observed the pressure variations of the tsunami at the instant of creation. The abrupt changes in water pressure at the seafloor clearly show the seafloor displacements of the earthquake, with sustained acoustic (pressure) waves bouncing up and down between the hard bottom and the sea surface (Li et al., 2009) while the tsunami wave evolves outward therefrom. These observations of the 2003 Tokachi-Oki earthquake and tsunami provided an important lesson: the sensors and cables of an observatory placed at the epicenter can survive the earthquake, allowing the near-real-time data to be used effectively for rapid warning of local tsunamis.

Another possible technology for detecting local tsunamis is high-frequency (HF) radar (Lipa et al., 2006). Coastal HF radar stations produce maps of the ocean surface currents using radar echoes from short period surface gravity waves. A tsunami wave, which exists at longer periods (1-30 minutes) than the waves (~10 seconds) that reflect the radar's microwave energy, will transport the shorter waves, adding to the ambient current and producing a signature detectable by the radar. The method has not been proven in the field, but theoretical and analytical studies are encouraging. The radars could provide accurate and rapid observations of tsunami waves before they make landfall and thereby aid in the formulation of better warning products.

Many radar stations installed along the coast are threatened by the Cascadia subduction zone (e.g., see http://bragg.coas.oregonstate.edu/). With software enhancements, these stations, and new ones in critical locations, could be key elements of a rapid warning system for near-field events. The radar stations are typically installed on high bluffs overlooking the shore, above any possible inundation. The potential for a broad user base of HF radar data in many locations would help justify the expense of installation and operations, resulting in enhanced sustainability.

Conclusion: Tsunami detection, warning, and preparedness activities for tsunamis arriving within minutes to an hour or so could benefit from existing, alternative technologies for rapid detection, especially considering the current sensor network's limitations for detecting tsunami earthquakes and tsunamis generated by submarine landslides.

Recommendation: For the purpose of developing more rapid and accurate warnings of local tsunamis, especially along the Washington and Oregon coasts, the committee recommends that the TWCs coordinate with the NEPTUNE-Canada and OOI observatory managers to ensure that their seismic and bottom pressure data are (or will be) made available in near-real time to the appropriate telecommunications gateways. Data interpretation tool(s), jointly applied to the seismic and bottom pressure data, will need to be developed to realize the most rapid tsunami detection possible.

Other NTHMP member states could seek similar opportunities to utilize existing and/or planned systems (including coastal HF radars) for the detection and warning of local tsunamis. It must be emphasized that investment for this adaptation would be minimal, because the observatories are being constructed and will be maintained with funds external to the U.S. Tsunami Program; thus, the benefit could be substantial.

RESEARCH OPPORTUNITIES AND NEW TECHNOLOGIES

The previous sections of this chapter have made it clear that present technologies and methodologies for evaluating the potential of earthquakes to produce dangerous tsunamis, and for detecting and forecasting those tsunamis, are far from the ideal of having an accurate and complete forecast of the expanding tsunami wave train within a few minutes of the initiating rupture. It is appropriate therefore to briefly review nascent technologies and methodologies that might be able to improve the ability of the U.S. TWCs and their international counterparts to provide quicker and more accurate tsunami warnings. Some of these technologies and methodologies, like the undersea, cabled observatories discussed in the previous section, are already available, simply waiting for the appropriate testing and software development to be integrated into the TWCs warning processes. Others require much more development before they will become useful.

Technologies such as satellite altimetry, passive microwave radiometry, ionospheric perturbation detection, and real-time kinematic-global positioning system (RTK-GPS) buoys have been proposed for detecting tsunamis in the wake of the Indian Ocean event of 2004. Although potentially promising, there has not been any demonstration of a viable operational alternative to the current systems, perhaps due to lack of funding. In general, most alternatives are not adequately sensitive to serve as a replacement for present technologies, with which small waves (<1 cm) can be observed and used for wave model inputs, fine-tuning of forecasts and warnings (including cancellation of warnings), and tsunami research. Nevertheless, continued research and development may prove fruitful. The descriptions below of some interesting technologies and methodologies are provided simply to indicate possibilities and should not be interpreted as endorsements of their utility by this committee.

Duration of High-Frequency P-Waves for Earthquake Moment Magnitude Estimation

Because of the difficulty of obtaining reliable estimates of seismic moments at the long periods relevant to tsunami generation, research is needed to explore the possibility of using other methods, possibly drawing on different technologies, in order to improve the accuracy of moment estimates, and the ability to detect unusual events, such as tsunami earthquakes.

One approach to the near-real-time investigation of large seismic sources consists of targeting their duration in addition to their amplitude. The comparison between the amplitude and duration reveal violations of scaling laws (e.g., slow events such as tsunami earthquakes). Following the 2004 Sumatra earthquake, Ni et al. (2005) noted that source duration can be extracted by high-pass filtering of the P-wave train at distant stations, typically between 2 and 4 Hz. Only P-waves escape substantial inelastic attenuation, so that this procedure eliminates spurious contributions by later seismic phases and delivers a "clean" record of the history of the source.

This approach has been pursued recently by Lomax et al. (2007) and Okal (2007a). In particular, the latter study has applied techniques initially developed in the field of seismic source discrimination (of manmade explosions as opposed to earthquakes) to characterize the duration of the source through the time $\tau_{1/3}$ over which the envelope of the high-frequency P-wave is sustained above one third of its maximum value. It is shown, for example, that this approach would have clearly recognized the 2004 Sumatra earthquake as a great earthquake, or the 2006 Java tsunami earthquake rupture as exceptionally slow. In addition, alternative methods for the rapid identification of source duration of major earthquakes are presently the topic of significant research endeavors, e.g., by Lomax et al. (2007) and Newman and Convers (2008).

The high-frequency band of the Sumatra earthquake was recorded in Japan using the Hi-Net seismic array comprising 700 borehole instruments at an approximate 20 km spacing. Ishii et al. (2005) used the data from the array to produce back-projected images of the earthquake rupture over approximately eight minutes across a 1,300 km long aftershock region including both the slip history and overall extent of the seismic zone. A comparison of the subsequent fault image, when compared to previous great earthquakes, supported the hypothesis that the moment magnitude of the earthquake was 9.3—the largest earthquake ever recorded with modern seismic instruments. The authors believe that such images of the aftershock zone could be made available within 30 minutes of the initiation of a similar event. Although networks or arrays like Hi-Net are rare, a similar or even more capable array is currently being implemented across the continental United States, funded by the NSF EarthScope program. Today's high-speed, high-capacity networks coupled with large-capacity computing facilities such as cloud computing provide the technologies for implementing an early warning system. The compressional wave velocity is high (>8 km/s) and will provide fault images more quickly than the hydrophone approaches discussed below. The technique used for acoustics, however, is similar to seismic back-projection.

> **Conclusion:** The P-wave duration and back projection methods appear robust and applicable to high-frequency records. These methods have some advantages over the W-phase approach because they can provide constraints on the rupture length and duration and do not rely on having seismometers with a stable long-period response.

> **Recommendation:** The committee recommends that NOAA and the TWCs consider the use of arrays and networks such as Hi-Net and EarthScope Array National Facility to determine rupture extent and moment of great earthquakes. The networking and computational requirements are significant and would need to be included in TWC upgrades in the future.

Hydroacoustic Monitoring of Underwater Geophysical Events

Sound wave ("hydroacoustic") signals can propagate a great distance within a waveguide in the ocean, termed the sound fixing and ranging channel ("SOFAR channel"). This propagation was discovered during World War II, and immediately following declassification scientists began exploring the possibility of using hydroacoustic signals generated by large earthquakes (the so-called T phases) for the purpose of tsunami warning (Ewing et al., 1950). With the development of the UN International Monitoring System of the CTBTO, several state-of-the-art hydrophone stations have been deployed in the world ocean, offering an opportunity for complementary use in the context of tsunami warning. Each station comprises three hydrophones separated by approximately 2 km to provide some directionality at low frequencies.

By placing hydrophone sensors within the SOFAR channel, a scientist can "listen" to seafloor seismic, tectonic, and volcanic events occurring at a great distance. The potential of using hydroacoustic techniques to monitor underwater landslides has yet to be fully explored, but it may represent the best approach for detecting unsuspected underwater landslides, as occurred in the 1998 Papua New Guinea (PNG) tsunami (Okal, 2003). However, that detection represents to this day a unique, unrepeated occurrence. Furthermore, the PNG landslide was identified as such because its hydroacoustic signal was too weak for its duration, in violation of earthquake scaling laws. At the same time, T phases can be used to complement the identification of anomalously slow events, such as tsunami earthquakes, because hydroacoustic signals include very high frequencies (3 Hz and above) and their energy bears the imprint of the earthquake at very short periods (Okal et al., 2003).

In this respect, hydroacoustic signals play a complementary role in tsunami warning because they travel slowly (1,500 m/s). However, de Groot-Hedlin (2005) and Tolstoy and Bohnenstiehl (2005) demonstrated that it was possible to use ocean hydrophones to track the rupture of the 2004 Sumatra event from the original epicenter to the termination more than 600 km to the north. The hydrophones were 2,800 and 7,000 km from the epicenter and acoustic propagation required 31-78 minutes while the fault itself ruptured for more than 8 minutes. The information would not be useful for alerting nearby communities but could have provided meaningful warnings for Sri Lanka and more distant countries.

Other properties of T phases can shed some interesting, but again complementary, light on properties of the seismic sources, for example, their duration, along lines similar to the $\tau_{1/3}$ method described earlier. Salzberg (2008) has also proposed to precisely constrain hypocentral depth using the decay of very high frequency (20-80 Hz) T phases from the parent earthquakes. Once such techniques reach an operational status, they could contribute to tsunami warning.

An additional aspect of SOFAR hydrophone sensors is that they can record pressure variations accompanying the passage of the tsunami, and in this sense could supplement the network of DART buoys, as their sensors (in both cases pressure detectors) essentially share the same technology, with the only difference being that the latter are deployed on the ocean bottom. However, within the context of the CTBTO, the Integrated Maritime Surveillance (IMS) sensors have been hard-wired with drastic high-pass filters (with a corner frequency of 10 Hz), and the main spectral components of the 2004 Sumatra tsunami (around 1 mHz) were recorded only as digital noise (Okal et al., 2007). The use of software rather than hardware filters for any future deployment of hydrophones in the SOFAR could be extremely valuable to the tsunami community. The cabled NSF OOI Regional Scale Nodes (RSNs) to be deployed off Washington and Oregon and the existing NEPTUNE-Canada network (see above) could support both bottom pressure gauges as well as hydrophones in the SOFAR channel for enhancing tsunami research and warning in the Cascadia area.

Continuous GPS Measurements of Crustal Movement

When combined with seismic data, continuous global positioning system (GPS) measurements of displacement have proven to be powerful in studying continental earthquakes; for example, in illuminating the processes of earthquake after-slip, creep, and viscoelastic deformation. Continuous GPS can provide a map of the three-dimensional deformation incurred at the surface in the proximity of the epicenter as a result of the earthquake rupture. It provides a resolution to the problem of the long-period component of the seismic source by simply allowing measurement during a time window long enough to be relevant to tsunami generation even for nearby sources.

GPS and broadband seismic measurements differ substantially in that GPS geodetic measurements provide distances between neighboring stations, while individual seismometers are affected by applied forces and signals are proportional to acceleration. Normally, the output of a seismometer is "shaped" to be proportional to velocity above some frequency (1/360 Hz for an STS-1; Appendix G). Because earthquakes cannot apply a constant force at zero frequency, it's not possible to directly infer displacements from a seismometer. Furthermore, a seismometer is limited by its mechanics and electronics to recording signals smaller than some threshold; arbitrarily large displacements can be measured by GPS.

Bock et al. (2000) demonstrated that GPS receivers can measure ground motion in real time as often as every few seconds. They tested the accuracy of these estimates over baselines as large as 37 km and found that the horizontal components have accuracies no worse than 15 mm; they anticipated that the baselines could be extended to at least 50 km with no further loss in accuracy. The vertical measurements were less useful with accuracies a factor 7-8 times worse. The accuracies have improved over the past decade with the advent of new receivers, new algorithms, and statistical analyses. GPS receivers of 10-50 Hz and methods are now practical and measurements routine (e.g., Genrich and Bock, 2006).

The application of near-real-time, continuous GPS measurements have made great strides as well. For example, Song (2007) used coastal GPS stations (E-W and N-S horizontal measurements) to infer displacements on the seafloor offshore using the location of the fault and inferring the vertical uplift from conservation of mass. Song tested the method against geodetic data from the 2005 Nias, 2004 Sumatra, and 1964 Alaska earthquakes. In the case of Nias and Sumatra, both continuous GPS data as well as campaign GPS data were available. He tested the model against satellite altimetry measurements of the tsunami wave using Topex, Jason, and Envisat data (altimetry profiles included time epochs of 1:55-2:10, 1:48-2:03, and 3:10-3:22 [hr:min after the origin time]). The Nias and Alaska events were also tested against available coastal tide gauge data. The methods were used again after the February 27, 2010, Chile earthquake and later verified by satellite altimetry from JASON-1 & 2 satellites operated by National Aeronautics and Space Administration (NASA) and the French Space Agency.

The successful use of GPS data for these four earthquakes makes a strong case for the use of continuous GPS stations to measure coastal ground displacements to infer the corresponding displacements offshore. In turn, these displacements can be used to predict tsunami generation including accurate wave heights as a function of time, range, and azimuth.

Near-field tsunamis are generated by the rupture of hundreds of kilometers of an offshore subduction fault. As in the case of the Sumatra earthquake this rupture can last as long as eight minutes and more. During this period of time, GPS data will mimic seismic data with oscillatory behavior that obscures the smaller, permanent displacements. The most distant part of the fault from a station can be at least as large as eight minutes of propagation time away, and the displacements generated by that distant source will take as long to propagate back to the station. By that time, however, the static offsets will begin to be apparent, allowing the inference of offshore displacements and realistic assignment of magnitudes (as little as 4-5 minutes after the initiation of faulting). The tsunami associated with the earthquake will not come ashore much earlier than 30 minutes following the beginning of the rupture, and technology-based warnings could be made in time to provide useful warnings. None of these operations lie even remotely outside the capabilities of modern networks, computational workflows, and computing capabilities.

Today there are thousands of GPS geodetic receivers located around the earth. Just in southern California, there are more than 250 continuously recording GPS geodetic stations that are available in near-real time from a variety of sources (e.g., http://sopac.ucsd.edu; Schmidt et al., 2008). Recently, NSF Geosciences elected to undertake the improvement and densification of seismic and geodetic stations in the Cascadia region including the enhancement of near-real time access to GPS (http://www.oceanleadership.org/2010/nsf-cascadia-initiative-workshop/). Sweeney at al. (2005) have demonstrated that centimeter-level horizontal accuracy can be achieved on the seafloor using GPS coupled to seafloor geodetic monuments using acoustic methods. These technologies might be extended to verify offshore displacements predicted by accurate coastal GPS stations.

Permanent GPS stations should be incorporated into the tsunami warning program and expanded, if needed, to provide tsunami prediction capabilities. Although Cascadia is one of the most critical sites for U.S. tsunami warning in the near-field regions, Alaska and the Caribbean are also critical sites. There are few new technologies that promise such revolutionary approaches for improving tsunami warning, especially in the near-field region.

Conclusion: GPS geodesy, exploiting near-real-time data telemetry from permanent geodetic stations, holds great promise for extending the current seismic networks to include capabilities for measuring displacements in the coastal environment for great and mega-earthquakes. Displacements onshore can potentially be used to infer offshore displacements in times as short as five minutes in an area such as the Cascadia Fault Zone.

Recommendation: NOAA should explore further the operational integration of GPS data into TWC operations from existing and planned GPS geodetic stations along portions of the coast of the United States potentially susceptible to near-field tsunami generation including Alaska, Cascadia, the Caribbean, and Hawaii. Where GPS geodetic coverage is not adequate NOAA should work with NSF and the states in extending coverage including the long-term operation and maintenance of the stations.

Observation of Tsunami Wave Trains with Satellite Altimeters

Satellite altimeter measurement of the ocean's surface height, in use since 1978, consists of measuring (with a precision of a few centimeters) the deformation of the surface of the ocean by precisely timing the reflection of a radar beam emitted and received at a satellite. Its capability to detect a tsunami was proposed following the 1992 Nicaragua tsunami (Okal et al., 1999), and it achieved a definitive detection following the 2004 Sumatra tsunami, with a signal of 70 cm in the Bay of Bengal (Scharroo et al., 2005; Ablain et al., 2006). (See also the preceding topic, "Continuous GPS Measurements of Crustal Movement.")

Although the method has obvious promising potential in the field of tsunami warning, two major problems presently hamper its systematic use: (1) delayed processing of the data, which in the case of the 2004 event was made available to the scientific community several weeks after the event, and (2) the presently sparse coverage of the earth's oceans by altimetry satellites. In lay terms, the satellite has to be over the right spot at the right time; in the case of the Sumatra tsunami, the passage of two satellites over the Bay of Bengal as the tsunami propagated across was a lucky coincidence. Thus, making satellite altimetry operational for tsunami warning requires geostationary satellites over the ocean basins of interest, or a dense array of low earth orbit (LEO) satellites, with either set-up providing data availability in near-real time. In fact, Iridium Communications, Inc. is designing its second generation of LEO communications satellites (called Iridium NEXT), which are expected to be fully deployed by 2016 and will carry scientific payloads such as altimeters for sea height determination, including observation of tsunamis (http://www.iridium.com/About/IridiumNEXT/HostedPayloads.aspx). The planned constellation of 66 satellites suggests that a tsunami created anywhere in the world could be observed close to the moment of inception. At the present time, however, the NEXT constellation is not being touted as a tool for operational tsunami warning.

Tsunami-Induced Sea-Surface Roughness and "Tsunami Shadows"

Godin (2004) theoretically justified so-called "tsunami shadow" observations (Walker, 1996), namely that the surface of the ocean exhibits a change of appearance during the propagation of a tsunami. In simple terms, the tsunami creates a coherent change in sea-surface slope, inducing turbulence in wind currents at the surface, which in turn results in enhanced roughness of the sea- air interface. Godin et al. (2009) further showed that the phenomenon was detectable in the form of anomalous scattering in the radar signal from the JASON satellite altimeter, during its transit over the wavefront of the 2004 Sumatra tsunami in the Bay of Bengal. This remarkable scientific confirmation and physical explanation of what had amounted to anecdotal reports provides some promise as a complementary means of near-real-time tsunami detection. In its reported form, the method suffers from the same limitations as satellite altimetry, namely the need to have a satellite at the right place at the right time. On the other hand, it may be feasible to develop a land-based detector of sea-surface roughness using over-the-horizon radar technology.

Direct Recording of Tsunami Waves by Island Seismometers

Another notable observation made in the wake of the 2004 Sumatra event was that the actual tsunami wave was detectable on horizontal long-period seismometers located on oceanic islands or on the shores of continental masses (e.g., Antarctica) (Yuan et al., 2005). Okal (2007b) later verified that such signals could be extracted from past events (e.g., Peru, 2001), and showed that the recordings expressed the response of the seismometer to the combined horizontal displacement and tilt of the ocean floor during the passage of the tsunami wave, the latter having such large wavelengths (typically 300 km) that the structure of a small island can be neglected. In particular it was verified that such records could be interpreted quantitatively on this basis, which amounts to saying that near-shore seismometers can play the role of tsunameters deployed on the high seas for tsunami detection. The present network of island seismic stations (see Figure 4.1) thus has the potential of increasing the density of the tsunami (sea level) detection network, at essentially no cost, since the stations already exist.

"Upward Continuation" of the Tsunami Wave and Its Detection in Space

Because of the finite density of the atmosphere, a tsunami wave does not stop at the surface of the sea, but induces a displacement of the atmosphere, in the form of a gravitational wave accompanying the tsunami during its propagation. The volumetric energy density of this upward continuation of the tsunami decreases with height, but because the atmosphere rarefies even faster, the amplitude of the resulting vibration will actually increase with height. A tsunami wave of amplitude 10 cm at the surface of the ocean will reach 1 km at the base of the ionosphere at an altitude of 150 km. This fascinating proposition was initially suggested by Peltier and Hines (1976) and confirmed by Artru et al. (2005) during the 2001 Peruvian tsunami. The detection methodology uses dense arrays of GPS receivers, because large-scale fluctuations of the ionosphere affect the propagation of the electromagnetic waves from the GPS satellites, thus distorting the signals recorded at the receivers. Occhipinti et al. (2006) have successfully modeled such records quantitatively and have shown that other space-based techniques involving reflection at the bottom of the ionosphere (e.g., over-the-horizon radar) could be useful for remote detection of a tsunami on the high seas without the need to instrument the ocean basin itself. The speed of propagation of the atmospheric gravity wave, however, is very low and presents an even greater complication than that described above for acoustic propagation in the ocean's SOFAR channel.

Conclusion: Novel and potentially useful approaches to the estimation of earthquake magnitude and tsunami detection are emerging. Some of these approaches could become operational in the not-too-distant future with proper support for research and testing.

CHAPTER FIVE

Long-Term Reliability and Sustainability of Warning Center Operations

SUMMARY

This chapter evaluates long-term prospects of the two Tsunami Warning Centers (TWCs) to providing reliable and internally consistent tsunami detection, decision support, and product generation, and for effectively supporting threat detection, warning management, and public response. The goal of the current geographically distributed organization of the TWCs (i.e., a center each in Hawaii and Alaska) with distinct areas of responsibility (AORs) is to provide the system with back-up in the case of critical failure at the other center. However, the two TWC technology suites differ considerably from each other, and those differences lead to technological incompatibilities and limited capabilities for back-up, redundancy, and checks and balances, which are important mission capabilities for the tsunami warning system.

In addition, as discussed in detail in Chapter 3, inconsistencies in warning products issued by the two TWCs and the current division of AOR results in messages that have caused confusion and the potential to cause confusion in the future, making the products less effective in eliciting the appropriate response. The committee recommends that the National Oceanic and Atmospheric Administration's (NOAA's) National Weather Service (NWS) develop tsunami warning system products that reflect best practices, as well as lessons learned from other operational real-time, large-scale, mission-critical distributed systems, and that comply with international information technology and software engineering product and process standards. In addition, the committee recommends that the TWCs undertake an external review by information technology (IT) specialists in the area of communication technology for the latest technology in message composition software and formats to ensure compatibility with current and next generation information and communication technology (web and cell-phone) for message dissemination.

Because the centers are under different management, use different analytical software and hardware, and appear to have distinct organizational cultures, the committee concludes that they do not function as redundant systems. The committee discusses several options for alternative organizational structures including operational convergence of the two centers, merging the two centers into a single center, and/or co-locating center(s) with other research or operational units and recommends that the decision to develop an organizational structure for the TWCs should be undertaken as part of the comprehensive, enterprise-wide, long-range planning effort recommended in this chapter.

The success of the TWC mission is critically dependent on technical infrastructure and human capital, both of which the committee assessed to be insufficiently supported. Because of the rapid evolution of IT and its importance in the overall process of detecting and warning, the committee found that the TWCs lack sufficient state-of-the-art technology; IT support, maintenance, assessment and planning processes; and IT personnel and leadership. The committee recommends that NOAA/NWS provide these capabilities to the TWCs and establish an external IT advisory body, with membership from the U.S. Geological Survey (USGS), other seismic network operators, social and information scientists, emergency managers, and other large-scale, safety-critical systems professionals to advise the TWCs.

Workforce development and recruitment can be challenging. Frequent, regular, and varied types of training as well as stronger connections with the external research community are required. TWC human capital requirements, training, re-training, development, and mentoring and requirements for professional exchanges should be included, reassessed, and updated as part of the recommended enterprise-wide tsunami planning effort, so that technology and human and organizational requirements can be considered and developed together by tsunami program members and their customers. Overall, the TWCs should adopt NOAA- and government-wide standards for recruiting, retaining, training and re-training, planning, developing, nurturing, and mentoring the critical human resources that are at the center of tsunami warning and detection system success.

Finally, the committee found that an organizational culture change within the NOAA/NWS Tsunami Program would be beneficial to advance operational excellence. Such a change should also lead to increased support to adopt national and international standards, processes, best practices and lessons learned for all functions, technologies, and processes and products; and result in ongoing, continuous process improvements.

As detailed in the chapter, some of the steps to improve long-term operations recommended by the committee include the following:

- NOAA/NWS should undertake a comprehensive, enterprise-wide, long-range planning effort for the TWCs. The goal of the planning effort would be to analyze TWC functions and requirements; articulate the technological, human, physical, and intellectual infrastructure required to meet the TWC requirements; and integrate the technology, applications, tools, processes, networks, leadership, policies, organizational structures, and human capital required to provide long-term reliable and sustainable global TWC operations. Such a technology planning effort should develop assessments of:
  - o technology, applications, tools, processes, networks, hardware, software, and systems;
  - o the requirements for human capital, training, re-training, development, mentoring, professional exchange, leadership, and policies; and
  - o the organizational structure(s) required to ensure that the two TWCs can function

as a single warning center and can provide the required support for reliable and sustainable global TWC operations.

- NOAA/NWS and the TWCs should adopt national, and where available, international, standards, best practices, and lessons learned for all functions, technology, processes, and products.

- If NOAA/NWS maintains the current organizational structure, then it should harmonize and standardize technologies, processes, and products between the two TWCs.

- NOAA/NWS and the TWCs should undertake ongoing, joint, or NOAA-wide benchmarking and continuous process improvement activities for their functional, technological, organizational, or human capital initiatives; report those activities internally and externally; and incentivize excellent performance as well as best practices.

- NOAA/NWS and the TWCs should develop measures of performance and benchmark individual, organizational, and technical performance against industry and agency metrics; identify areas for improvement; set short- and long-term performance goals; develop reward and incentive systems for such goals; and celebrate TWC and agency accomplishments as performance improves, in order to raise the level of TWC performance to that expected of a high-reliability organization.

- NOAA/NWS and the TWCs should increase their use of internal and external review processes, as detailed below.

THE TSUNAMI WARNING CENTERS

The Pacific Tsunami Warning Center (PTWC) was established as the Honolulu Observatory in 1949 after the April 1, 1946, tsunami generated in the Aleutian Islands[1] caused casualties and damage on the Hawaiian Islands. Following the 1960 Chile tsunami, the Honolulu Observatory expanded its AOR to cover all nations along the Pacific basin (Intergovernmental Oceanographic Commission, 1965). After the 2004 Indian Ocean tsunami, the PTWC's responsibility was again expanded to include the Indian Ocean and Caribbean Sea nations.

The West Coast/Alaska Tsunami Warning Center (WC/ATWC) was established as the Palmer Observatory after the great 1964 Alaskan earthquake, which devastated parts of Anchorage. In the 1980s and 1990s, its AOR was expanded to include tsunami warnings for California, Oregon, Washington, and British Columbia if potential tsunamigenic earthquakes were detected in their coastal areas. This delineation was changed in 1996 to include all Pacific-basin tsunamigenic sources for California, Oregon, Washington, British Columbia, and Alaska. After the 2004 Indian Ocean tsunami, the WC/ATWC's responsibility expanded to include the U.S. Atlantic and Gulf coasts, Puerto Rico, the Virgin Islands, and the Atlantic coast of Canada. The PTWC has the following areas of responsibility (Figure 5.1):

- the State of Hawaii
- Guam, American Samoa, and other U.S. Pacific assets

---

[1] Weinstein, 2008.

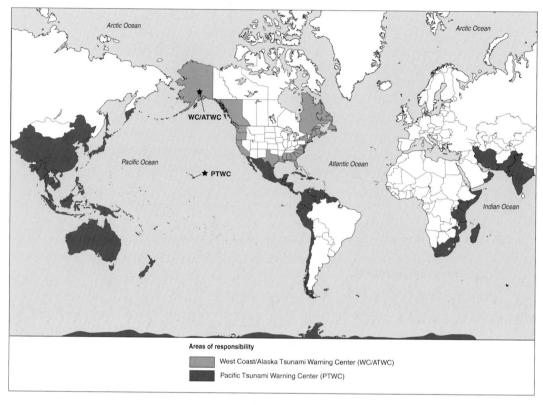

FIGURE 5.1 Areas of responsibility (AORs) for the PTWC (dark gray) and the WC/ATWC (light gray). SOURCE: Government Accountability Office, 2010.

- many Pacific rim countries (as an operational center of the Pacific Tsunami Warning and Mitigation System [PTWS] of the Intergovernmental Oceanographic Commission [IOC] occasioned by the 1960 Pacific-wide tsunami generated in Chile)
- Indian Ocean countries (as an interim center for the Indian Ocean Tsunami Warning and Mitigation System [IOTWS] of the IOC since 2005)
- Caribbean countries except the U.S. commonwealth of Puerto Rico and the U.S. and British Virgin Islands (as interim center for the Intergovernmental Coordination Group for the Tsunami and Other Coastal Hazards Warning System for the Caribbean and Adjacent Regions [ICG/CARIBE EWS] since 2006)

The WC/ATWC has the following areas of responsibility (Figure 5.1):

- all U.S. states except Hawaii
- Canadian coastal regions
- Puerto Rico and the Virgin Islands

TSUNAMI WARNING CENTER FUNCTIONS

The Tsunami Warning Centers have a widespread client base: emergency managers, the scientific community, and the public. They are responsible for gathering information from sensor and observational systems; detecting events of interest (tsunamigenic earthquakes) and determining magnitudes for those events; developing decision support information for operational and scientific decision makers; and providing and disseminating warning and notification products to the public and other entities. The TWCs are not designed for or capable of detecting landslide-induced tsunamis such as those that might occur in Alaska, Puget Sound, or in some of the Hawaiian Islands. Operational components of each TWC (Figure 5.2) include:

- *Earth Data Observations,* which allow the detection of earthquakes and tsunami occurrence (described in Chapter 4).
- *Data and Information Collection* of seismic and sea-level data, impact reports from agencies and the public, and data and information sharing with other centers (described in Chapter 4).

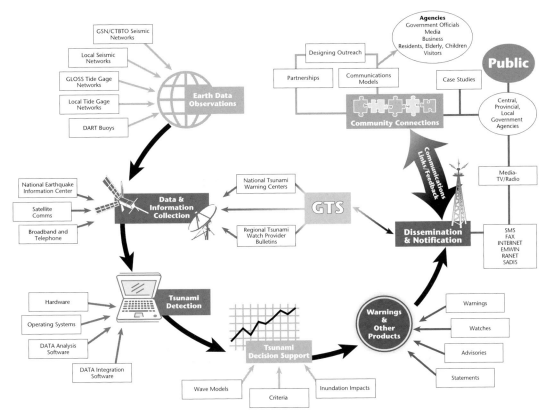

FIGURE 5.2 Key operational components of the tsunami warning centers. SOURCE: U.S. Indian Ocean Tsunami Warning System Program, 2007.

- *Tsunami Detection Systems* that rely upon scientific expertise and practical experience to process and analyze the gathered data in support of tsunami detection and impact projections, and hardware and software that support the analysis. Critical to effective detection system performance are appropriate personnel, training, policies, procedures, and an organizational culture that values constant and frequent organizational learning and that reinforces the core goals of excellence and high performance in the warning system.

- *Tsunami Warning Center Decision Support Systems* that assist the TWC watchstander[2] in determining what type of warning product should be issued. Tools available to support the decision making process include criteria and bulletin thresholds, software support to detect seismic events in real time, and computer models of tsunami wave heights and coastal inundation models for impact assessments. Critical needs for this component include frequent and varied training for operational watchstanders (e.g., simulations, walk throughs, case studies, table-top exercises); organizationally supported interactions between watchstanders in the two warning centers; and ongoing and prioritized research and development to support operations and implement new technology.

- *Warning and Other Products,* which are standardized messages issued to the public and other customers. The following four products are described in detail in this chapter:
  - A *Tsunami Warning*, issued when a potential tsunami with significant widespread inundation is imminent or expected. Initial warnings are normally based only on seismic information.
  - A *Tsunami Watch,* issued to alert emergency managers and the public of an event that could later impact the watch area. As updated information becomes available (e.g., from sea level networks), the watch may be changed to a warning or advisory.
  - A *Tsunami Advisory*, issued for the threat of a potential tsunami causing strong currents or waves that would be dangerous to those in or near the water. Significant widespread inundation is not expected for those areas under an advisory.
  - A *Tsunami Information Statement*, issued to alert emergency managers and the public that an earthquake has occurred. In most cases, information statements are issued to indicate that there is not a threat of a destructive tsunami affecting the TWC's AOR, and are used to avoid evacuations in coastal areas that may have felt the earthquake. In some cases, the information statement is also used to indicate the need to stay tuned for more information as it becomes available.
  - *Dissemination and Notification Plans,* which are produced in advance to ensure that TWC customers are able to receive and understand the warning products. It is critical that dissemination systems are tested and that roles and responsibilities for different actors are clearly defined and articulated.

---

[2] A watchstander is a person who stands a shift (a watch) in an operational command center, such as those centers that monitor earthquakes, storms, emergency events, electrical and power failures, and the operation of various plants, factories, emergency services operations, equipment, and vessels.

- *Community Connections* that educate the public about tsunami safety and preparedness and inform the public about the TWC's role in tsunami warning. Effective community connections include partnerships with media and the community that develop community preparedness and other resilience initiatives (U.S. Indian Ocean Tsunami Warning System Program, 2007) (community preparedness efforts are discussed in detail in Chapter 3).

Tsunami Detection

Tsunami detection requires information gathering, data analysis and assessment, and decision making and communication. Earthquake detection functions required at both the TWCs are similar and comprise the following basic steps (see Weinstein, 2008; Whitmore et al., 2008):

(1) Seismic data analysis systems automatically and rapidly evaluate the location, size, and focal mechanism of an earthquake to determine if it has significant potential to trigger a tsunami. Watchstanders reassess the event by analyzing select seismic data and may empirically adjust the moment magnitude determination;

(2) Watchstanders determine if the magnitude is above certain thresholds and based on this analysis, generate and disseminate initial messages (with either a watch, warning, or information bulletin);

(3) For each significant earthquake, watchstanders estimate corresponding tsunami arrival times and heights for selected critical locations;

(4) Once sea level data are acquired, watchstanders reassess the threat, including scaling earlier, computed tsunami forecast scenarios to fit the sea level observations from the Deep-ocean Assessment and Reporting of Tsunamis (DART) systems; if needed (e.g., if sea level data are lacking) watchstanders continue post-processing the seismic data to refine the threat assessment;

(5) Watchstanders generate and disseminate follow-up informational messages with the additional detailed information available from sea level and seismic data analysis (National Oceanic and Atmospheric Administration, 2008a, c);

(6) Watchstanders iterate steps 5 and 6 until an appropriate time interval (based on the expected tsunami propagation speed and modeled duration of inundation) has passed and all watches or warnings for the AOR shorelines can be lifted.

The TWCs consider "significant" earthquakes as events exceeding certain predetermined magnitude thresholds, which also depend on the distance offshore. These events automatically trigger the audio alarm systems of the two TWCs, prompting on-duty watchstanders to initiate a detailed investigation of the earthquake and its potential to trigger tsunamis. Typically, watchstanders at both TWCs are required to make initial and independent estimates of the earthquake location and magnitudes within 5 to 15 minutes of the earthquake origin time (National Oceanic and Atmospheric Administration, 2008a, c). The location determines which TWC will be responsible for officially determining earthquake parameters. Based on the committee's review of the warning messages, magnitude determinations included in the warning messages sometimes differ between the two TWCs. The warning message is communicated to the National Earthquake Information Center (NEIC, see also Box 5.1) but the TWCs typically attempt to determine the seismic location and magnitude more rapidly than NEIC. Subsequent messages from the TWCs may include corrections to the magnitude determination issued in the first message. Given the benefit of consistency in warning message dissemination (discussed in Chapter 3), avoiding such inconsistencies in earthquake parameters between different warning products would be beneficial to the warning process.

Although many of the required functions are similar, the detection procedures followed by each of the TWCs differ and the centers use different operating systems and graphical user interfaces (Table 5.1). The WC/ATWC uses personal computer (PC) workstations running Microsoft Windows and all seismic, sea level, geographic information system (GIS), forecasting, and product-generation software run on a network of 10 operational PCs with complete hardware and communications redundancy. Development, testing, and training operations are performed on three other computers. In contrast, the PTWC utilizes seven redundant Sun/Solaris workstations, and functions are migrating to the open-source Linux operating system. The development of incompatible hardware platforms was based on the availability of local knowledge at each TWC, rather than on a formal information architecture development or planning process.

Each TWC uses different software for its real-time monitoring and analysis systems: the PTWC runs Antelope[3] to manage seismic data streams. This software package is currently used at several other earth science programs (e.g., Earthscope Array National Facility, Saudi Arabia National Seismic Network, Alaska Seismic Network, Singapore National Telemetry Network, Dominican Republic Seismic Network, Ocean Observatories Initiative Cyberinfrastructure component). The WC/ATWC and the PTWC use software based on Earthworm (version 7.1), an open source package for regional seismic networks developed by the USGS in 1993. As of 2007, 40 observatories[4] in addition to the WC/ATWC and the PTWC were using Earthworm despite the lack of active software development or maintenance.

Because there is no common software architecture, each TWC operates with a different set of standards and procedures. The processes followed by scientists receiving sensor information differ and are supported by different analytical tools in each TWC. The analyses, results, messages, thresholds, and notification processes supported by each TWC's hardware and software technology are also therefore different, and the customer-facing software interfaces and reports provided to the public and the media from each TWC differ; the result of the differences suggests that the two TWCs are not part of the same organization, with the same mission. The differences introduce another set of operational challenges, as, based on the committee's observations, it appears that a scientist working at one TWC would have difficulty covering a shift at the other TWC without significant additional training. Despite these differences, how-

---

[3] http://www.brtt.com/.
[4] The list was obtained from http://folkworm.ceri.memphis.edu/ew-doc/.

TABLE 5.1 Comparison of Tsunami Warning Center Technology and Management Products and Processes

Technology Product, Process	West Coast/Alaska Tsunami Warning Center	Pacific Tsunami Warning Center
Hardware Platforms		
Hardware Platforms	10 PC workstations and servers with hardware and communications redundancy	7 Sun workstations 2 PC workstations from PMEL to run SIFT
Software Platforms		
Operating System	Windows XP for EarlyBird	SUN Solaris and Linux
Applications		
• Seismic processing, analysis	Data acquisition: Earthworm and Nanometrics for WC/ATWC network data Data processing: standard Earthworm architecture with specialized tsunami analysis modules known as EarlyBird	Data acquisition: Earthworm, EdgeCWB, and Antelope for PTWC network data Data processing: for non-Hawaii events, standard Earthworm architecture with local developed user interfaces and some use of EarlyBird modules. For Hawaii events, locally developed analysis system
• Sea level, tidal data analysis	Data ingest through Global Telecommunications System (GTS) and Earthworm using locally developed decode software Data analysis using TideView	Data analysis using TideTool

continued

TABLE 5.1 Continued

Technology Product, Process	West Coast/Alaska Tsunami Warning Center	Pacific Tsunami Warning Center
• Mapping software	EarthVu GIS—locally developed based on Geodyssey Limited's Hipparchus system. Also use Generic Mapping Tool (GMT) and developing with ESRI and Google Maps	GMT open source software maintained at the University of Hawaii
• Messaging software	Written internally Aging—software support difficult Watchstander previews are not supported	
	Coded in C	Coded in FORTRAN
• Geographic information system	EarthVu	
• Tsunami forecasting systems	Use Alaska Tsunami Forecast Model (ATFM) through EarthVu GIS interface Use PMEL's SIFT server software and hardware	
	Precomputed model database WC/ATWC developing ATFM version 2, an upgrade to ATFM	Developing forecast system capable of producing international forecasts known as real-time inundation forecasting of tsunamis Will use ATFM version 2 once available
• Product formats	Standard NWS, HTML, RSS, CAP/XML, SMS	Standard NWS, RSS, CAP/XML, SMS
• Configuration management tools	Subversion Configuration Management System Configuration Management Plan (CMP) in use	GIT
• Security, quality assurance, file sharing software	Recently completed a Certificate and Accreditation audit (May 2010). Authority to operate granted with moderate threat level.	Ready to undergo Certificate and Accreditation audit
Programming Languages	C Some Java-based code	Procedural—C, FORTRAN Tcl/TK for scripting
Databases	MySQL	
Web Infrastructure	HTML, PHP, MapTools	PHP
Web 2.0, 3.0 Planning, Products	Not described	Initial explorations

continued

TABLE 5.1 Continued

Technology Product, Process	West Coast/Alaska Tsunami Warning Center	Pacific Tsunami Warning Center
Technology Life Cycle Processes		
System Planning Processes	Not observed	
System Development Processes	Software is internally developed, during periods of operational slack, by watchstanding scientists. IT support staff express reservations about stability of hardware, software infrastructure.	
System Deployment Processes	Performed by watchstanding scientists, during periods of operational slack	
	Release deployment follows CMP's instructions and requirements	
System Testing Processes	Development and test systems provide testbed before systems are operationally deployed. CMP provides testing requirements.	Not described
System Support	Watchstanding scientists maintain the pieces of code that they are responsible for developing.	
	IT support staff report software is unstable and aging.	
System Procurement Processes	Informal Dated Inadequately budgeted	
System Maintenance	Regularly performed Informal processes	
System Configuration Management	No formal processes described Updates difficult IT support staff performing updates and installing changes impacted by unstable and aging software.	No formal processes described
Organizational Learning Processes, Process Improvements, Incorporation of Lessons Learned, Dissemination of Best Practices	Not observed	

continued

TABLE 5.1 Continued

Technology Product, Process	West Coast/Alaska Tsunami Warning Center	Pacific Tsunami Warning Center
Technology Management Processes		
System Back-up and Redundancy	Regularly scheduled Common processes between TWCs	
System Security, Information Assurance, Quality Control	Not observed	
Personnel, Watchstanding Redundancy, Back-up	Personnel can perform back-up activities and issue the other center's messages (federal mandate requirement satisfied). Made difficult due to incompatible operating systems, applications, and processes Training required before back-up and redundancy is operational	
Code Sharing, Code Integration between TWCs	Not possible due to incompatible operating systems, applications	
Documentation	Limited	
System Performance Benchmarking, Reporting, Metrics	System accuracy and response time reported monthly Center uptime (partially dependent on IT system uptime) also reported monthly	Not available or developed Not part of an articulated IT performance evaluation process
Adherence to/ Compliance with International Process, Product Standards	Not apparent Software risks mitigated by use of legacy personnel performing timely repairs	

SOURCE: Committee member.

ever, the TWCs are able to perform federally mandated back-up activities for each other and are able to issue each other's messages.

The TWCs' technology products and processes can benefit from the experiences and lessons learned from other distributed, safety-critical, large-scale systems with significant data and communication requirements, including aviation, medicine, transportation, chemical processes, offshore oil and gas, and, within NOAA, the NCEP, as well as the NEIC and the NWS' Hurricane Center and Severe Storm Center (see Box 5.1). The TWCs' software and hardware technology planning, development, operations, and maintenance processes and products should be aligned with those of other safety-critical systems and NOAA departments in order to (1) enhance compatibility, modeling, and analysis with other large-scale environmental systems; (2) improve data and analysis sharing to enhance lessons learned; (3) increase compliance with international standards; (4) reduce hardware and software vulnerability and dependence; and (5) decrease training and maintenance costs.

Given their shared mission, TWC functions (including but not limited to tsunami detection) could be supported by a common technology suite having common operational requirements, processes, training, policies, and procedures. The committee notes that a number of highly reliable, mission-critical, large-scale systems currently support real-time distributed operations across a broad geographic area, including, within NOAA, the NWS with global telecommunications service (GTS), the NCEP, and the Hurricane and Severe Storm labs; and outside NOAA, the medical (Physicians On-Line/Medscape, 2006), geophysical (Gulbrandsøy et al., 2002; Herring, 2002), financial (Society for Worldwide Interbank Financial Telecommunications, 2009), and energy distribution systems (Andersen and Drejer, 2005). These systems provide examples of how to operate and maintain a distributed, complex operational system. Technology industry best practices adopted by these organizations include (1) adherence to platform-independent hardware and software architectures, applications, and interfaces; (2) the use of international hardware, software planning, development, operations, and maintenance product and process standards, including the Software Engineering Institute's Capability Maturity Model and the software development life cycle (Carnegie Mellon Software Engineering Institute, 2010); (3) regular and systematic use of continuing process and product improvement models for hardware, software, personnel and organizational planning, development, operations, and maintenance; and (4) evaluation and assessment of technology and organizational processes and products as part of a continuous process improvement and learning organizational culture (Senge, 1990).

> **Conclusion:** The two TWC technology suites differ considerably from each other, with different hardware platforms, software suites, products, processes, and interfaces to the public and their users. Differences in TWC technologies, processes, products, and interfaces lead to technological incompatibilities and limited capabilities for back-up, redundancy, and checks and balances, which are important mission capabilities for the tsunami warning system.

Tsunami Decision Support

Detection of a significant earthquake leads to a series of critical decisions by TWC watch-standers: What type of product needs to be issued? When and how can the initial message be refined, based on real-time observations? Basic tools available in the decision making process include:

- Threshold criteria for an initial warning, which are based on historical events;
- Software for monitoring relevant data and visualizing the tsunami event; and
- Pre-run models that provide scenarios of inundation and impacts, also based on historical events.

The type of notification product issued from each TWC is based on an earthquake's location, magnitude, and depth. These "bulletin threshold" criteria differ for different regions of an AOR, as seen in Table 5.2. However, some regional differences in the threshold criteria could result in confusion. For example, in the case of a 6.5 magnitude earthquake in the vicinity of Puerto Rico, the WC/ATWC would issue a warning for Puerto Rico, but based on PTWC back-up criteria from the WC/ATWC user's guide, the PTWC would issue an information statement for the Caribbean Sea (National Oceanic and Atmospheric Administration, 2009). It is not clear what factors have contributed to this difference in threshold criteria between Puerto Rico and rest of the Caribbean Sea. As previously discussed, issuing messages with potentially conflicting or confusing content—as would be the result from issuing a warning for Puerto Rico and an information statement for the remaining Caribbean Sea—is counter to the science of effective warning messaging. Thus, harmonizing threshold criteria to yield consistent warning products for regions in close proximity would reduce a potential source of confusion.

As discussed in Chapter 4, the initial warning is based on seismic information alone and relies on fairly conservative thresholds. Therefore, in the case of most earthquakes—except for "mega-" and "tsunami" earthquakes (see Appendix G and H)—these thresholds result in the tendency to over-warn. Given the potential consequences of failing to warn, that is, the loss of lives, over-warning is the preferred outcome for the initial product, which can be refined based on the subsequent detection of sea level changes. However, as previously noted, tsunami earthquakes (i.e., slow earthquakes—see Appendix H) might not be recognized quickly enough based on the current analysis and might not result in the correct warning decision. Even this early decision process could benefit from refinement as more advanced methods have been developed and are discussed in greater detail in Chapter 4.

Product Generation and Warnings

Once a watchstander decides to issue a message, s/he fills out a pre-generated checklist that is incorporated into a pre-written or "canned" message. After the product is generated, it is carefully reviewed by the watchstander prior to issuance. Checklists and tsunami warn-

TABLE 5.2 Bulletin Thresholds to Provide PTWC Backup

PTWC Backup									
Hawaii ^ Hawaii	Pacific ^ Hawaii	WCATWC Hawaii	All ^ Pacific	Indian Ocean Ocean	Ocean Ocean	Caribbean Sea Sea	Atlantic Ocean Ocean	Caribbean/ Atlantic	
									3.5
TIS SEHW70				<100 KM deep	Inland or >100 deep	<100 KM deep	<100 KM deep	Inland or >100 deep	4
									4.5
									5
									5.9
									6
								(Caribbean > 6.0)	6.1
									6.2
									6.3
									6.4
									6.5
									6.6
	TIS WEHW42	TIS WEHW42	TIS WEPA42	TIS WEIO23	TIS WEIO23	TIS WECA 43	TIS WECA43		6.7
									6.8
Warning County WEHW40								(Atlantic > 6.5)	6.9
									7
									7.1
				Local Watch WEIO21		Local Watch WECA41			7.2
		Advisory WEHW40					TIS WECA43		7.3
									7.4
									7.5
	Adw/Wrn** 1000Km WEHW40		Warning** 1000Km WEPA40	Regional Watch WEIO21		Regional Watch WECA41			7.6
									7.7
									7.8
Warning State WEHW40	Advisory*/ Warning 3W/3W WEHW40	Watch/ Warning WEHW40	Warning 3W/3W WEPA40	All IO Nations Watch WEIO21		All Carib. Nations Watch WECA41	All Carib. Nations Watch WECA41		7.9
									8.5
									9
									9.5
									10

** countries within 1000 km of epicenter
* if North of 35 degrees, Hawaii in Watch
PTWC issues two messages for most Pacific events (Hawaii and international)
^ Top shows source zone and line below is where the product is sent

SOURCE: http://wcatwc.arh.noaa.gov/ops/opsmanual.pdf; West Coast/Alaska Tsunami Warning Center, NOAA.

ing products are generated differently by each TWC's suite of software, although both are in-house products (Table 5.1; for additional details, see Chapter 3). Messaging software is written in either the formula translation/translator (FORTRAN) or C programming languages, so the watchstander needs to recompile and regenerate the message if an error is detected. This inability to preview and edit a warning product has the potential to introduce delays in the initial warning phase, and the reliance on dated software technology deters easy interfacing with current network and mobile data structures. Addressing these problems is difficult for the

TWCs, as IT software development and maintenance activities are collateral duties for scientific personnel and watchstanders, most of whom have formal training or education in geophysical sciences, not software engineering. Despite this lack of formal education and/or training, IT support, maintenance, and development of operational software consumes a significant portion of a scientific person's daily tasks.

Given these issues and the increasing prevalence of social networking and mobile technologies in emergency, crisis, and disaster management, messaging software that permits an interface with the social media and compatibility with newer generation software and web products/mobile interfaces (e.g., Extensible Markup Language (XML), Simple Messaging System (SMS) formats) would provide several benefits (Sutton et al., 2008).

The committee reviewed the "canned messages" that are composed and delivered by the TWCs and discovered that many documented principles for effective warning messages (see Chapter 3 for detailed discussion) have not been applied. In addition, the generation of two sets of warning products from the two TWCs can be a major source of confusion among the emergency management community and the public as illustrated by the June 14 case study (Appendix F). On June 14, 2005, after an earthquake within the Gorda Plate west of the northern California and southern Oregon coast, the TWCs issued two different warning messages. While the WC/ATWC issued a warning for its AOR, which included California and Oregon, the PTWC subsequently issued a bulletin for its AOR (including Mexico) stating that there was no tsunami warning in effect (for its AOR). Both messages were correctly stating the threat of their respective AORs, but they led local officials to believe that the PTWC message canceled the previous message from the WC/ATWC; thus, reducing the effectiveness of the message. This is of particular concern in areas where the separation in AOR is not intuitive. For example, if an earthquake with the potential to generate a tsunami occurs in the Caribbean, the WC/ATWC would issue a message for Puerto Rico and the U.S. Virgin Islands, while the PTWC would issue a notification for the rest of the Caribbean. Similarly, after a major earthquake in the Pacific, the WC/ATWC issues a warning message for Alaska, the Canadian Pacific coast, Washington, Oregon, and California, but not Mexico or Hawaii.

Although these messages are not intended to replace messages from local/state officials to the public with instructions about appropriate response actions, products from both the TWCs are distributed to members of the public via the internet, social networking tools, or the media (TV and radio) in addition to the other official channels (see Chapter 3 for details). Consequently, the messages are immediately disseminated to the public via multiple and diverse channels—as was the case during the Samoan tsunami (Appendix I) and the Chilean tsunami (Appendix J). Obtaining messages from multiple channels is positively correlated with protective action taking, if the messages are consistent. Inconsistencies in message content create the potential for confusion and can result in greater uncertainty about protective actions needed. Therefore, it is central to the success of the TWCs that they further improve the consistency and clarity between their messages to prevent any confusion resulting from the distinct AOR. Alternatively, the issuance of a single message after internal consultation between the TWCs ought to be considered.

Conclusion: The TWCs' warning products could be much more effective if their content and their delivery incorporated the social science on composing effective warning messages and were compatible with current software, hardware, and social media.

Conclusion: Based on the latest science the committee reviewed, the greatest likelihood for achieving appropriate response before and during an impending disaster results from a consistent warning message from multiple sources. However, the current division in AOR between the two centers and the different products issued by the two centers has resulted in confusion in the past (see case study on June 14) and has great potential to cause confusion again in the future. Despite improvements after the June 14 event, there are still *inconsistencies* in warning products from the two TWCs; and the current division of AORs results in messages that do not clearly communicate who needs to take protective actions and who does not.

Recommendation: NOAA/NWS should harmonize and standardize checklists, tsunami warning products, and decision support tools, and standard TWC software tools and applications should be used in the TWCs, following current software engineering practices and taking advantage of current programming language best practices.

Recommendation: The TWCs should consider alternative warning message composition software (considering software technology and product generation that result in current generation software and web products [e.g., XML, SMS formats]) and should improve protocols by undertaking an external review by IT specialists in the area of communication technology to identify the latest technology in message composition software and formats, and to ensure compatibility with current and next generation information and communication (web and cell-phone) technology for message dissemination.

TECHNOLOGY NEEDS

The TWC technological infrastructure needs to support all of the tsunami warning system functions, including detection, decision support, product and warning message generation, and dissemination and community outreach. As mentioned in previous sections, each TWC has different, non-interoperable hardware platforms and software suites. These choices, made without consultation between the TWCs, are a result of the historical evolution of these centers and appear to be based on skill sets of each center's available personnel rather than on any systemwide architecture plan. Although IT development, support, and maintenance activities are critical to support the TWCs' functions, the activities are often ad hoc and informal and are carried out as a collateral duty by geoscientists. Little IT planning has taken place at either TWC, and planning activities are not articulated in a systemwide enterprise or system architecture.

TWC staff maintains and develops software in their spare time or when there are no pressing operational requirements. Staff is not formally trained in technology life cycle processes or products, leading to nonstandard hardware and software that does not represent current development practices. In addition, some of the TWC software is aging, was developed using inappropriate or aging programming languages, was developed using nonstandard single person coding models, and is difficult to support and maintain. Failure risks have been mitigated by both relatively infrequent significant tsunami events and a small legacy TWC staff who maintain the aging systems as needed.

NOAA has developed an IT Convergence Plan to harmonize software and hardware suites between the TWCs for FY2009-2012. The goal is to create a single, platform-independent technology architecture to be deployed at each TWC and to develop a shared tsunami portal (National Oceanic and Atmospheric Administration, 2009). This plan, which would be funded by telecommunications surplus funds, was cited several times by TWC management and staff as a future source of IT planning and direction. Upon review of the Concept of Operations and Operational Requirements for the IT Convergence Program (National Oceanic and Atmospheric Administration, 2009), the committee believes that although the IT Convergence plan offers the possibility of streamlined and state-of-the-art knowledge transfer for software development, it will not address the need for an enterprise-wide technology architecture and process planning effort. Additionally, plan requirements were developed without consultation of TWC users, emergency managers, academia, the public, or other tsunami program stakeholders.

The current software suites are tightly linked to their hardware platforms; migration to platform-independent architectures could offer the benefit of reduced system maintenance requirements and could enhance information sharing among the TWCs, universities, watch-standers, and the real-time software development community. A long-range technology planning effort for the TWCs that is consistent with international technology process and product standards (U.S. Navy Space and Aviation Warfare Command, 1998; International Standards Organization, 2009; Carnegie Mellon Software Engineering Institute, 2010; Institute of Electrical and Electronic Engineers, 2010) could effectively develop an enterprise-wide technology architecture that supports tsunami warning system missions and its accompanying development, deployment, support, and maintenance operations. The committee learned that no metrics for system up-time or program faults are collected, resulting in a lack of necessary benchmarks for system performance evaluation and improvement. Senior IT leadership and adequate multi-year funding are required to support this long-range effort. It should also entail:

- Adopting process, documentation, hardware, and software standards;
- Modernizing the current software system so that a single common extensible core runs at both centers (IEEE Std 610.12-1990);
- Creating development and planning processes that allow both the TWCs to perform joint planning and execution, including better transparency and communication; and
- Providing better oversight to ensure that the centers are adhering to standards and minimizing duplication of effort.

Conclusion: The committee believes that the Convergence Plan is well-motivated but also notes that the Plan is a single project effort and is not part of a systematic, comprehensive IT system plan and enterprise architecture, which the tsunami program needs.

Each TWC has different web design processes, resulting in dissimilar websites.[5] Although both websites utilize a similar NWS banner, different user interfaces contribute to an impression that the TWCs are separate, unrelated entities.

Dissemination and notification technologies could involve a variety of message dissemination pathways, including satellite-based paths, text messages, faxes, mass media, public announcement systems (e.g., sirens, alarms), telephones, Internet, and syndicated news feeds. In addition, development and use of Web 2.0 and 3.0 technologies (e.g., mashups, blogs, microblogs, social networking sites) offer new opportunities to alert the public and interface with peers and the scientific community. These technologies are only in their infancy at the PTWC and not described in any WC/ATWC planning documents.

As discussed in Chapter 4, these newer technologies proliferate, and the public will increasingly use them for obtaining real-time updates and information, forcing both the TWCs to develop strategies for communication and warning technology. Evaluation and assessment of the utility of traditional websites and Web 2.0 and 3.0 technologies for crisis and disaster management is under development in the information systems, hazard awareness, and crisis and disaster management communities (Boiney et al., 2008; Buscher et al., 2008; Hughes et al., 2008; Shankar, 2008). This literature could provide important direction to the tsunami warning community.

Conclusion: The centers rely on technical infrastructure and human capital, both of which the committee assessed to be insufficiently supported. The TWCs lack sufficient support for the essentials that complex, large-scale, high-reliability systems with mission-critical requirements demand: state-of-the-art technology; IT support, maintenance, assessment and planning processes; IT personnel; and dedicated senior IT leadership.

Conclusion: Despite the importance of technology to fulfilling the TWC mission, technology development, deployment, support, maintenance, back-up, recovery, and configuration management are collateral duties for most TWC staff members. Neither management nor staff members at the TWCs have formal training in technology, software engineering, design, maintenance, or IT support, yet almost all staff members have significant technology responsibilities.

Recommendation: The tsunami warning program should undertake a comprehensive, enterprise-wide,[6] long-range technology planning effort, consistent with international technology process and product standards, in order to develop both an enterprise-

---

[5] http://wcatwc.arh.noaa.gov/ and http://www.prh.noaa.gov/ptwc/.
[6] Enterprise-wide refers to organization-wide or across a large system or enterprise.

wide technology architecture for TWC operations and the accompanying enterprise-wide technology support processes.

Recommendation: Given the importance of technology, particularly IT, in the overall process of detecting and warning, and the rapid evolution of IT, NOAA/NWS should provide the TWCs with stronger IT commitment and leadership, and greater resources for software and hardware personnel, planning, development, operations, maintenance, and continuous process and product improvement.

Recommendation: IT staff should be provided to the TWCs so that IT hardware and software design, development, and maintenance are not a collateral duty of a watchstanding scientist, as is the case presently. An external IT Advisory Board, with membership from the USGS, other seismic network operators, human factors, information technology, and other large-scale, safety-critical systems professionals should be established to advise the TWCs. The Board should meet on at least an annual basis and provide TWC management and operational personnel with guidance and expertise in building, developing, maintaining, and nurturing a highly effective, large-scale distributed, tsunami warning system.

Six elements are critical to the enterprise-wide planning effort: highly effective leadership; a common set of functional, operational, and organizational processes; adherence to international standards; assessment processes that lead to continuous improvement; effective and compelling communication; and adequate and consistent funding to ensure that the processes, people, organizational structures, and policies effectively support the tsunami mission. Specifically, the committee recommends the following improvements:

Senior IT leadership to guide the organization and ensure that the TWC technology architecture supports TWC operational requirements for reliability and sustainability;

- *A common set of functional, operational and organizational processes, including*
  - *Articulated technology development, procurement, deployment, maintenance, support, security, and configuration management processes;*
  - *Planning processes* that are compliant with software engineering and computer science standard processes and lessons learned from other large-scale, mission-critical systems, and which are effectively carried out by TWC IT management and staff personnel, IT support personnel, in regular consultation with TWC customers and emergency management personnel;
  - *An enterprise-wide IT requirements development process* that transparently identifies all TWC requirements with a single planning process and a single enterprise-wide IT architecture; and
  - *IT development, deployment, maintenance, support, quality assurance, security, and configuration management processes* that are adequately planned and budgeted,

and undertaken by IT and software engineering professionals following current IT development and deployment practices;

- *Adherence to international standards,* including adoption of current international software engineering, process, documentation, hardware, and software standards throughout the tsunami system;
- *Assessment processes* that lead to continuous improvement, including
  - Yearly systemwide performance monitoring, management, testing and bench-marking; yearly assessment, evaluation, accountability audits, and reporting of the TWC technology infrastructure's support for the long-term reliability and sustainability of TWC operations;
  - Common, systemwide performance metrics and benchmarks; and
  - Incentives to ensure compliance with standards and high levels of performance;
- *Mechanisms to communicate best practices,* lessons learned and to enhance organizational learning; and
- *Adequate, substantial, multi-year, dedicated funding* for all elements of the technology plan.

Multi-year funding to support the technology planning, development, deployment, maintenance, support, and security operations will need to be appropriately budgeted over the technology life cycle. As part of this planning and modernization process the committee identifies an additional urgent need:

- Attention should be paid to the use of traditional, nontraditional, and next generation technology in support of community outreach and dissemination. IT for community outreach and dissemination efforts should be included as part of the long-range technology planning process and should be incorporated as an ongoing component of TWC planning processes.

Recommendation: NOAA/NWS and the TWCs should adopt national, and where applicable, international, standards, best practices, and lessons learned for all functions, technology, processes, and products. Specifically, the TWCs should develop platform-independent hardware and software architectures, applications, and interfaces; and employ international hardware and software planning, development, operations, and maintenance product and process standards, including the Software Engineering Institute's Capability Maturity Model and the software development life cycle (Carnegie Mellon Software Engineering Institute, 2010).

Recommendation: The TWCs should also regularly and systematically apply continuing process and product improvement models for hardware and software planning, development, operations, and maintenance; organizational processes; and should develop a learning organizational culture.

HUMAN RESOURCES

Each TWC is staffed by a center director, a tsunami warning science officer (who serves as the deputy director), an information technology officer, nine science duty staff (geophysicists and physical oceanographers), a senior electronics technician, an electronics technician, and an administrative assistant (Charles McCreery, presentation to the committee, 2008). The nine science duty staff members perform watchstanding duties in addition to their research and development duties (National Oceanic and Atmospheric Administration, 2008b, c). Because of the watchstander's critical role in maintaining situational awareness and issuing correct notification and warning products, the committee reviewed this position's shift schedules, training, support, and responsibilities as part of its assessment of the TWC' long-term sustainability.

When the TWCs are fully staffed, the nine watchstanders serve on rotating two-person, eight-hour shifts that provide 24/7 coverage. Both TWCs have identical watch schedules (0800-1600, 1600-2400, 2400-0800 local time), although the two watch centers are in different time zones. Watches can be rotated to cover busy periods, vacations, and other TWC needs. Two watchstanders are always present at the WC/ATWC. In contrast, the PTWC has only one on-duty watchstander in the watch station at all times, while the second watchstander is on a 90-second response standby, allowing him/her to sleep or to be outside the watch station. Each watchstander is responsible for checking all workstations every four hours to ensure functionality. Unless they need to respond to an event, watchstanders spend approximately six hours on software development and two for operational activities (Paul Whitmore, West Coast and Alaska Tsunami Warning Center, personal communication). When an alarm is sounded, the watchstanders leave their other duties to respond. Because they are most often attending to other duties, there is an indeterminate period of time required for watchstanders to acclimate themselves to an alarm.

The WC/ATWC rotates watches every two weeks, which can result in sleep disorders (Sack et al., 2007) and work/life balance issues, but provides equitable sharing of night shift watchstanding duties. Resulting sleep disorders, issues with work/life balance, as well as several month-long shortages of two full-time watchstanders at the WC/ATWC in 2008 suggest that staff fatigue may be an issue.

TWC staff have varying levels of engagement with the external research community. The PTWC's proximity to Honolulu (23 miles away) is conducive to interactions with the civil defense and academic communities at the University of Hawaii at Manoa, while WC/ATWC's distance from Anchorage (43 miles away) and its location in a small town of 7,000 residents (Palmer, Alaska) does not lend itself as easily to such interactions.

Conclusion: Because of the importance of technical and scientific know-how within the TWC program, opportunities for interactions between TWC staff and the external scientific and professional communities are important, need to be encouraged and institutionalized within the tsunami program, and require adequate resources.

Such interactions might include attendance at professional conferences; participation in seminars, workshops, or other structured learning opportunities; scientific and personnel exchanges and sabbaticals; study away opportunities at related scientific venues; and grants, fellowships, and stipends to further professional study.

The watchstander has a critical role in tsunami decision support by maintaining situational awareness and issuing the correct products and information. Although visualization software assists by monitoring seismic and sea level data and mapping event locations, it is the watchstanders' training, experience, and scientific expert judgment that are essential in making the appropriate decisions when creating notification products for emergency managers, local government, and the general public.

Tsunami detection and warning requires frequent, effective, and purposeful communication and interactions between watchstanders, staff, and management in the TWCs, and with operational decision makers and the public. To enhance the effectiveness of TWC decision making and the TWC staff's ability to inform decision making processes of their customers, frequent, regular, and varied types of training for operational watchstanders (e.g., simulations, walk throughs, case studies, table-top exercises) are needed. In addition, scheduled and organizationally supported interactions between watchstanders and management in distributed watch centers are beneficial to the TWCs' reliability of operations; these activities could include seminars, personnel and information exchanges, technical meetings, and scheduled joint work sessions.

Conclusion: Given the highly technical and specialized skill sets required for tsunami watchstanding, workforce development and recruiting can be challenging. The success of the TWC mission is critically dependent on human resources for tsunami warning and detection.

Conclusion: Given the importance of technology in the overall process of tsunami detection and warning, and the rapid evolution of IT, stronger IT commitment and leadership and greater human resources devoted to IT are required in the TWCs. The TWCs lack senior IT leadership to support TWC operations, guide the enterprise-wide technology planning efforts, and provide guidance in adopting enterprise-wide technology processes.

Conclusion: The TWCs require frequent, regular, and varied types of training for operational watchstanders (e.g., simulations, walk throughs, case studies, table-top exercises); frequent, regular, and organizationally supported interactions between watchstanders in distributed watch centers; and ongoing, funded, and prioritized research and development to support operations, which requires an explicit process for implementing new technology into operations.

Recommendation: Because of the importance of technical and scientific expertise to the TWCs' functions, TWC human capital requirements and TWC recruiting, training, retraining, development, mentoring, and professional exchange needs should be included, re-assessed, and updated as part of the NOAA/NWS enterprise-wide technology planning effort, and should be consistent with NOAA- and government-wide standards, so that technological, human, and organizational requirements can be considered and developed together by Tsunami Program members and their customers.

As part of the process of maintaining and developing expertise in the TWCs, opportunities for interactions between TWC staff and external scientific and professional communities could be funded, encouraged, and institutionalized within the Tsunami Program. Such interactions might include drafting and implementing a formal plan for maintaining and increasing scientific currency; attendance at professional conferences; participation in seminars, workshops, or other structured learning opportunities; scientific and personnel exchanges and sabbaticals; study away opportunities at related scientific venues; and grants, fellowships, and stipends to further professional study.

ORGANIZATIONAL STRUCTURES

Tsunami detection and warning is currently undertaken by multiple, distributed members linked together to achieve a goal: management of and response to a tsunami disaster. As indicated in the first chapter, the tsunami warning system needs to exhibit properties of a high-reliability organization (HRO), sharing common processes that are supported by distributed information technology (Davidow and Malone, 1992; Mowshowitz, 1997; Jarvenpaa and Leidner, 1999; Kock, 2000). In the event of a tsunami, tsunami warning system managers must assemble effective, functioning response organizations in periods of less than 24 hours and then adjust the organizational structure to the needs of the response (Tuler, 1988; Bigley and Roberts, 2001).

A challenge for tsunami warning systems is therefore to develop effective organizational structures that provide reliable and sustainable operations in non-tsunami periods as well as during catastrophic incidents. HROs typically have flexible and redundant organizational structures that permit organizational slack, allow testing of different response modes and techniques, and provide members the opportunity to develop communication, decision making, and organizational culture that are essential to cooperative, interdependent operations (Weick, 1987; Weick et al., 1999). Organizational structures that provide back-up, redundancy, skill overlap, checks and balances, and one-over-one reviews are critical to the development of effective HROs (Grabowski and Roberts, 1997, 1999; Jarvenpaa and Leidner, 1999), and are thus key structures for highly reliable tsunami operations. Both structures and actors need to be tested regularly in rehearsal and simulation of tsunami events because tsunamis are relatively rare. Such regular rehearsals prevent the potential loss of institutional memory about appropriate organizational response.

During this review, the committee found the current organizational structure—two TWCs managed by distinct regional weather service offices—associated with several benefits and risks:

Benefits

- Having independent research and development efforts at the two centers increases the competition and the potential for innovative approaches developed at either center.
- If identical IT systems, software, and analytical methods were used, then the two centers would offer redundancy in the system, which is especially important given that each center's location is vulnerable to natural hazards.
- If identical IT systems, software, and analytical methods were used, then the two centers could leverage resources for IT development and modernization.
- Having two centers in regions that are at great risk of tsunamis offers opportunities to engage with and to educate the emergency management community, the public, and media.

Risks

- Because the two TWCs are managed by two different regional NWS offices, the two centers display different organizational cultures.
- Two centers with distinct AORs and different message thresholds greatly increase the potential for confusion and hampers effective decision making during an emergency.
- The current geographic separation in the AOR is not intuitive and can result in difficulties with regard to interpreting who is under a tsunami warning or not.
- The current physical settings and the organizational structure within the NWS provides minimal integration with the tsunami and earthquake research community or other operational forecast and warning centers within NOAA (e.g., all other centers are managed by NCEP).
- Because both the TWCs are located in remote locations, and neither is co-located at another NOAA, seismic or mission-critical center, opportunities to leverage lessons learned and best practices and to adopt standard processes and procedures are limited.
- NOAA's already limited technical, professional, and economic resources are stretched to support both centers' infrastructure, IT, and engineering maintenance and upgrades.
- Maintaining and modernizing software and hardware systems is difficult because of limited staffing support. Because the TWCs are currently not operationally redundant, they lack the benefit from effectively leveraging limited staff resources (e.g., they must hire more highly specialized IT personnel to update both systems simultaneously).
- Supporting two robust and redundant communication networks incurs additional expenses.
- The two TWCs are designed to be back-ups for each other, but they do not operate as such, creating an illusion of redundancy that could prove dangerous and costly, because adequate resources may not have been deployed to provide needed back-up and redundancy.

Ideally, the TWC organizational structures allow for distributed groups of scientists to manage their joint processes, work together effectively in support of their shared mission, and avoid unnecessary redundancy. Support for TWC functions can be provided within an HRO framework, which opens several options for organizational structures. As described above, there are many differences between the centers' operations, technology infrastructure, and human resources. The incompatibilities in the technology infrastructure run counter to the goal of redundancy in operations. In fact, the inconsistencies in outcomes and messages reduce the benefit of having two centers provide redundancy. Although the TWC organizational structures could permit some organizational redundancy, that redundancy can cause difficulties if duplicate tasks are executed in geographically dispersed operational settings by organizational members who follow different procedures, use different warning thresholds, and communicate inconsistent messages to a public within confusing AORs.

> **Conclusion:** The goal of the current geographically distributed organization of the TWCs is to provide the system with back-up in the case of critical failure at the other center. However, based on the June 14, 2005, event analysis, the review of the literature on high-reliability organizations, and the current geographically distinct boundaries in AORs, the committee concludes that this redundancy is currently more likely to cause confusion than provide benefits. In addition, because the centers are under different management, use different analytical software and hardware, and appear to have distinct organizational cultures, the committee concludes that they function as separate rather than redundant systems.

> **Conclusion:** Even if the IT convergence plan is fully executed, the issues arising from producing different products and messages remain and increase the risk of confusing the public and state and local emergency managers.

> **Conclusion:** The current organizational model is problematic and reduces the ability of the TWCs to provide timely, accurate, and consistent warning products.

The committee concludes that significant changes would need to occur in the management, operations, software and hardware architecture, and organizational culture for the two TWCs to become functionally redundant systems. As a result, the committee discussed alternative organizational options, in addition to the TWCs' current organizational structure:

(1) Based on the committee's assessment of the relatively slow transition and incorporation of research advances into operations at both TWCs (see Chapters 3 and 4), the committee recognizes various benefits from co-locating the two centers (or a center) with other academic or scientific institutions or with other centers responsible for detection and warning (e.g., NCEP, the NEIC, PMEL, etc). For example, NCEP is co-located with the research community to increase the exchange of ideas between the scientific and operational staff. Alternatively, colocating the TWC(s) with the NEIC would give

NOAA access to seismological expertise at this operational center run by the USGS and would pair the critical assets of two agencies with complementary missions.

(2) Alternatively, because of issues arising from parallel or inconsistent procedures and potential confusion in the public from different warning products from the two centers, the committee suggests that a centrally managed center located in a single geographic location (such as the Hurricane Center) is a potential solution to the inconsistency in methods, architecture, culture, and messages. A single center might also allow limited resources to be pooled and might ease the difficulties in 24/7 staffing of the centers.

There are many current examples of highly reliable, mission-critical, large-scale systems that support real-time distributed operations using various organizational forms across a broad geographical area. In the NOAA/NWS community, these include the NWS Hurricane Center, the Severe Storm Lab, and the Storm Prediction Center at NCEP. There are other examples outside NOAA including the seismic, oceanographic, meteorological, undersea, cyberspace, and space systems communities, for example (National Aeronautics and Space Administration, 2005; National Science Foundation, 2006). Each of these HROs provide opportunities to leverage lessons learned and best practices for distributed tsunami warning operations.

Recommendation: Organizational structures for the two TWCs should be evaluated and fully described as part of the enterprise-wise technology planning effort previously described. Whether there should be a single or multiple TWCs, or whether the TWC operations should be consolidated in a different location, should be addressed in the enterprise-wide, long-range planning effort.

In evaluating the TWC organizational structure and locale(s), consideration should be given to the proximity to the research community, its user community, and the vulnerability to hazards. HROs achieve high levels of performance when their organizational structures support the decision making, communications, organizational culture(s), and trust required for success, and facilitate provision of the requisite information and knowledge sources to system users, participants, and customers (Grabowski and Roberts, 1999; Bigley and Roberts, 2001). Although developing a strong, unified organizational culture in a warning system can be difficult when members are geographically dispersed, (Grabowski et al., 2007), it is crucial in order to avoid dysfunctionality and miscommunication (Porter, 1993; Stephenson, 1995).

CONCLUSIONS

Effective tsunami detection, warning, and preparedness require that multiple, distributed tasks that are linked together to achieve the goal of reducing loss of life and economic assets using common processes are supported by highly reliable, distributed information technology. In a tsunami event, all of the distributed efforts must come together to produce an effective response and function as if it were a single organization (Tuler, 1988; Bigley and Roberts, 2001).

In addition, tsunami warning and preparedness efforts must develop effective organizational structures that provide reliable and sustainable operations in non-tsunami periods, especially given the short time available to respond in a crisis (only minutes in the case of a near-field tsunami). Maintaining awareness in such a rapidly evolving crisis (i.e., situational awareness) poses a major challenge.

The committee found that personnel at the TWCs are highly committed to the TWC functions, from detection through community outreach. TWC management and staff are highly supportive of the organization's missions and are articulate advocates for the importance of the mission and TWC functions. The TWC mission is well understood by TWC management and staff, is clearly communicated throughout the tsunami warning system, and is well documented. TWC personnel were found to be knowledgeable and effective in providing TWC functions, systems, and operations. However, aside from notable motivation and good communication, the committee found many shortcomings of the TWCs in terms of functions, technology, human capital, organizational structures, and, as a result, many opportunities for significant improvements of the centers' operations.

The most fundamental changes required are an improved organizational model and an organizational culture that fosters expectations of performance excellence, and consistently measures and rewards that performance across the tsunami warning system. If the organizational structures are adjusted to meet the required characteristics of an HRO, the TWC(s) could provide the function of an incident command center (ICC). Its central role as such an ICC would be to maintain situational awareness during a tsunami event, continuously updating its threat assessment, understanding the time horizon for informing critical decisions of the emergency management community, repeatedly informing critical users (public officials, emergency managers, media, etc.) of the evolving threat assessment, and monitoring the information broadcast in the public arena to correct misinformation.

At the same time that an improved TWC structure could function as the ICC during crisis mode, an improved tsunami program could provide long-term strategic planning and guidance, support for preparedness efforts, and coordination and support for educational efforts to ensure consistent content in the educational materials. If the TWCs become well integrated with other components of the tsunami program, assets at the TWCs could be used to more rapidly transition research at other program units into operational use. As previously discussed, a critical goal of the Tsunami Program should be to ensure that staffing and funding resources are allocated appropriately.

> **Conclusion:** An organizational culture change is needed within the NOAA/NWS Tsunami Program that supports and celebrates operational excellence; adopts national and international standards, processes, best practices, and lessons learned for all functions, technologies, processes, and products; and engages in ongoing, continuous process improvements.

> **Conclusion:** The committee found that the TWCs do not sufficiently engage in ongoing, joint or agency-wide, continuous process improvement activities for their functional,

technological, organizational, or human capital initiatives, such as those followed by other high-reliability, safety-critical distributed systems. Such activities provide critical benchmarks against which to measure performance over time, and could provide opportunities to reward and incentivize desired performance and behavior across the tsunami warning system.

Recommendation: NOAA/NWS and the TWCs should undertake ongoing, joint or NOAA-wide, continuous process improvement activities for their functional, technological, organizational, and human capital initiatives, including the following:

- developing measures of performance and benchmarking individual, organizational, and technical performance against industry and agency metrics,
- identifying areas for improvement,
- setting short- and long-term performance goals,
- developing reward and incentive systems for such goals, and
- celebrating TWC and agency accomplishments as performance improves, in order to raise the level of TWC performance to that expected of a high-reliability organization.

References

Abe, K. 1979. Size of great earthquakes of 1837–1974 inferred from tsunami data. *Journal of Geophysical Research* 84(B4):1561-1568.

Ablain, M., J. Dorandeu, P.Y. Le Traon, and A. Sladen. 2006. High resolution altimetry reveals new characteristics of the December 2004 Indian Ocean tsunami. *Geophysical Research Letters* 33:L21602.

Alexandra, K., G. Cain, and P. Iwasaki. 2009. *Tsunami Education: A Blueprint for Coastal Communities*. Report of the Pacific Tsunami Museum and County of Hawaii Planning Program, Honolulu, Hawaii.

Ambraseys, N.N. 1960. The seismic sea wave of July 9, 1956, in the Greek archipelago. *Journal of Geophysical Research* 65(4):1257-1265.

Andersen, P. and I. Drejer. 2005. Distributed innovation in integrated production systems: The case of offshore wind farms. In *The DRUID 10th Anniversary Summer Conference 2005 on Dynamics of Industry and Innovation: Organizations, Networks and Systems*, Copenhagen, Denmark.

Artru, J., V. Ducic, H. Kanamori, P. Lognonne, and M. Murakami. 2005. Ionospheric detection of gravity waves induced by tsunamis. *Geophysical Journal International* 160(3):840-848.

Atwater, B.F. 1987. Evidence for great Holocene earthquakes along the outer coast of Washington State. *Science* 236(4804):942-944.

Atwater, B.F., S. Musumi-Rokkaku, K. Satake, Y. Tsuji, K. Ueda, and D.K. Yamaguchi. 2005. *The Orphan Tsunami of 1700: Japanese Clues to a Parent Earthquake in North America*. U.S. Geological Survey Professional Paper 1707, U.S. Geological Survey, Reston, Virginia. [Online]. Available: http://pubs.usgs.gov/pp/pp1707/ [2010, September 14].

Audet, C. and J.E. Dennis, Jr. 2006. Mesh adaptive direct search algorithms for constrained optimization. *SIAM Journal on Optimization* 17(2):188-217.

Balaban, V. 2006. Psychological assessment of children in disasters and emergencies. *Disasters* 30(2):178-198.

Balk, D., Y. Gorokhovich, and M. Levy. 2005. *Estimation of Coastal Populations Exposed to 26 December 2004 Tsunami*. Center for International Earth Science Information Network, Columbia University, Palisades, New York.

Barberopoulou, A., M. Legg, B. Uslu, and C.E. Synolakis. 2008. *Tsunami Hazards in San Diego Bay, California*. American Geophysical Union Fall Meeting, San Francisco, California.

Barberopoulou, A., J.C. Borrero, B. Uslu, K. Kalligeris, J.D. Goltz, R.I. Wilson, and C.E. Synolakis. 2009. New maps of California to improve tsunami preparedness. *Eos, Transactions, American Geophysical Union* 90(16):137-144.

Barkan, R., U.S. ten Brink, and J. Lin. 2009. Far field tsunami simulations of the 1755 Lisbon earthquake: Implications for tsunami hazard to the U.S. East Coast and the Caribbean. *Marine Geology* 264(1-2):109-122.

Bement, A.L., Jr. 2005. *NSF and Congress Testimony Before the U.S. Senate Committee on Commerce, Science, and Transportation*. U.S. Senate, Washington, DC.

Bernard, E.N. 2005. The U.S. National Tsunami Hazard Mitigation Program: A successful state-federal partnership. *Natural Hazards* 35(1):5-24.

Bernard, E.N., L. Dengler, and S. Yim. 2007. *National Tsunami Research Plan: Report of a Workshop Sponsored by NSF/NOAA*. NOAA Technical Memorandum OAR PMEL-133, National Oceanic and Atmospheric Administration, Silver Spring, Maryland.

Bigley, G.A. and K.H. Roberts. 2001. Structuring temporary systems for high reliability. *Academy of Management Journal* 44:1281-1300.

Birkman, J., N. Setiadi, and N. Gebert. 2008. Socio-economic vulnerability assessment at the local level in context of tsunami early warning and evacuation planning in the city of Padang, West Sumatra. In *International Conference on Tsunami Warning*, Bali, Indonesia.

Bock, Y., R.M. Nikolaidis, P.J. de Jonge, and M. Bevis. 2000. Instantaneous geodetic positioning at medium distances with the Global Positioning System. *Journal of Geophysical Research* 105(B12):28223-28253.

Boiney, L.G., B. Goodman, R. Gaimari, J. Zarrella, C. Berubi, and J. Hitzeman. 2008. Taming multiple chat room collaboration: Real-time visual cues to social networks and emerging threads. In *Proceedings of the 5th International Conference on Information Systems for Crisis Response and Management 2008 (ISCRAM '08)*, Fiedrich, F. and B. Van de Walle (Eds.). International Society for Information Systems for Crisis Response and Management, Seattle, Washington.

Bonk, W., R. Lachman, and M. Tatsuoka. 1960. *A Report of Human Behavior During the Tsunami of May 23, 1960.* The Hawaii Division of the Hawaiian Academy of Science, Hilo, Hawaii.

Borrero, J., S. Cho, J.E. Moore, II, H.W. Richardson, and C. Synolakis. 2005. Could it happen here? *Civil Engineering* 75(4):54-65.

Bostrom, A., S.P. French, and S.J. Gottlieb. 2008. *Risk Assessment, Modeling and Decision Support: Strategic Decisions.* Springer-Verlag, Berlin, Germany.

Bourgeois, J., C. Petroff, H. Yeh, V.V. Titov, C.E. Synolakis, B. Benson, E. Norabuena, J. Kuroiwa, and J. Lander. 1999. Geologic setting, field survey and modeling of the Chimbote northern Peru, tsunami of 21 February 1996. *Pure and Applied Geophysics* 154(3-4):513-540.

Briggs, M.J., C.E. Synolakis, G.S. Harkins, and D.R. Green. 1995. Laboratory experiments of tsunami runup on a circular island. *Pure and Applied Geophysics* 144(3-4):569-593.

Bucknam, R.C., E. Hemphill-Haley, and E.B. Leopold. 1992. Abrupt uplift within the past 1700 years at southern Puget Sound, Washington. *Science* 258(5088):1611-1614.

Burby, R., L. Steinberg, and V. Basolo. 2003. The tenure trap: The vulnerability of renters to joint natural and technological disasters. *Urban Affairs Review* 39(1):32-58.

Burningham, K., J. Fielding, and D. Thrush. 2008. "It'll never happen to me": Understanding public awareness of local flood risk. *Disasters* 32(2):216-38.

Buscher, M., P.H. Mogensen, and M. Christensen. 2008. When and how (not) to trust IT? Supporting virtual emergency teamwork. In *Proceedings of the 5th International Conference on Information Systems for Crisis Response and Management 2008 (ISCRAM '08)*, Fiedrich, F. and B. Van de Walle (eds.). International Society for Information Systems for Crisis Response and Management, Seattle, Washington.

Carnegie Mellon Software Engineering Institute. 2010. *Capability Maturity Model Integration (CMMI)*. [Online]. Available: http://www.sei.cmu.edu/cmmi/ [2009, August 14].

Carrier, G.F. and H.P. Greenspan. 1958. Water waves of finite amplitude on a sloping beach. *Journal of Fluid Mechanics* 4(1):97-109.

Carver, G. and G. Plafker. 2008. Paleoseismicity and neotectonics of the Aleutian subduction zone: An overview. *Geophysical Monograph* 179:43-63.

Centrec Consulting Group, LLC. 2007. *An Investigation of the Economic and Social Value of Selected NOAA Data and Products for Geostationary Operational Environmental Satellites (GOES)*. A report submitted to the NOAA's National Climatic Data Center, Centrec Consulting Group, Savoy, Illinois.

Charland, J. and G. Priest. 1995. *Inventory of Critical and Essential Facilities Vulnerable to Earthquake or Tsunami Hazards on the Oregon Coast: Portland, Oregon.* Oregon Department of Geology and Mineral Industries, Portland, Oregon.

Connor, D. 2005. *The City of Seaside's Tsunami Awareness Program: Outreach Assessment—How to Implement an Effective Tsunami Preparedness Outreach Program.* Oregon Department of Geology and Mineral Industries, Portland, Oregon.

Cooney, S.M., M. Huser, S. Small, and C. O'Connor. 2007. Evidence based-programs: An overview. In *What Works, Wisconsin Research to Practice Series 6*, University of Wisconsin, Madison, Wisconsin.

Cox, D.C. and J.F. Mink. 1963. The tsunami of 23 May 1960 in Hawaiian Islands. *Bulletin of the Seismological Society of America* 53(6):1191-1209.

Crawford, G.L. 2005. NOAA Weather Radio (NWR)—A coastal solution to tsunami alert and notification. *Natural Hazards* 35(1):163-171.

Cutter, S.L. 2003. The vulnerability of science and the science of vulnerability. *Annals Association American Geographers* 93(1):1-12.

Daily Telegraph, The. 2005. *Girl, 10, Used Geography Lesson to Save Lives.* [Online]. Available: http://www.telegraph.co.uk/news/1480192/Girl-10-used-geography-lesson-to-save-lives.html [2010, May 4].

Dall'Osso, F., A. Cavalletti, and P. Polo. 2006. *Risk Assessment and Evaluation ArcGIS Toolbox User's Manual, CRATER Coastal Risk Analysis of Tsunamis and Environmental Remediation*. Italian Ministry for the Environment and Territory, Rome, Italy and Asian Disaster Preparedness Center, Pathumthani, Thailand.

Davidow, W.H. and M.S. Malone. 1992. *The Virtual Corporation: Structuring and Revitalizing the Corporation for the 21st Century*. Edward Burlingame Books/Harper Business Press, New York.

de Groot-Hedlin, C.D. 2005. Estimation of the rupture length and velocity of the Great Sumatra earthquake of Dec 26, 2004 using hydroacoustic signals. *Geophysical Research Letters* 32:L11303.

De Holan, P.M. and N. Phillips. 2004. Remembrance of things past? The dynamics of organizational forgetting. *Management Science* 50(11):1603-1613.

Dengler, L. 1998. *Strategic Implementation Plan for Tsunami Mitigation Projects*. Mitigation Subcommittee of the National Tsunami Hazard Mitigation Program, NOAA Technical Memorandum ERL PMEL-113 (PB99-115552), Pacific Marine Environmental Laboratory, National Oceanic and Atmospheric Administration, Seattle, Washington.

Dengler, L. 2005. The role of education in the National Tsunami Hazard Mitigation Program. *Natural Hazards* 35(1):141-153.

Dengler, L. 2009. Frequency of tsunami alert bulletins in California. *American Geophysical Union* 90(52):OS43A-1372.

Dengler, L., B. Uslu, A. Barberopoulou, J. Borrero, and C.E. Synolakis. 2008. The vulnerability of Crescent City, California, to tsunamis generated by earthquakes in the Kuril Islands region of the northwestern Pacific. *Seismological Research Letters* 79(5):608-619.

Dow, K. 1992. Exploring differences in our common future(s): The meaning of vulnerability to global environmental change. *Geoforum* 23(3):417-436.

Dudley, W. 1999. The Pacific Tsunami Museum: A memorial to those lost to tsunami, and an education center to prevent further casualties. *Science of Tsunami Hazards* 17(2):127-134.

Dunbar, P.K. and C.S. Weaver. 2008. *U.S. States and Territories National Tsunami Hazard Assessment: Historical Record and Sources for Waves*. National Tsunami Hazard Mitigation Program, National Oceanic and Atmospheric Administration, Silver Spring, Maryland, and U.S. Geological Survey, Reston, Virginia.

Earthquake Engineering Research Institute. 2010. *Learning from Earthquakes: Samoa Earthquake and Tsunami of September 29, 2009*. Earthquake Engineering Research Institute, Oakland, California.

Eaton, J.P., D.H. Richter, and W.U. Ault. 1961. The tsunami of May 23, 1960, on the Island of Hawaii. *Bulletin of the Seismological Society of America* 51(2):135-157.

Ewing, W.M., I. Tolstoy, and F. Press. 1950. Proposed use of the T phase in tsunami warning systems. *Bulletin of the Seismological Society of America* 40(1):53-58.

Farber, D.A., R.G. Bea, K.H. Roberts, E. Wenk, and K. Inkabi. 2006. Reinventing flood control. *Tulane Law Review* 81(4):1085-1127.

Fischhoff, B., P. Lichtenstein, S. Read, and B. Combs. 1978. How safe is safe enough? A psychometric study of attitudes towards technological risks and benefits. *Policy Sciences* 9(2):127-152.

Fischhoff, B., S.R. Watson, and C. Hope. 1984. Defining risk. *Policy Sciences* 17(2):123-139.

Federal Emergency Management Agency. 2008. *Guidelines for Design of Structures for Vertical Evacuation from Tsunamis*. Federal Emergency Management Agency, Washington, D.C.

Geist, E.L. 2005. *Local Tsunami Hazards in the Pacific Northwest from Cascadia Subduction Zone Earthquakes*. [Online]. Available: http://pubs.usgs.gov/pp/pp1661b [2010, August 26].

Geist, E.L., F.I. González, and U. ten Brink. 2005. *Workshop on Optimizing the DART Network for Tsunami Forecasting*. [Online]. Available: http://soundwaves.usgs.gov/2005/10/meetings.html [2010, February, 3].

Geist, E.L. and T. Parsons. 2006. Probabilistic analysis of tsunami hazards. *Natural Hazards* 37(3):277-314.

Geist, E.L., V.V. Titov, A. Kelly, and H. Gibbons. 2007. *Tsunami-Forecasting System Tested by Recent Subduction-Zone Earthquakes*. [Online]. Available: http://soundwaves.usgs.gov/2007/04/research2.html [2010, February, 3].

Geist, E.L., P.J. Lynett, and J.D. Chaytor. 2009. Hydrodynamic modeling of tsunamis from the Currituck landslide. *Marine Geology* 264(1-2):41-52.

Genrich, J.F and Y. Bock. 2006. Instantaneous geodetic positioning with 10-50 Hz GPS measurements: Noise characteristics and implications for monitoring networks. *Journal Geophysical Research* 111(B3):B03403.

Godin, O.A. 2004. Air-sea interaction and feasibility of tsunami detection in the open ocean. *Journal of Geophysical Research* 109:C05002.

Godin, O.A., V.G. Irisov, R.R. Leben, B.D. Hamlington, and G.A. Wick. 2009. Variations in sea surface roughness induced by the 2004 Sumatra-Andaman tsunami. *Natural Hazards and Earth Science Systems* 9(4):1135-1147.

González, F.I., H.M. Milburn, E.N. Bernard, and J.C. Newman. 1998. Deep-ocean assessment and reporting of tsunamis (DART): Brief overview and status report. In *Proceedings of the International Workshop on Tsunami Disaster Mitigation*, Tokyo, Japan.

González, F.I., V.V. Titov, H.O. Mofjeld, A.J. Venturato, and J.C. Newman. 2001. The NTHMP inundation mapping program. In *Proceedings of the International Tsunami Symposium 2001 (ITS 2001)*, Seattle, Washington.

González, F.I., E.L. Geist, C.E. Synolakis, D. Arcas, D. Bellomo, D. Carlton, T. Horning, B. Jaffe, J. Johnson, U. Kanoglu, H. Mojfeld, J. Newman, T. Parsons, R. Peters, C. Peterson, G. Priest, V.V. Titov, A. Venturato, J. Weber, F. Wong, and A. Yalciner. 2006. *Seaside, Oregon tsunami pilot study—Modernization of FEMA flood hazard maps*. Tsunami Pilot Working Group, NOAA, USGS, and FEMA, Washington, DC.

Government Accountability Office. 2006. *U.S. Preparedness: Federal and State Partners Collaborate to Help Communities Reduce Potential Impacts, But Significant Challenges Remain*. Government Accountability Office, Washington, DC.

Government Accountability Office. 2010. *U.S. Tsunami Preparedness: NOAA Has Expanded Its Tsunami Programs But Improved Planning Could Enhance Effectiveness*. Government Accountability Office, Washington, DC.

Grabowski, M.R. and K.H. Roberts. 1996. Human and organizational error in large-scale systems. *IEEE Transactions on Systems, Man & Cybernetics* 26(1):2-16.

Grabowski, M.R. and K.H. Roberts. 1997. Risk mitigation in large scale systems: Lessons from high reliability organizations. *California Management Review* 39(4):152-162.

Grabowski, M.R. and K.H. Roberts. 1999. Risk mitigation in virtual organizations. *Organization Science* 10(6):704-721.

Grabowski, M.R., P. Ayyalasomayajula, J.R. Merrick, J.H. Harrald, and K.H. Roberts. 2007. Leading indicators of safety in virtual organizations. *Safety Science* 45(10):1013-1043.

Gregg, C.E., B. Houghton, D. Paton, D. Swanson, and D. Johnston. 2004. Community preparedness for lava flows from Mauna Loa and Hualalai volcanoes, Kona, Hawaii. *Bulletin of Volcanology* 66(6):531-540.

Gregg, C.E., B.F. Houghton, D. Paton, D.M. Johnston, D.A. Swanson, and B.S. Yanagi. 2007. Tsunami warnings: Understanding in Hawai'i. *Natural Hazards* 40(1):71-87.

Gregory, R. and K. Wellman. 2001. Bringing stakeholder values into environmental policy choices: A community-based estuary case study. *Ecological Economics* 39(1):37-52.

Grindlay, N.R., P. Mann, J.F. Dolan, and J.P. van Gestel. 2005. Neotectonics and subsidence of the northern Puerto Rico—Virgin Islands margin in response to the oblique subduction of high-standing ridges. *Geological Society of America Bulletin Special Paper* 385:31-60.

Gulbrandsøy, K., V. Hepsø, and A. Skavhaug. 2002. Virtual collaboration in oil and gas organizations. *ACM SIGGROUP Bulletin* 23(3):42-47.

Haflidason, H., R. Lien, H.P. Sejrup, C.F. Forsberg, and P. Bryn. 2005. The dating and morphometry of the Storegga Slide. *Marine and Petroleum Geology* 22(1-2):123-136.

Hammack, J.L. and H. Segur. 1978. Modelling criteria for long water waves. *Journal of Fluid Mechanics* 84(2):359-373.

Hawaii Research and Economic Analysis Division. 1996. Methodology for estimating the economic loss of a tsunami false alert. In *Tsunami Alert Economic Loss Estimation*, Oboshi, P.I. and R.C. Price (Eds.). Hawaii Research and Economic Analysis Division, Department of Business, Economic Development and Tourism, State of Hawaii, Honolulu, Hawaii.

Heaton, T.H. and S.H. Hartzell. 1987. Earthquake hazards on the Cascadia subduction zone. *Science* 236(4798):162-168.

Herring, E. 2002. *A European Union*. [Online]. Available: http://www.chevron.com/news/publications/oct02/docs/pdf_european.pdf [2009, January 12].

Hewitt, K. 1997. *Regions of Risk: A Geographical Introduction to Disasters*. Addison Wesley Longman, Essex, United Kingdom.

Hollnagel, E., C.P. Nemeth, and S. Dekker. 2008. *Resilience Engineering Perspectives: Remaining Sensitive to the Possibility of Failure, Volume 1*. Ashgate Publishing, Hampshire, England, United Kingdom.

Hughes, A., L. Palen, J. Sutton, S. Liu, and S. Vieweg. 2008. "Site-seeing" in disaster: An examination of on-line social convergence. In *Proceedings of the 5th International ISCRAM Conference*, International Society for Information Systems for Crisis Response and Management, Washington, DC.

Imeura, H., M.H. Pradono, A. bin Husen, T. Jauhari, and M. Sugimoto. 2008. Information dissemination for reality-based tsunami disaster education. In *The 14th World Conference on Earthquake Engineering*, Beijing, China.

Institute of Electrical and Electronic Engineers. 2010. *IEEE Standard Glossary of Data Management Terminology Standard 610.5-1990*. [Online]. Available: http://www.techstreet.com/standards/IEEE/610_5_1990?product_id=222904 [2010, August 26].

Intergovernmental Oceanographic Commission. 1965. *Report of the Working Group Meeting on the International Aspects of the Tsunami Warning System in the Pacific, Honolulu, Hawaii, April 27-30, 1965*. Intergovernmental Oceanographic Commission, Honolulu, Hawaii.

Intergovernmental Oceanographic Commission. 2008. *Exercise Pacific Wave 08: A Pacific-Wide Tsunami Warning and Communication Exercise Pacific, 28-30 October 2008*. United Nations Educational, Scientific and Cultural Organization, Paris, France.

Intergovernmental Oceanographic Commission and International Hydrographic Organization. 2005. General bathymetric chart of the oceans (GEBCO). In *Twenty-first Meeting of the GEBCO Guiding Committee 11-12 July, 2005 and Twenty-first meeting of the Sub-Committee on Digital Bathymetry 7-8 July, 2005*, Aguascalientes, Mexico.

Intergovernmental Oceanographic Commission, International Strategy for Disaster Reduction, and World Meteorological Organization. 2005. *Assessment of Capacity Building Requirements for an Effective and Durable Tsunami Warning and Mitigation System in the Indian Ocean: Consolidated Report for 16 Countries Affected by the 26 December 2004 Tsunami*. UNESCO-IOC Information document No. 1219, United Nations Educational, Scientific and Cultural Organization, Paris, France.

International Community on Information Systems for Crisis Response and Management. 2008. *ISCRAM2008 Proceedings*. [Online]. Available: http://www.iscram.org/index.php?option=content&task=view&id=2236&Itemid=2 [2010, September].

International Community on Information Systems for Crisis Response and Management. 2009. *6th International Conference on Information Systems for Crisis Response and Management*. [Online]. Available: http://www.iscram.org/index.php?option=content&task=view&id=2264&Itemid=2 [2010, September].

International Standards Organization. 2009. *Ergonomics of Human System Interactions*. [Online]. Available: http://www.iso.org/iso/catalogue_detail.htm?csnumber=37031 [2009, August 19].

International Strategy for Disaster Reduction. 2004. *Living with Risk: A Global Review of Disaster Reduction Initiatives 2004 Version*. [Online]. Available: http://www.unisdr.org/eng/about_isdr/bd-lwr-2004-eng.htm [2010, August 26].

International Tsunami Information Center. 2004. *Tsunami Newsletter, 36(1)*. International Tsunami Information Center, Honolulu, Hawaii.

Ishii, M., P.M. Shearer, H. Houston, and J.E. Vidale. 2005. Extent, duration and speed of the 2004 Sumatra-Andaman earthquake imaged by the Hi-Net array. *Nature* 435(7044):933-936.

Ismail, F., A. Hakam, O. Nur, and B. Adji. 2008. Study on tsunami evacuation route in Padang City, West Sumatra. In *International Conference on Tsunami Warning*, Bali, Indonesia.

Jankaew, K., B.F. Atwater, Y. Sawai, M. Choowong, T. Charoentitirat, M.E. Martin, and A. Prendergast. 2008. Medieval forewarning of the 2004 Indian Ocean tsunami in Thailand. *Nature* 455:1228-1231.

Jarvenpaa, S.L. and D.E. Leidner. 1999. Communication and trust in virtual teams. *Organization Science* 10(6):791-815.

Johnson, D.L. 2005. *NOAA Service Assessment: West Coast Tsunami Warning, June 14*. West Coast/Alaska Tsunami Warning Center, National Oceanic and Atmospheric Administration, Palmer, Alaska.

Johnston, J. 2003. Personal accounts from survivors of the Hilo tsunamis of 1946 and 1960: Toward a disaster communication model. Unpublished master's thesis, University of Hawaii, Honolulu, Hawaii.

Johnston, D., D. Paton, G.L. Crawford, K. Ronan, B. Houghton, and P. Burgelt. 2005. Measuring tsunami preparedness in coastal Washington, United States. *Natural Hazards* 35(1):173-184.

Johnston, D., J. Becker, C.E. Gregg, B. Houghton, D. Paton, and G. Leonard. 2007. Developing warning and disaster response capacity in the tourism sector in coastal Washington, USA. *Disaster Prevention and Management* 16(2):210-216.

Kanamori, H. 1972. Mechanisms of tsunami earthquakes. *Physics of the Earth and Planetary Interiors* 6(5):346-359.

Kanamori, H. and M. Kikuchi. 1993. The 1992 Nicaragua earthquake: A slow tsunami earthquake associated with subducted sediments. *Nature* 361:714-716.

Kano, M., M.M. Wood, D.S. Mileti, and L.B. Bourque. 2008. *Public Response to Terrorism: Findings from the National Survey of Disaster Experiences and Preparedness.* Southern California Injury Prevention Research Center, University of Southern California, Los Angeles, California.

Katada, T., N. Kuwasawa, H. Yeh, and C. Pancake. 2006. Integrated simulation of tsunami hazards. In *100th Anniversary Earthquake Conference including the 8th U.S. National Conference on Earthquake Engineering (8NCEE), the SSA Centennial Meeting, and the OES Disaster Resistant California Conference*, San Francisco, California.

Keating, B. 2006. Status of tsunami science research and future directions of research. *Science of Tsunami Hazards* 24(5):385-395.

Kock, N. 2000. Benefits for virtual organizations from distributed groups. *Communications of the ACM* 43(11):107-112.

Koike, N., Y. Kawata, and F. Imamura. 2003. Far-field tsunami potential and a real-time forecast system for the Pacific using the inversion method. *Natural Hazards* 29(3):423-436.

Koshimura, S., T. Katada, H.O. Mofjeld, and Y. Kawata. 2006. A method for estimating casualties due to the tsunami inundation flow. *Natural Hazards* 39(2):265-274.

Kowalik, Z. and P.M. Whitmore. 1991. An investigation of two tsunamis recorded at Adak, Alaska. *Science of Tsunami Hazard* 9(2):67-83.

Lachman, R., M. Tatsuoka, and W.J. Bonk. 1961. Human behavior during the tsunami of May 1960. *Science* 133(3462):1405-1409.

Lammel, G., H. Klupfel, and K. Nagel. 2008. Preliminary result of a large scale microscopic evacuation simulation for the city of Padang in the case of tsunami. In *International Conference on Tsunami Warning*, Bali, Indonesia.

LaPorte, T.R. and P. Consolini. 1991. Working in practice but not in theory: Theoretical challenges in high reliability organizations. *Journal of Public Administration Research and Theory* 1(1):19-47.

Laska, S. and B. Morrow. 2006. Social vulnerabilities and Hurricane Katrina: An unnatural disaster in New Orleans. *Marine Technology Society Journal* 40(4):16-26.

Lavigne, F., R. Paris, D. Grancher, P. Wassmer, D. Brunstein, F. Vautier, F. Leone, F. Flohic, B. De Coster, T. Gunawan, C. Gomez, A. Setiawan, R. Cahyadi, and R. Fachrizal. 2009. Reconstruction of tsunami inland propagation on December 26, 2004 in Banda Aceh, Indonesia, through field investigations. *Pure and Applied Geophysics* 166(1-2):259-281.

Lay, T., H. Kanamori, C.J. Ammon, M.N., S.N. Ward, R.C. Aster, S.L. Beck, S.L. Bilek, M.R. Brudzinski, R.B., H.R. DeShon, G. Ekström, K. Satake, and S. Sipkin. 2005. The Great Sumatra-Andaman earthquake of 26 December 2004. *Science* 308(5725):1127-1133.

Lee, H.J. 2009. Timing of occurrence of large submarine landslides on the Atlantic Ocean margin. *Marine Geology* 264(1-2):53-64.

Levitt, B. and J.G. March. 1988. Organizational learning. *Annual Review of Sociology* 14(1):319-340.

Lewis, D. 2007. Statewide seismic needs assessment: Implementation of 2005 Oregon Senate Bill 2 relating to public safety, seismic safety, and seismic rehabilitation of public buildings: Report to the seventy-fourth Oregon Legislative Assembly. In *Oregon Department of Geology and Mineral Industries Report to the Seventy-Fourth Oregon Legislative Assembly*, Oregon Department of Geology and Mineral Industries, Portland, Oregon.

Li, W., H. Yeh, K. Hirata, and T. Baba. 2009. Ocean-bottom pressure variations during the 2003 Tokachi-Oki Earthquake. In *Nonlinear Wave Dynamics*, Lynett, P. (Ed.). World Scientific Publishing Co., Singapore.

Lipa, B.J., D.E. Barrick, J. Bourg, and B.B. Nuden. 2006. HF radar detection of tsunamis. *Journal of Oceanography* 62(5):705-716.

Liu, P.L.F., Y-C. Cho, M. Briggs, U. Kanoglu, and C.E. Synolakis. 1995. Solitary wave runup on a conical island. *Journal of Fluid Mechanics* 302:259-285.

Liu, P.L.F., H. Yeh, and C.E. Synolakis (Eds.). 2008. Advanced numerical models for simulating tsunami waves and runup. In *Advances in Coastal and Ocean Engineering, Volume 10*. World Scientific Publishing Co., Singapore.

Locat, J., H. Lee, U.S. ten Brink, D. Twichell, E.L. Geist, and M. Sansoucy. 2009. Geomorphology, stability and mobility of the Currituck slide. *Marine Geology* 264(1-2):28-40.

Lomax, A., A. Michelini, and A. Piatanesi. 2007. An energy-duration procedure for rapid determination of earthquake magnitude and tsunamigenic potential. *Geophysical Journal International* 170(3):1195-1209.

López-Venegas, A.M., U.S. ten Brink, and E.L. Geist. 2008. Submarine landslide as the source for the October 11, 1918 Mona Passage tsunami: Observations and modeling. *Marine Geology* 254:35-46.

Madsen, O.S. and C.C. Mei. 1969. The transformation of a solitary wave over an uneven bottom. *Journal of Fluid Mechanics* 39(4):781-791.

March, J.G., L. Sproull, and M. Tamuz. 1991. Learning from samples of one or fewer. *Organization Science* 2(1):1-13.

McAdoo, B., L. Dengler, G. Prasetya, and V.V. Titov. 2006. *Smong*: How an oral history saved thousands on Indonesia's Simeulue Island during the December 2004 and March 2005 tsunamis. *Earthquake Spectra* 22(S3):S661-S669.

McCaffrey, R. 2008. Global frequency of magnitude 9 earthquakes. *Geology* 36(3):263-266.

McCann, W.R. 1985. On the earthquake hazards of Puerto Rico and the Virgin Islands. *Bulletin of the Seismological Society of America* 75(1):251-262.

McCann, W.R., L. Feldman, and M. McCann. Undated. Catalog of felt earthquakes for Puerto Rico and neighboring islands 1492-1899 with additional information for some 20th century earthquakes. Unpublished material provided as a PDF file by W.R. McCann.

McGuire, L., E. Ford, and C. Okoro. 2007. Natural disasters and older U.S. adults with disabilities: Implications for evacuation. *Disasters* 31(1):49-56.

McGuire, R.K. 2008. Seismic risk mitigation decisions under uncertainty. In *Risk Assessment, Modeling and Decision Support: Strategic Decisions*. Bostrom, A., S.P. French, and S.J. Gottlieb (Eds.). Springer-Verlag, Berlin, Germany.

McGuire, R.K. and D. Becker. 2004. *Seismic Hazard and Risk Analysis*. Earthquake Engineering Research Institute, Oakland, California.

McMillan, A. and I. Hutchinson. 2002. When the mountain dwarfs danced: Aboriginal traditions of paleoseismic events along the CSZ of western North America. *Ethnohistory* 49(1):41-68.

Meinig, C., S.E. Stalin, A.I. Nakamura, and H. Milburn. 2005. *Real-Time Deep-Ocean Tsunami Measuring, Monitoring, and Reporting System: The NOAA DART II Description and Disclosure*. National Ocean and Atmospheric Administration, Silver Spring, Maryland.

Mercado-Irizarry, A. and P.L.F. Liu (Eds.). 2006. *Caribbean Tsunami Hazard*. World Scientific, Hackensack, New Jersey.

Mileti, D.S. 1995. Factors related to flood warning response. In *The U.S.-Italy Research Workshop on the Hydrometeorology, Impacts, and Management of Extreme Floods*. Perugia, Italy.

Mileti, D.S. 1999. *Disasters by Design: A Reassessment of Natural Hazards in the United States*. The Joseph Henry Press, Washington, DC.

Mileti, D.S. and D. Sorenson. 1990. *Communication of Emergency Public Warnings: A Social Science Perspective and State-of-the-Art Assessment*. Oak Ridge National Laboratory, Oak Ridge, Tennessee.

Mileti, D.S. and C. Fitzpatrick. 1992. Causal sequence of risk communication in the Parkfield earthquake prediction experiment. *Risk Analysis* 12(3):393-400.

Mileti, D.S., C. Fitzpatrick, and B.C. Farhar. 1992. Fostering public preparations for natural hazards: Lessons from the Parkfield earthquake prediction. *Environment* 34(3):16-39.

Miller, M., D. Paton, and D. Johnston. 1999. Community vulnerability to volcanic hazard consequences. *Disaster Prevention and Management* 8(4):255-260.

Mofjeld, H.O. 2009. Tsunami measurements. In *The Sea, Volume 15: Tsunamis*. Bernard, E.N. and A.R. Robinson (Eds.). Harvard University Press, Cambridge, Massachusetts.

Morgan, J. 1984. A tsunami avoidable susceptibility index. *Science of Tsunami Hazards* 2(1):3-12.

Morrow, B. 1999. Identifying and mapping community vulnerability. *Disasters* 23(1):1-18.

Mowshowitz, A. 1997. Virtual organization. *Communications of the ACM* 40(9):30-37.

Muir-Wood, R. and A. Mignan. 2009. A phenomenological reconstruction of the Mw9 November 1st 1755 Lisbon earthquake. In *The 1755 Lisbon Earthquake, Revisited*, Mendes-Victor, L.A., C.S. Oliveira, J. Azebedo, and A. Ribeiro (Eds.). Springer, Dordrecht, The Netherlands.

Nakaseko, T., F. Kimata, S. Tanaka, and M. Takahashi. 2008. Tsunami warning and evacuation system in Nishiki of central Japan. In *International Conference on Tsunami Warning*, Bali, Indonesia.

Nanayama, F., K. Satake, R. Furukawa, K. Shimokawa, B.F. Atwater, K. Shigeno, and S. Yamaki. 2003. Unusually large earthquakes inferred from tsunami deposits along the Kuril trench. *Nature* 424:660-663.

National Aeronautics and Space Administration. 2005. *Live Demonstration of 21st Century National-Scale Team Science.* [Online]. Available: http://earthobservatory.nasa.gov/Newsroom/view.php?id=27764 [2009, August 17].

National Oceanic and Atmospheric Administration. 2008a. *WC/ATWC User's Guide.* West Coast/Alaska Tsunami Warning Center, National Oceanic and Atmospheric Administration, Palmer, Alaska.

National Oceanic and Atmospheric Administration. 2008b. *NOAA's Tsunami Program 2008-2017 Strategic Plan.* National Oceanic and Atmospheric Administration, Silver Spring, Maryland.

National Oceanic and Atmospheric Administration. 2008c. *Operations, Systems, and Procedures Manual.* Pacific Tsunami Warning Center, National Oceanic and Atmospheric Administration, Ewa Beach, Hawaii.

National Oceanic and Atmospheric Administration. 2009. *Concept of Operations and Operational Requirements: Tsunami Operations Software and Operational Testbed.* National Oceanic and Atmospheric Administration, Silver Spring, Maryland.

National Research Council. 1996a. *Understanding Risk: Informing Decisions in a Democratic Society.* National Academy Press, Washington, DC.

National Research Council. 1996b. *National Science Education Standards.* National Academy Press, Washington, DC.

National Research Council. 2006. *Facing Hazards and Disasters: Understanding Human Dimensions.* The National Academies Press, Washington, DC.

National Research Council. 2007. *Tools and Methods for Estimating Populations at Risk from Natural Disasters and Complex Humanitarian Crises.* The National Academies Press, Washington, DC.

National Science and Technology Council. 2005. *Tsunami Risk Reduction for the United States: A Framework for Action.* Subcommittee on Disaster Reduction and United States Group on Earth Observations, National Science and Technology Council, Washington, DC.

National Science Foundation. 2003. *Review of the Global Seismographic Network.* Prepared for the Instrumentation and Facilities Program, Earth Sciences Division, National Science Foundation, IRIS Consortium, Washington, DC.

National Science Foundation. 2006. *NCAR Supercomputing Data Project.* [Online]. Available: www.ncar.ucar.edu/nsa/NSA.../Buja_NSA_20060530_public.ppt [2009, August 17].

National Tsunami Hazard Mitigation Program. 2007. *Documents and Presentations from NTHMP 5-year Assessment.* [Online]. Available: http://nthmp.tsunami.gov/assessments.html [2010, August 25].

National Tsunami Hazard Mitigation Program. 2008. *Tsunami Data Management: An Initial Report on the Management of Environmental Data Required to Minimize the Impact of Tsunamis in the United States.* National Oceanic and Atmospheric Administration Data Management Committee Special Report, National Tsunami Hazard Mitigation Program, National Oceanic and Atmospheric Administration, Silver Spring, Maryland.

National Tsunami Hazard Mitigation Program. 2009a. *National Tsunami Hazard Mitigation Program 2009-2013 Strategic Plan.* [Online]. Available: http://nthmp.tsunami.gov/documents/NTHMPStrategicPlanvf-081309.doc [2010, August 25].

National Tsunami Hazard Mitigation Program. 2009b. *Exercise LANTEX 09 Participant Handbook: A Northwest-Atlantic Tsunami Warning Exercise April 2, 2009.* [Online]. Available: http://nthmp.tsunami.gov/documents/LANTEX09HandbookV3.pdf [2010, August 25].

Newman, A.V. and J.A. Convers. 2008. An energy rate magnitude for large earthquakes. *Eos, Transactions, American Geophysical Union* 89:S13B-1804.

Ni, S., H. Kanamori, and D. Helmberger. 2005. Seismology: Energy radiation from the Sumatra earthquake. *Nature* 434:582.

Normark, W.R., M. McGann, and R. Sliter. 2004. Age of Palos Verdes submarine debris avalanche, southern California. *Marine Geology* 203(3-4):247-259.

O'Loughlin, K.F. and J.F. Lander. 2003. *Caribbean Tsunamis: A 500-Year History from 1498-1998.* Kluwer Academic, Dordrecht, Boston, Massachusetts.

Occhipinti, G., P. Lognonne, E.A. Kherani, H. Hebert, F. Crespon. 2006. *The Indian Ocean Tsunami 2004 in the Ionosphere: Observations and Modeling.* American Geophysical Union Fall Meeting 2006, San Francisco, California.

Okal, E.A. 1988. Seismic parameters controlling far-field tsunami amplitudes: A review. *Natural Hazards* 1(1):67-96.

Okal, E.A. 2003. *T* waves from the 1998 Papua New Guinea earthquake and its aftershocks: Timing the tsunamigenic slump. *Pure and Applied Geophysics* 160(10-11):1843-1863.

Okal, E.A. 2007a. Performance of robust source estimators for last year's large earthquakes. *Eos, Transactions, American Geophysical Union* 88:S44A-01.

Okal, E.A. 2007b. Seismic records of the 2004 Sumatra and other tsunamis: A quantitative study. *Pure and Applied Geophysics* 164(2-3):325-353.

Okal, E.A., A. Piatanesi, and P. Heinrich. 1999. Tsunami detection by satellite altimetry. *Journal of Geophysical Research* 104(B1):599-615.

Okal, E.A. and A.V. Newman. 2001. Tsunami earthquakes: The quest for a regional signal. *Physics of the Earth and Planetary Interiors* 124(1-2):45-70.

Okal, E.A., P.J. Alasset, O. Hyvernaud, and F. Schindelé. 2003. The deficient *T* waves of tsunami earthquakes. *Geophysical Journal International* 152(2):416-432.

Okal, E.A., J. Talandier, and D. Reymond. 2007. Source quantification of the 2004 Sumatra earthquake from hydrophone records of its tsunami. *Pure and Applied Geophysics* 164(2-3):309-323.

Papathoma, M., D. Dominey-Howes, Y. Zong, and D. Smith. 2003. Assessing tsunami vulnerability, an example from Herakleio, Crete. *Natural Hazards and Earth Systems Sciences* 3(5):377-389.

Park, J., R. Butler, J. Berger, P. Davis, H. Benz, C.R. Hutt, C.S. McCreery, T. Ahern, G. Ekström, K. Anderson, and R. Aster. 2005. Performance review of the global seismographic network for the Sumatra-Andaman megathrust earthquake. *Seismological Research Letters* 76(3):331-343.

Paton, D., B.F. Houghton, C.E. Gregg, D.A. Gill, L.A. Ritchie, D. McIvor, P. Larin, S. Meinhold, J. Horan, and D.M. Johnston. 2008. Managing tsunami risk in coastal communities: Identifying predictors of preparedness. *The Australian Journal of Emergency Management* 23(1):4-9.

Peltier, W.R. and C.O. Hines. 1976. On the possible detection of tsunamis by a monitoring of the ionosphere. *Journal of Geophysical Research* 81(12):1995-2000.

Physicians On-Line/Medscape. 2006. Medscape. [Online]. Available: http://www.medscape.com/ [2010, September 1].

Porter, A.L. 1993. Virtual companies reconsidered. *Technology Analysis and Strategic Management* 5(4):413-420.

Post, J., M. Mück, K. Zosseder, S. Wegscheider, T. Steinmetz, T. Riedlinger, G. Strunz, H. Mehl, S. Dech, J. Birkmann, N. Gebert, H. Anwar, and H. Harjono. 2008. Tsunami risk assessment for local communities in Indonesia to provide information for early warning and disaster management. In *International Conference on Tsunami Warning*, Bali, Indonesia.

Priest, G.R. 1997. *Tsunami Hazard Map of the Yaquina Bay Area, Lincoln County, Oregon: Interpretive Map Series*. Oregon Department of Geology and Mineral Industries, Portland, Oregon.

Priest, G.R., C. Goldfinger, K. Wang, R.C. Witter, Y. Zhang, and A.M. Baptista. 2009. Confidence levels for tsunami-inundation limits in northern Oregon inferred from a 10,000-year history of great earthquakes at the Cascadia subduction zone. *Natural Hazards* DOI 10.1007/s11069-009-9453-5.

Quarantelli, E.L. 1984. Perceptions and reactions to emergency warnings of sudden hazards. *Ekisticts* 309:511-515.

Reid, H.F. and S. Taber. 1919. The Puerto Rico earthquakes of October-November, 1918. *Bulletin of the Seismological Society of America* 9:95-127.

Reid, H.F. and S. Taber. 1920. The Virgin Islands earthquakes of 1867-1868. *Bulletin of the Seismological Society of America* 10(1):9-30.

Renn, O. 2008. *Risk Governance: Coping with Uncertainty in a Complex World*. Earthscan, London, England, United Kingdom.

Roberts, K.H. 1990. Some characteristics of high reliability organizations. *Organization Science* 1(2):160-177.

Roberts, K.H., P. Madsen, and V. Desai. 2005. The space between in space transportation: A relational analysis of the failure of STS107. In *Organization at the Limit: Lessons from the Columbia Disaster*, Farjoun, M. and W. Starbuck (Eds.). Blackwell Publishing, Malden, Massachusetts.

Sack, R.L., D. Auckley, R.R. Auger, M.A. Carskadon, K.P. Wright, Jr., M.V. Vitiello, and I.V. Zhdanova. 2007. Circadian-rhythm sleep disorders: Part I, basic principles, shift work and jet lag disorders. *Sleep* 30(11):1460-1483.

Salzberg, D.H. 2008. A hydro-acoustic solution to the local tsunami warning problem. In *American Geophysical Union Fall Meeting*, San Francisco, California.

Sandwell, D.T. and W.H.F. Smith. 2009. Global marine gravity from retracked Geosat and ERS-1 altimetry: Ridge segmentation versus spreading rate. *Journal of Geophysical Research* 114:B01411.

Satake, K., K. Wang, and B.F. Atwater. 2003. Fault slip and seismic moment of the 1700 Cascadia earthquake inferred from Japanese tsunami descriptions. *Journal of Geophysical Research* 108(B11):2535.

Scharroo, R., W.H.F. Smith, V.V. Titov, and D. Arcas. 2005. Observing the Indian Ocean tsunami with satellite altimetry. *Geophysical Abstracts* 7: 230.

Schein, E.A. 1992. *Organizational Culture and Leadership, 2nd Edition*. Jossey-Bass, San Francisco, California.

Schein, E.A. 1996. Three cultures of management: The key to organizational learning. *Sloan Management Review* 38(1):9-20.

Schmidt, M., H. Dragert, Y. Lu, and K. Wang. 2008. Status of the Canadian pilot project for a GPS-augmented tsunami warning system. *American Geophysical Union* 89(53):G43A-0644.

Scientific Committee on Oceanic Research. 2001. *Improved Global Bathymetry Final Report of SCOR Working Group 107*. Intergovernmental Oceanographic Commission Technical Series 63, UNESCO, New York.

Senge, P.M. 1990. *The Fifth Discipline: The Art and Practice of the Learning Organization*. Doubleday, New York.

Shankar, K. 2008. Wind, water and wifi: New trends in community informatics and disaster management. *The Information Society* 24(2):116-120.

Shearer, P.M. and R. Bürgmann. 2010. Lessons learned from the 2004 Sumatra-Andaman megathrust rupture. *Annual Review of Earth and Planetary Sciences* 38:103-131.

Shiro, B.R., S.K. Koyanagi, P.G. Okubo, C.J. Wolfe. 2006. *Towards the Establishment of the Hawaii Integrated Seismic Network for Tsunami, Seismic, and Volcanic Hazard Mitigation*. American Geophysical Union, Fall Meeting 2006, San Francisco, California.

Slovic, P., B. Fischhoff, and S. Lichtenstein. 1980. Facts and fears: Understanding perceived risk. In *Societal Risk Assessment: How Safe Is Safe Enough?*, Schwing, R.C. and W.A. Albers (Eds.). General Motors Corporation, Research Laboratories, Plenum Press, New York.

Society for Worldwide Interbank Financial Telecommunications. 2009. [Online]. Available: http://www.SWIFT.com [2009, August 19].

Song, Y.T. 2007. Detecting tsunami genesis and scales directly from coastal GPS stations. *Geophysical Research Letters* 34(19):L19602.

Spillane, M.C., E. Gica, V.V. Titov, and H.O. Mofjeld. 2008. *Tsunameter Network Design for the U.S. DART Arrays in the Pacific and Atlantic Oceans*. NOAA Technical Memorandum OAR PMEL-143, Pacific Marine Environmental Laboratory, National Oceanic and Atmospheric Administration, Seattle, Washington.

Stephenson, K. 1995. The formation and incorporation of virtual entrepreneur groups. *Entrepreneurship: Theory and Practice* 19(3):35-52.

Suter, L., T. Birkland, and R. Larter. 2009. Disaster research and social network analysis: Examples of the scientific understanding of human dynamics at the National Science Foundation. *Population Research and Policy Review* 28(1):1-10.

Sutton, J. and K. Tierney. 2006. *Disaster Preparedness: Concepts, Guidance, and Research*. The Fritz Institute Conference on Assessing Disaster Preparedness, Sebastopol, California.

Sutton, J., L. Palen, and I. Shklovski. 2008. Backchannels on the front lines: Emergent uses of social media in the 2007 California wild fires. In *Proceedings of the 5th International Systems Conference for Crisis Response and Management*, International Society for Information Systems for Crisis Response and Management, Washington, DC.

Sweeney, A.D., C.D. Chadwell, J.A. Hildebrand, and F.N. Spiess. 2005. Centimeter-level positioning of seafloor acoustic transponders from a deeply-towed interrogator. *Marine Geodessy* 28(1):39-70.

Synolakis, C.E. 1987. The runup of solitary waves. *Journal of Fluid Mechanics* 185:523-545.

Synolakis, C.E., P.L.F. Liu, G. Carrier, and H. Yeh. 1997. Tsunamigenic sea-floor deformations. *Science* 278(5338):598-600.

Synolakis, C.E, J.P. Bardet, J. Borrero, H. Davies, E.A. Okal, E. Silver, J. Sweet, and D. Tappin. 2002. Slump origin of the 1998 Papua New Guinea tsunami. *Proceedings of the Royal Society A* 458:763-769.

Synolakis, C.E., E.N. Bernard, V.V. Titov, U. Kanoglu, and F.I. González. 2007. *Standards, Criteria, and Procedures for NOAA Evaluation of Tsunami Numerical Models*. NOAA Technical Memorandum OAR PMEL-135, Pacific Marine Environmental Laboratory, National Oceanic and Atmospheric Administration, Seattle, Washington.

Tang, L., C.D. Chamberlin, E. Tolkova, M. Spillane, V.V. Titov, E.N. Bernard, and H.O. Mofjeld. 2006. *Assessment of Potential Tsunami Impact for Pearl Harbor, Hawaii.* NOAA Technical Memorandum OAR PMEL-131, Pacific Marine Environmental Laboratory, National Oceanic and Atmospheric Administration, Seattle, Washington.

Tang, L., V.V. Titov, Y. Wei, H.O. Mofjeld, M. Spillane, D. Arcas, E.N. Bernard, C.D. Chamberlin, E. Gica, and J. Newman. 2008. Tsunami forecast analysis for the May 2006 Tonga tsunami. *Journal of Geophysical Research* 113:C12015.

Tang, L., V.V. Titov, and C.D. Chamberlin. 2009. Development, testing, and applications of site-specific tsunami inundation models for real-time forecasting. *Journal of Geophysical Research* 114(C12025):1-22.

ten Brink, U.S. 2009. Tsunami hazard along the U.S. Atlantic coast. *Marine Geology* 264:1-3.

ten Brink, U.S., W.W. Danforth, C.F. Polloni, B. Andrews, P. Llanes, S. Smith, E. Parker, and T. Uozumi. 2004. New seafloor map of the Puerto Rico trench helps assess earthquake and tsunami hazards. *Eos, Transactions, American Geophysical Union* 85(37):349-360.

ten Brink, U.S., E.L. Geist, and B.D. Andrews. 2006. Size distribution of submarine landslides and its implication to tsunami hazard in Puerto Rico. *Geophysical Research Letters* 33:L11307.

Theilen-Willige, B. 2006. Emergency planning in northern Algeria based on remote sensing data in respect to tsunami hazard preparedness. *Science of Tsunami Hazards* 25(1):3-17.

Titov, V.V. 2009. Tsunami forecasting. In *The Sea, Volume 15: Tsunamis*, Bernard, E.N. and A.R. Robinson (Eds.). Harvard University Press, Cambridge, Massachusetts.

Titov, V.V. and F.I. González. 1997. *Implementation and Testing of the Method of Splitting Tsunami (MOST) Model.* NOAA Technical Memorandum ERL PMEL-122, Pacific Marine Environmental Laboratory, National Oceanic and Atmospheric Administration, Seattle, Washington.

Titov, V.V., H.O. Mofjeld, F.I. González, and J.C. Newman. 1999. *Offshore Forecasting of Hawaiian Tsunamis Generated in Alaska-Aleutian Subduction Zone.* NOAA Technical Memorandum ERL PMEL-114, Pacific Marine Environmental Laboratory, National Oceanic and Atmospheric Administration, Seattle, Washington.

Titov, V.V., F.I. González, E.N. Bernard, M. Eble, H.O. Mofjeld, J. Newman, and A.J. Venturato. 2005. Real-time tsunami forecasting: Challenges and solutions. *Natural Hazards* 35(1):41-58.

Tolstoy, M. and D.R. Bohnenstiehl. 2005. Hydroacoustic constraints on the rupture duration, length and speed of the Great Sumatra-Andaman Earthquake. *Seismological Research Letters* 76(4):419-425.

Trabant, P., P. Watts, F.L. Lettieri, and G.A. Jamieson. 2001. East breaks slump, northwest Gulf of Mexico. In *Proceedings of Offshore Technology Conference*, Houston, Texas.

Tuler, S. 1988. Individual, group and organizational decision making in technological emergencies. *Industrial Crisis Quarterly* 2(2):109-138.

Turner, B.L., R.E. Kasperson, P.A. Matson, J.J. McCarthy, R.W. Corell, L. Christensen, N. Eckley, J.X. Kasperson, A. Luers, M.L. Martello, C. Polsky, A. Pulsipher, and A. Schiller. 2003. Framework for vulnerability analysis in sustainability science. *Proceedings of the National Academy of Sciences of the United States of America* 100(14):8074-8079.

Twichell, D.C., J.D. Chaytor, U.S. ten Brink, and B. Buczkowski. 2009. Morphology of late Quaternary submarine landslides along the U.S. Atlantic continental margin. *Marine Geology* 264(1-2):4-15.

U.S. Department of Commerce Office of Inspector General. 2008. *The National Data Buoy Center Should Improve Data Availability and Contracting Practices.* Final Inspection Report No. IPE-18585/May 2008, Office of Inspector General, U.S. Department of Commerce, Washington, DC.

U.S. Department of Transportation. 2006. *Pacific Peril 2006: Lessons Learned.* U.S. Department of Transportation, Washington, DC.

U.S. Indian Ocean Tsunami Warning System Program. 2007. *Tsunami Warning Center Reference Guide.* U.S. Agency for International Development, Bangkok, Thailand.

U.S. Navy Space and Aviation Warfare Command. 1998. *Overview of IEEE-EIA 12207, Standard for Information Technology.* [Online]. Available: http://sepo.spawar.navy.mil/12207.doc [2010, August 26].

Uslu, B., J.C. Borrero, L.A. Dengler, and C.E. Synolakis. 2007. Tsunami inundation at Crescent City, California generated by earthquakes along the Cascadia Subduction Zone. *Geophysical Research Letters* 34:L20601.

Walker, D.A. 1996. Observations of tsunami "shadows": A new technique for assessing tsunami wave heights? *Science of Tsunami Hazards* 14:3-11.

Ward, S.N. 1980. Relationships of tsunami generation and an earthquake source. *Journal of Physics of the Earth* 28:441-474.

Wei, Y., E.N. Bernard, L. Tang, R. Weiss, V.V. Titov, C. Moore, M. Spillane, M. Hopkins, and U. Kâno Iu. 2008. Real-time experimental forecast of the Peruvian tsunami of August 2007 for U.S. coastlines. *Geophysical Research Letters* 35:L04609.

Weichselgartner, J. 2001. Disaster mitigation: The concept of vulnerability revisited. *Disaster Prevention and Management* 10(2):85-94.

Weick, K.E. 1987. Organizational culture as a source of high reliability. *California Management Review* 29(2):112-127.

Weick, K.E. 1990. The vulnerable system: An analysis of the Tenerife air disaster. *Journal of Management* 16(3):571-593.

Weick, K.E. 1993. The collapse of sensemaking in organizations: The Mann Gulch disaster. *Administrative Science Quarterly* 38(4):628-652.

Weick, K.E. 1998. Improvisation as a mindset for organizational analysis. *Organization Science* 9(5):543-555.

Weick, K.E. 2003. Organizational design and the Gehry experience. *Journal of Management Inquiry* 12:93-97.

Weick, K.E. and K.H. Roberts. 1993. Collective mind in organizations: Heedful interrelating on flight decks. *Administrative Science Quarterly* 38(3):357-381.

Weick, K.E., K.M. Sutcliffe, and D. Obstfeld. 1999. Organizing for high reliability: Processes of collective mindfulness. In *Research in Organizational Behavior*, Staw, B. and R. Sutton (Eds.). JAI Press, Greenwich, Connecticut.

Weinstein, S.A. 2008. *PTWC Operations, Systems and Procedures Manual, Edition 1.0.* Pacific Tsunami Warning Center, National Oceanic and Atmospheric Administration, Ewa Beach, Hawaii.

Wesson, R.L., O.S. Boyd, C.S. Mueller, and A.D. Frankel. 2008. Challenges in making a seismic hazard map for Alaska and the Aleutians. In *Active Tectonics and Seismic Potential of Alaska, Geophysical Monograph* 179:385-397.

White, G.F. and J.E. Haas. 1975. *Assessment of Research on Natural Hazards.* MIT Press, Cambridge, Massachusetts.

Whitmore, P.M. 2003. Tsunami amplitude prediction during events: A test based on previous tsunamis. *Science of Tsunami Hazards* 21:135-143.

Whitmore, P.M. and T.J. Sokolowski. 1996. Predicting tsunami amplitudes along the North American coast from tsunamis generated in the Northwest Pacific Ocean during tsunami warnings. *Science of Tsunami Hazards* 14(3):147-166.

Whitmore, P.M., T.J. Sokolowski, S. Tsuboi, and B. Hirshorn. 2002. Magnitude-dependent correction for M_{wp}. *Science of Tsunami Hazards* 20(4):187-192.

Whitmore, P.M., H. Benz, M. Bolton, G.L. Crawford, L. Dengler, G. Fryer, J. Goltz, R. Hansen, K. Kryzanowski, S. Malone, D. Oppenheimer, E. Petty, G. Rogers and J. Wilson. 2008. NOAA/WestCoast and Alaska Tsunami Warning Center Pacific Ocean response criteria. *Science of Tsunami Hazards* 27(2):1-21.

Wilson, R.I., L. Dengler, M.R. Legg, K. Long, and K.M. Miller. 2010. *The 2010 Chilean Tsunami on the California Coastline.* State of California, Sacramento, California.

Winerman, L. 2009. Crisis communication. *Nature* 457(22):376-378.

Wisner, B., P. Blaikie, T. Cannon, and I. Davis. 2004. *At Risk: Natural Hazards, People's Vulnerability and Disasters, 2nd Edition.* Routledge, New York.

Wood, N. 2007. *Variations in City Exposure and Sensitivity to Tsunami Hazards in Oregon.* U.S. Geological Survey Scientific Investigations Report 2007-5283, U.S. Geological Survey, Reston, Virginia.

Wood, N. 2009. Tsunami exposure estimation with land-cover data: Oregon and the Cascadia subduction zone. *Applied Geography* 29(2):158-170.

Wood, N. and J. Good. 2004. Vulnerability of a port and harbor community to earthquake and tsunami hazards: The use of GIS in community hazard planning. *Coastal Management* 32(3):243-269.

Wood, N. and J. Good. 2005. Perceptions of earthquake and tsunami issues in U.S., Pacific Northwest port and harbor communities. *International Journal of Mass Emergencies and Disasters* 23(3):103-138.

Wood, N. and C. Soulard. 2008. *Variations in Community Exposure and Sensitivity to Tsunami Hazards on the Open-Ocean and Strait of Juan de Fuca Coasts of Washington.* U.S. Geological Survey Scientific Investigations Report 2008-5004, U.S. Geological Survey, Reston, Virginia.

Wood, N., J. Good, and R. Goodwin. 2002. Vulnerability assessment of a port and harbor community to earthquake and tsunami hazards: Integrating technical expert and stakeholder input. *Natural Hazards Review* 3(4):148-157.

Wood, N., A. Church, T. Frazier, and B. Yarnal. 2007. *Variations in Community Exposure and Sensitivity to Tsunami Hazards in the State of Hawai`i.* U.S. Geological Survey Scientific Investigation Report 2007-5208, U.S. Geological Survey, Reston, Virginia.

Wood, N., C.G. Burton, and S.L. Cutter. 2010. Community variations in social vulnerability to Cascadia-related tsunamis in the U.S. Pacific Northwest. *Natural Hazards* 52(2):369-389.

Yamazaki, Y., Z. Kowalik, and K.F. Cheung. 2009. Depth-integrated, non-hydrostatic model for wave breaking and run-up. *International Journal for Numerical Methods in Fluids* 61(5): 473-497.

Yeh, H., F. Imamura, C.E. Synolakis, Y. Tsuji, P.L.F. Liu, and S. Shi. 1993. The Flores Island tsunamis. *Eos, Transactions, American Geophysical Union* 74(33):371-373.

Yeh, H., P.L.F. Liu, M. Briggs, and C.E. Synolakis. 1994. Propagation and amplification of tsunamis at coastal boundaries. *Nature* 372:353-355.

Yeh, H., P.L.F. Liu, and C.E. Synolakis (Eds.). 1996. *Long-Wave Runup Models.* World Scientific Publishing Co., Singapore.

Yeh, H., I. Robertson, and J. Preuss. 2005. *Development of Design Guidelines for Structures That Serve as Tsunami Vertical Evacuation Sites.* Washington Division of Geology and Earth Resources, Olympia, Washington.

Yeh, H., T. Fiez, and J. Karon. 2009. *A Comprehensive Tsunami Simulator for Long Beach Peninsula Phase-1: Framework Development.* Washington State Military Department, Tacoma, Washington.

Yuan X., R. Kind, and H.A. Pedersen. 2005. Seismic monitoring of the Indian Ocean tsunami. *Geophysical Research Letters* 32:L15308.

APPENDIX A

Examples of Tsunami Sources That Threaten the United States

Estimates of tsunami losses and heights are from the NOAA tsunami database.[1]

Source	Tsunamis	Unknowns

FAULTS—Seismic slip on faults generates tsunamis directly by displacing the floors of water bodies. The slip can also generate tsunamis indirectly through shaking that triggers slides; a category of source treated separately below. Tsunamis most commonly result from slip on the subduction-zone faults that convey one tectonic plate beneath another.

Source	Tsunamis	Unknowns
Aleutian-Alaskan subduction zone—Along about 2,500 km of its length, ruptured almost completely in a series of earthquakes between 1938 and 1965.[2]	The zone's largest 20th-century tsunamis, both on nearby coasts and on distant ones, were generated during the Aleutian earthquake of 1946 and the Alaskan earthquake of 1964. The far-field part of the 1946 tsunami, chiefly generated directly by faulting,[3] caused most of Hawaii's recorded tsunami deaths. Similarly, the greatest tsunami in Washington, Oregon, and California written history originated off Alaska with tectonic displacement during the 1964 earthquake. Judging from geologic records of predecessors to the 1964 earthquake during the last 6,000 years (Fig. 3-3c),[4] ocean-wide tsunamis from the 1964 source recur at irregular intervals averaging close to 600 years.	How often do Aleutian sources spawn tsunamis comparable in far-field size to the tsunamis of 1946 and 1964? How much are recurrence intervals lengthened by aseismic slip in the fault-rupture areas? Will the next large tsunami from the 1964 source recur sooner than average because the 1964 earthquake ended a recurrence interval close to 900 years, about 300 years longer than average? How persistent are the lateral limits of Aleutian-Alaskan fault ruptures of the 20th century as boundaries that define individual tsunami source areas?[5]
Cascadia subduction zone—1,100 km long. Confirmed as a tsunami hazard by geophysical and geological research in the 1980s and 1990s.[6]	The main nearby tsunami source for Washington, Oregon, and northern California. Also among the main distant sources for Hawaii.[7] Intervals between the zone's great earthquakes (of estimated magnitude 8.0 or 9.0) average close to 500 years and range from a few centuries to a millennium (Fig. 3-3d).[8, 9] The most recent of Cascadia's great earthquakes, of estimated magnitude 8.7-9.2,[10] spawned an ocean-wide tsunami in A.D. 1700 (Fig. 2f).	What proportion of Cascadia's great earthquakes produce unusually large tsunamis by attaining magnitude 9.0?[9, 11, 12] How do those proportions vary along the length of the subduction zone? What partial-length ruptures should be assumed by tsunami modelers?[13] What parts of the zone are likely to augment tsunamis on nearby shores by producing greater than average deformation of the ocean floor?[12]

Source	Tsunamis	Unknowns
Caribbean subduction zone—Faults from oblique convergence between the North America and Caribbean plates near Puerto Rico and the Virgin Islands	A tsunami in 1946, which caused an estimated 1,790 deaths in the Dominican Republic, resulted from a thrust earthquake on or near the plate boundary.[14] A tsunami in 1867, with some 30 fatalities in the Virgin Islands, was generated during an earthquake southeast of Puerto Rico in the Anegada Trough (Fig. 3-1b).	What is the tsunami potential of the plate boundary north and northeast of Puerto Rico, and of a probable backthrust south of the island (Muertos Trough)?[15] What far-field tsunami hazard does the plate boundary pose to the U.S. Atlantic seaboard?[16]
Subduction zone off south-central Chile—Source of largest known earthquake, of 1960, and of a predecessor in 1837[17]	The 1837 and 1960 tsunamis each took some 60 lives in Hawaii. The 1960 tsunami also produced strong currents in Los Angeles–Long Beach Harbor. In the source area of the 1960 tsunami, a swath of ocean floor almost 100 km by 800 km probably rose 2 m or more during the 1960 mainshock.[18] Tsunamis like the big one in 1960 may have recurred at roughly four-century intervals, on average, during the last 2,000 years (Fig. 3-3b). [19]	What factors enabled this subduction zone to produce the outsize earthquake and tsunami of 1960,[20] and what do these factors imply for tsunami hazards from subduction zones—including the Kuril, Japan, and Mariana examples below—that are not known to have produced earthquakes of magnitude 9.0 yet may be capable of doing so?[21, 22]
Subduction zone along the Kuril Trench—Produced earthquake of Mw 8.3 in 2006	The tsunami from the 2006 earthquake caused an estimated $700,000 in damage in Crescent City, California.[23]	How large were the unusually large Kuril earthquakes inferred from geological signs of tsunamis and postseismic uplift in Hokkaido?[24, 25]
Subduction zone along the Japan Trench—No measured earthquake larger than Mw 8.3[22]	In simulations with unit sources having 1 m of seismic slip on fault-rupture patches 50 km by 100 km, Crescent City's greatest tsunami threat from the western Pacific is the subduction zone along the Japan Trench.[23]	Is the Japan Trench limited to earthquakes as large as those in its written historical record?[26]
Mariana subduction zone—No measured earthquake larger than Mw 7.2[21] or 7.7[22]	Simulated for earthquakes as large as Mw 9.3 to make hazard assessments for nearby Guam[27] and distant Pearl Harbor.[7]	What is the maximum plausible earthquake from the Mariana subduction zone, classically considered a place where plates are weakly coupled and the interplate thrust earthquakes consequently of modest size?[28]

Source	Tsunamis	Unknowns
Source of the 1755 Lisbon tsunami—Offshore faulting related to collision of the Nubian (African) and Eurasian plates[29]	The 1755 tsunami was noted in the Caribbean, from Barbados northwestward to Cuba. Its maximum estimated Caribbean height is 7 m. The tsunami is unknown from ports along the U.S. Atlantic seaboard, probably because of shielding by submarine hills that directed the transatlantic waves northwestward toward Newfoundland and southwestward toward the Caribbean and Brazil.[30]	How did the 1755 tsunami affect Puerto Rico and the Virgin Islands? Does it account for overwash of Anegada, in the British Virgin Islands northeast of Puerto Rico (Fig. 3-3e)? How often can tsunamis like the one in 1755 be expected?
Seattle fault—One of several faults capable of displacing waters of Puget Sound.[31]	Six-meter uplift along the Seattle fault generated a tsunami in Puget Sound during an earthquake about 1,100 years ago.[32] The same earthquake set off slides in Lake Washington.[33]	How often does the Seattle fault produce earthquakes like the one 1,100 years ago? Do tsunamis result from slip limited to the Seattle fault's backthrusts, which have a post-glacial history of repeated earthquakes?[34]

SLIDES—Most slides that set off tsunamis have been triggered by earthquake or, less commonly, by volcanic eruption. Several grand examples:

- Lituya Bay, Alaska, 1958—An earthquake-induced rockslide in 1958 set off a giant wave that trimmed trees to an altitude of 525 m.[35]
- Sunda Strait, Indonesia, 1883—The explosion of Krakatau triggered a tsunami that killed an estimated 35,000 persons.[36]
- North Sea, 8,000 years ago—The Storegga slide displaced 2,400-3,200 km^3 of ocean-bottom materials[37] and generated waves known from tsunami deposits in Norway and Scotland.[38, 39]
- Big Island of Hawaii, 120,000 years ago—Flank collapse produced tsunami run-ups to heights of hundreds of meters.[40] A catastrophic ancestor to the local Hawaiian tsunamis that killed 46 persons in 1846 and 2 in 1975.[41]

Slide-generated tsunamis rarely amount to much on distant shores. Compared with the areas of ocean floor displaced by faulting during great subduction zone earthquakes, their source areas are usually compact. Slides therefore yield tsunami waves of short period that diminish rapidly with distance. This decrease helps limit the hazards to the U.S. Atlantic coast from flank collapse in the Canary Islands, off West Africa.[42]

Alaskan slides during the 1964 earthquake—Slides at Chenaga,[43] Kenai Lake,[44] Seward,[45] Valdez,[46] and Whittier[47]	The separate tsunamis from the Chenaga, Seward, Valdez, and Whitter slides together account for 79 of the 106 Alaskan deaths from tsunamis that the 1964 earthquake triggered (Fig. 3-2e). Most of the slides resulted from shaking-induced failures of deltas.	What do these slides imply for Puget Sound deltas as potential tsunami sources?[31]

Source	Tsunamis	Unknowns
Slides off Puerto Rico—Aided by a wealth of steep slopes (Fig. 3-1b) and by active faults associated with the nearby plate boundary[48]	The 1918 tsunami, which caused roughly 40 deaths, may have resulted from an earthquake-induced slide.[49] The slide extends from a headscarp at 1,200 m depth to a terminus at 4,200 m in Mona Passage, the strait between Puerto Rico and Hispaniola. It likely displaced 10 km^3 of water.	What slides are poised to generate tsunamis elsewhere on the steep submarine slopes off Puerto Rico?[50]
Volcanic debris flows—Hot and cold debris flows into Cook Inlet and Bristol Bay, Alaska; debris avalanche at Mount St. Helens	A tsunami in Cook Inlet resulted from a debris avalanche off erupting Augustine Volcano in 1883. Sedimentary deposits suggest that Augustine and Redoubt Volcanoes triggered additional Cook Inlet tsunamis in the last 4,000 years,[51] and that a caldera-forming eruption of Aniakchak Volcano generated a tsunami 3,500 years ago in Bristol Bay.[52] The debris avalanche at the outset of the May 1980 eruption of Mount St. Helens, upon entering Spirit Lake, set off a tsunami that reached heights of 250 m above the former lake level.[53]	How many of Augustine's debris avalanches, a dozen of which have reached Cook Inlet in the last 2,000 years alone, sent tsunamis onto now-populated parts of the Kenai Peninsula?[54]
Slides off southern California—Include the Palos Verdes slide, cf 0.8 km^3 with a headscarp 5 km off the coast near Los Angeles (Fig. 3-1c)[55]	The Palos Verdes slide serves as a poster child for southern California's near-field tsunami hazard. Other potential sources include a submarine slide near Santa Barbara and offshore faults with known or inferred Quaternary displacement.[56-61] The Palos Verdes slide occurred close to 7,500 years ago.[55]	How do southern California's nearby sources of tsunamis compare, in probability and size, with its distant causes of lesser inundation?

Source	Tsunamis	Unknowns
Slides off the edge of the U.S. Atlantic continental shelf— Cover one-third of the continental slope and rise off New England, one-sixth off the Middle Atlantic, and one-eighth off the Southeast[62]	Submarine slides are "considered the primary source of potential tsunamis along the U.S. Atlantic coast."[15] The Currituck slide, with an estimated volume of 165 km^3, is among the largest of these.[63] Its simulated tsunamis originate with peak-to-trough amplitudes of several tens of meters. The waves crest about 6 m above sea level as they overtop the sandy barrier between the Atlantic Ocean and Currituck Sound, North Carolina.[64]	How probable are these slides today? Most of the slides off the U.S. Atlantic coast occurred at least 5,000 years ago, the notable exception being Canada's Grand Banks slide, which generated a tsunami that took 28 lives in Newfoundland in 1929.[65] The Currituck slide dates to roughly 25,000-50,000 years ago.[66] Probabilities aside, simulating slides like Currituck requires uncertain estimates of slide size, speed, and duration, all factors in the slide's effectiveness at generating a tsunami.[64]
Slumps and slides beneath the Gulf of Mexico—Some generated by rise of salt domes,[67] others at scarps in carbonate rocks,[68] still others by ice-age lowering of sea level[69]	No confirmed tsunamis. Tsunami hazard inferred from a slump with a volume of 50-60 km^3 in the northwestern Gulf of Mexico.[70]	As with the slides off the U.S. Atlantic coast, are the Gulf of Mexico examples mainly relics from times of lowered sea level?[69]
Slides ascribed to human activity— Includes construction at Skagway, Alaska,[71] and fluctuation of the level of a reservoir in northeast Washington State[72]	Skagway: Wave heights said 5-6 m high in inlet and 9-11 m high at shore; one fatality.[71] Northeast Washington: Waves up to 20 m high from shores of the reservoir behind Grand Coulee Dam,[72] smaller examples from summer 2009.[73]	Causes considered for the Skagway slide include natural failure as well as dock construction.[71]

REFERENCES

1. National Oceanic and Atmospheric Administration. 2010. *NOAA/WDC Historical Tsunami Database at NGDC*. [Online]. Available: http://www.ngdc.noaa.gov/hazard/tsu_db.shtml. [2010, March 17].

2. Wesson, R.L., O.S. Boyd, C.S. Mueller, and A.D. Frankel. 2008. Challenges in making a seismic hazard map for Alaska and the Aleutians: Active tectonics and seismic potential of Alaska. *Geophysical Monograph* 179:385-397.

3. Okal, E.A. and H. Hebert. 2007. Far-field simulation of the 1946 Aleutian tsunami. *Geophysical Journal International* 169:1229-1238.

4. Carver, G. and G. Plafker. 2008. Paleoseismicity and neotectonics of the Aleutian Subduction Zone–An overview. In *Active Tectonics and Seismic Potential of Alaska*, Freymueller, J.T., P.J. Haeussler, R. Wesson, and G. Ekstrom (Eds). American Geophysical Union, Washington, DC.

5. Shennan, I., R. Bruhn, and G. Plafker. 2009. Multi-segment earthquakes and tsunami potential of the Aleutian megathrust. *Quaternary Science Reviews* 28(1-2):7-13.

6. Atwater, B.F., M.R. Satoko, S. Kenji, T. Yoshinobu, U. Kazue, and D.K. Yamaguchi. 2005. *The Orphan Tsunami of 1700: Japanese Clues to a Parent Earthquake in North America*. University of Washington Press, Seattle, Washington.

7. Tang, L., C. Chamberlin, E. Tolkova, M. Spillane, V.V. Titov, E.N. Bernard, and H.O. Mofjeld. 2006. *Assessment of Potential Tsunami Impact for Pearl Harbor, Hawaii*. NOAA Technical Memorandum OAR PMEL-131 36, National Oceanic and Atmospheric Administration, Silver Spring, Maryland.

8. Atwater, B.F., M.P. Tuttle, E.S. Schweig, C.M. Rubin, D.K. Yamaguchi, and E. Hemphill-Haley. 2004. Earthquake recurrence inferred from paleoseismology. *Developments in Quaternary Science* 1:331-350.

9. Goldfinger, C., K. Grijalva, R. Bürgmann, A.E. Morey, J.E. Johnson, C.H. Nelson, J. Gutiérrez-Pastor, A. Ericsson, E. Karabanov, J.D. Chaytor, J. Patton, and E. Grácia. 2008. Late Holocene rupture of the northern San Andreas Fault and possible stress linkage to the Cascadia Subduction Zone. *Bulletin of the Seismological Society of America* 98(2):861-889.

10. Satake, K., K. Wang, and B.F. Atwater. 2003. Fault slip and seismic moment of the 1700 Cascadia earthquake inferred from Japanese tsunami descriptions. *Journal of Geophysical Research* 108(B11):1-17.

11. Nelson, A.R., H.M. Kelsey, and R.C. Witter. 2006. Great earthquakes of variable magnitude at the Cascadia subduction zone. *Quaternary Research* 65(3):354-365.

12. Priest, G.R., C. Goldfinger, K. Wang, R.C. Witter, Y. Zhang, and A.M. Baptista. 2009. Confidence levels for tsunami-inundation limits in northern Oregon inferred from a 10,000-year history of great earthquakes at the Cascadia subduction zone. *Natural Hazards* 1-47.

13. Uslu, B., J.C. Borrero, L.A. Dengler, and C.E. Synolakis. 2007. Tsunami inundation at Crescent City, California generated by earthquakes along the Cascadia subduction zone. *Geophysical Research Letters* 34:L20601.

14. Dolan, J.F. and D.J. Wald. 1998. The 1943-1953 north-central Caribbean earthquakes: Active tectonic setting, seismic hazards, and implications for Caribbean-North America plate motions. *Geological Society of America Special Paper* 326:143-169.

15. ten Brink, U.S. 2009. Tsunami hazard along the U.S. Atlantic coast. *Marine Geology* 264:1-3.

16. Geist, E.L. and T Parsons. 2009. Assessment of source probabilities for potential tsunamis affecting the U.S. Atlantic coast. *Marine Geology* 264:98-108.

17. Barrientos, S.E. 2007. Earthquakes in Chile. In *The Geology of Chile*, Moreno, T. and W. Gibbons (Eds.). The Geological Society, London, England, United Kingdom.

18. Moreno, M.S., J. Bolte, J. Klotz, and D. Melnick. 2009. Impact of megathrust geometry on inversion of coseismic slip from geodetic data: Application to the 1960 Chile earthquake. *Geophysical Research Letters* 36:L16310.

19. Cisternas, M., B.F. Atwater, F. Torrejón, Y. Sawai, G. Machuca, M. Lagos, A. Eipert, C. Youlton, I. Salgado, T. Kamataki, M. Shishikura, C.P. Rajendran, J.K. Malik, Y. Rizal, and M. Husni. 2005. Predecessors of the giant 1960 Chile earthquake. *Nature* 437:404-407.

20. Kanamori, H. 2006. Lessons from the 2004 Sumatra-Andaman earthquake: Extreme natural hazards. *Philosophical Transactions of the Royal Society A Mathematical, Physical and Engineering Sciences* 364(1845):1927-1945.

21. Stein, S. and E.A. Okal. 2007. Ultralong period seismic study of the December 2004 Indian Ocean earthquake and implications for regional tectonics and the subduction process: The 2004 Sumatra-Andaman earthquake and the Indian Ocean tsunami. *Bulletin of the Seismological Society of America* 97(1A):S279-S295.

22. McCaffrey, R. 2008. Global frequency of magnitude 9 earthquakes. *Geology* 36(3):263-266.

23. Dengler, L., B. Uslu, A. Barberopoulou, J. Borrero, and C.E. Synolakis. 2008. The vulnerability of Crescent City, California, to tsunamis generated by earthquakes in the Kuril Islands region of the northwestern Pacific. *Seismological Research Letters* 79(5):608-619.

24. Nanayama, F., K. Satake, R. Furukawa, K. Shimokawa, B.F. Water, K. Shingeno, and S. Yamaki. 2003. Unusually large earthquakes inferred from tsunami deposits along the Kuril Trench. *Nature* 424:660-663.

25. Sawai, Y., K Sataki, T. Kamataki, H. Nasu, M. Shishikura, B.F. Atwater, B.P. Horton, H.M. Kelsey, T. Nagumo, and M. Yamaguchi. 2004. Transient uplift after a 17th-century earthquake along the Kuril subduction zone. *Science* 306(5703):1918-1920.

26. Hashimoto, C., A. Noda, T. Sagiya, and M. Matsu'ura. 2009. Interplate seismogenic zones along the Kuril-Japan trench inferred from GPS data inversion. *Nature Geoscience* 2:141-144.

27. Arcas, D., B. Uslu, V.V. Titov, and C. Chamberlin. 2008. Tsunami hazard assessment in Guam. *EOS, Transactions, American Geophysical Union* 89:OS42B-06.

28. Uyeda, S. and H. Kanamori. 1979. Back-arc opening and the mode of subduction. *Journal of Geophysical Research* 84(B3):1049-1061.

29. Baptista, M.A. and J.M. Miranda. 2009. Evaluation of the 1755 earthquake source using tsunami modeling: The 1755 Lisbon earthquake: Revisited. *Geotechnical, Geological and Earthquake Engineering* 7:425-432.

30. Barkan, R., U.S. ten Brink, and J. Lin. 2009. Far field tsunami simulations of the 1755 Lisbon earthquake: Implications for tsunami hazard to the U.S. East Coast and the Caribbean. *Marine Geology* 264:109-122.

31. González, F.I., B.L. Sherrod, B.F. Atwater, A.P. Frankel, S.P. Palmer, M.L. Holmes, R.E. Karlin, B.E. Jaffe, V.V. Titov, H.O. Mofjeld, and A.J. Venturato. 2003. *Puget Sound Tsunami Sources—2002 Workshop Report*. U.S. National Tsunami Hazard Mitigation Program, National Oceanic and Atmospheric Administration, Silver Spring, Maryland.

32. Bucknam, R.C., E. Hemphill-Haley, and E.B. Leopold. 1992. Abrupt uplift within the past 1700 years at southern Puget Sound, Washington. *Science* 258(5088):1611-1614.

33. Jacoby, G.C., P.L. Williams, and B.M. Buckley. 1992. Tree ring correlation between prehistoric landslides and abrupt tectonic events in Seattle, Washington. *Science* 258(5088):1621-1623.

34. Kelsey, H.M., B.L. Sherrod, A.R. Nelson, and T.M. Brocher. 2008. Earthquakes generated from bedding plane-parallel reverse faults above an active wedge thrust, Seattle fault zone. *Geological Society of America Bulletin* 120:1581-1597.

35. Miller, D.J. 1960. Giant waves in Lituya Bay, Alaska. *U.S. Geological Survey Professional Paper* 0354 C:51-86.

36. Simkin, T. and R.S. Fiske (Eds.). 1983. *Krakatau 1883: The Volcanic Eruption and Its Effects*. Smithsonian Institution Press, Washington, DC.

37. Haflidason, H., R. Lien, H.P. Sejrup, C.F. Forsberg, and P. Bryn. 2005. The dating and morphometry of the Storegga Slide. *Marine and Petroleum Geology* 22(1-2):123-136.

38. Bondevik, S., J. Mangerud, S. Dawson, A. Dawson, and O. Lohne. 2003. Record-breaking height for 8000-year-old tsunami in the North Atlantic. *EOS Transactions American Geophysical Union* 84(31):289-293.

39. Smith, D.E., S. Shi, R.A. Cullingford, A.G. Dawson, S. Dawson, C.R. Firth, I.D.L. Foster, P.T. Fretwell, B.A. Haggart, L.K. Holloway, and D. Long. 2004. The Holocene Storegga Slide tsunami in the United Kingdom. *Quaternary Science Reviews* 23(23-24):2291-2321.

40. McMurtry, G.M., G.J Fryer, D.R. Tappin, I.P. Wilkinson, M. Williams, J. Fietzke, D. Garbe-Schoenberg, and P. Watts. 2004. Megatsunami deposits on Kohala Volcano, Hawaii, from flank collapse of Mauna Loa. *Geology* 32(9):741-744.

41. Goff, J., W.C. Dudley, M.J. de Maintenon, G. Cain, and J.P. Coney. 2006. The largest local tsunami in 20th century Hawaii. *Marine Geology* 226(1-2):65-79.

42. Gisler, G., R. Weaver, and M.L. Gittings. 2006. SAGE calculations of the tsunami threat from La Palma. *Science of Tsunami Hazards* 24(4):288-301.

43. Plafker, G., R. Kachadoorian, E.B. Eckel, and L.R. Mayo. 1969. Effects of the earthquake of March 27, 1964 on various communities. *U.S. Geological Survey Professional Paper* P 0542-G:G1-G50.

44. McCulloch, D.S. 1966. Slide-induced waves, seiching, and ground fracturing caused by the earthquake of March 27, 1964, at Kenai Lake, Alaska. *U.S. Geological Survey Professional Paper* P 0543-A:A1-A41.

45. Lemke, R.W. 1967. Effects of the earthquake of March 27, 1964, at Seward, Alaska. *U.S. Geological Survey Professional Paper* 0542-E:E1-E43.

46. Coulter, H.W. and R.R. Migliaccio. 1966. Effects of the earthquake of March 27, 1964 at Valdez, Alaska. *U.S. Geological Survey Professional Paper* 0542-C:C1-C36.

47. Kachadoorian, R. 1965. Effects of the earthquake of March 27, 1964, at Whittier, Alaska. *U.S. Geological Survey Professional Paper* 0542-B:B1-B21.

48. ten Brink, U.S., W. Danforth, C. Polloni, B. Andrews, P. Lianes, S. Smith, E. Parker and T. Uozumi. New seafloor map of the Puerto Rico trench helps assess earthquake and tsunami hazards. *EOS Transactions, American Geophysical Union* 85(37):349-360.

49. Lopez-Venegas, A.M., U.S. ten Brink, and E.L. Geist. 2008. Submarine landslide as the source for the October 11, 1918 Mona Passage tsunami: Observations and modeling. *Marine Geology* 254:35-46.

50. ten Brink, U.S., E.L. Geist, P. Lynett, and B. Andrews. 2006. Submarine slides north of Puerto Rico and their tsunami potential. In *Caribbean Tsunami Hazards*, Mercado, A. and P. Liu (eds.). World Scientific Publishers, Singapore.

51. Beget, J., C. Gardner, and K. Davis. 2008. Volcanic tsunamis and prehistoric cultural transitions in Cook Inlet, Alaska: Volcanoes and human history. *Journal of Volcanology and Geothermal Research* 176(3):377-386.

52. Waythomas, C.F. and C.A. Neal. 1998. Tsunami generation by pyroclastic flow during the 3500-year B.P. caldera-forming eruption of Aniakchak Volcano, Alaska. *Bulletin of Volcanology* 60(2):110-124.

53. Voight, B., H. Glicken, R.J. Janda, and P.M. Douglass. 1981. Catastrophic rockslide avalanche of May 18: The 1980 eruptions of Mount St. Helens, Washington. *U.S. Geological Survey Professional Paper* 1250:347-377.

54. Waythomas, C.F., P. Watts, and J.S. Walder. 2006. Numerical simulation of tsunami generation by cold volcanic mass flows at Augustine Volcano, Alaska. *Natural Hazards and Earth System Sciences (NHESS)* 6:671-685.

55. Normark, W.R., M. McGann, and R. Sliter. 2004. Age of Palos Verdes submarine debris avalanche, southern California. *Marine Geology* 203(3-4):247-259.

56. McCulloch, D.S. 2004. Evaluating earthquake hazards in the Los Angeles region: An earth-science perspective. In *Evaluating Tsunami Potential*, Ziony, J.I. (Ed.). U.S. Geological Survey Professional Paper 1360:375-413, Washington, DC.

57. Borrero, J.C., M.R. Legg, and C.E. Synolakis. 2004. Tsunami sources in the Southern California Bight. *Geophysical Research Letters* 31:L13211.

58. Lee, H.J., H.G. Greene, B.D. Edwards, M.A. Fisher, and W.R. Normark. 2009. Submarine landslides of the Southern California Borderland. In *Earth Science in the Urban Ocean: The Southern California Continental Borderland*, Special Paper Geological Society of America 454:251-269, Boulder, Colorado.

59. Ryan, H.F., M.R. Legg, J.E. Conrad, and R.W. Sliter. 2009. Recent faulting in the Gulf of Santa Catalina: San Diego to Dana Point. In *Earth Science in the Urban Ocean: The Southern California Continental Borderland*, Special Paper Geological Society of America 454:291-315, Boulder, Colorado.

60. Fisher, M.A., C.C. Sorlien, and R.W. Sliter. 2009. Potential earthquake faults offshore Southern California, from the eastern Santa Barbara Channel south to Dana Point. In *Earth Science in the Urban Ocean: The Southern California Continental Borderland*, Special Paper Geological Society of America 454:271-290, Boulder, Colorado.

61. Barberopoulou, A., J.C. Borrero, B. Uslu, N. Kalligeris, J.D. Goltz, R.I. Wilson, and C.E. Synolakis. 2009. New maps to improve California tsunami preparedness. *EOS, Transactions, American Geophysical Union* 90(16):137-138.

62. Twichell, D.C., J.D. Chaytor, U.S. ten Brink, and B. Buczkowski. 2009. Morphology of late Quaternary submarine landslides along the U.S. Atlantic continental margin. *Marine Geology* 264:4-15.

63. Locat, J., H.J. Lee, U.S. ten Brink, D. Twichell, E.L. Geist, and M. Sansoucy. 2009. Geomorphology, stability and mobility of the Currituck slide. *Marine Geology* 264(1-2):28-40.

64. Geist, E.L., P.J. Lynett, and J.D. Chaytor. 2009. Hydrodynamic modeling of tsunamis from the Currituck landslide. *Marine Geology* 264(1-2):41-52.

65. Lee, H.J. 2009. Timing of occurrence of large submarine landslides on the Atlantic Ocean margin. *Marine Geology* 264(1-2):53-64.

66. Prior, D.B., E.H. Doyle, and T. Neurauter. 1986. The Currituck Slide, Mid-Atlantic continental slope: Revisited. *Marine Geology* 73(1-2):25-45.

67. Tripsanas, E.K., W.R. Bryant, and B.A. Phaneuf. 2004. Slope-instability processes caused by salt movements in a complex deep-water environment, Bryant Canyon area, northwest Gulf of Mexico. *AAPG Bulletin* 88(6):801-823.

68. Mullins, H.T., A.F. Gardulski, and A.C. Hine. 1986. Catastrophic collapse of the West Florida carbonate platform margin. *Geology* 14(2):167-170.

69. Lowrie, A., C.B. Lutken, and T.M. McGee. 2004. Multiple outer shelf deltas and downslope massive mass-wastings characterize the Mississippi Canyon, northern Gulf of Mexico. *Transactions Gulf Coast Association of Geological Societies* 54:383-392.

70. Trabant, P., P. Watts, F.L. Lettieri, and G.A. Jamieson. 2001. East Breaks slump, Northwest Gulf of Mexico. In *Proceedings Offshore Technology Conference*, Houston, Texas.

71. Rabinovich, A.B., R.E. Thomson, E.A. Kulikov, B.D. Bornhold, and I.V. Fine. 1999. The landslide-generated tsunami November 3, 1994, in Skagway Harbor, Alaska: A case study. *Geophysical Research Letters* 26:3009-3012.

72. Jones, F.O., D.R. Embody, W.L. Peterson, and R.M. Hazlewood. 1961. Landslides along the Columbia River valley, northeastern Washington. *U.S. Geological Survey Professional Paper* 367:1-98.

73. http://www.krem.com/topstories/stories/krem2-082509-landslide__.116caba52.html.

Review of the Tsunami Warning and Forecast System and Overview of the Nation's Tsunami Preparedness

STATEMENT OF TASK

The committee will review progress toward tsunami preparedness in response to "Tsunami Risk Reduction for the United States" (National Science and Technology Council, December 2005) and the Tsunami Warning and Education Act (P.L. 109-424, December 2006). The National Science and Technology Policy report, spurred by the 2004 Indian Ocean tsunami, called for a broad range of federal, state, and local efforts to reduce future losses from tsunamis in the United States. P.L. 109-424 authorized improvements to tsunami warning systems, community-based hazard mitigation programs, public education, scientific research, and international coordination. The committee's task is divided into two parts as described below.

In the first part, the committee will produce an interim report to fulfill the congressional request in P.L. 109-424. The committee will review the National Oceanic and Atmospheric Administration (NOAA) Tsunami Program to assess progress and improvements made since 2005 to strengthen the existing U.S. tsunami detection, forecast, and warning system. The committee will examine the effectiveness of this system for both near- and far-field tsunamigenic events, including:

- Modeling of tsunami generation, propagation, and inundation
- Forecast accuracy, warning notification, and dissemination
- Reliability of observing and monitoring networks
- Strategies to ensure long-term operational reliability and sustainability
- Data quality control, management, archiving, and dissemination
- Data acquisition, processing, and assessment for warning generation
- Further modernization and geographic coverage needs
- Probabilistic assessments of tsunami hazard that include data on the sizes and recurrence intervals of submarine earthquakes and landslides near U.S. shores
- Level of coordination and integration with:
  - State and local level tsunami programs for facilitating mitigation
  - U.S. ocean and coastal observation systems, including the Integrated Ocean Observing System

   ○ Global observing systems, including the Global Earth Observing System of Systems
  • Priority areas of targeted research and development in the United States to improve performance and guide modernization efforts

In its assessment, the committee will consider measures in the tsunami detection, forecast, and warning program implemented and planned by NOAA under the Tsunami Warning and Education Act (P.L. 109-424).

In the second part, the committee will provide a general overview of national preparedness, based on existing compilations and national assessments including topics such as the following:

- Adequacy of federal coordination and integration with state and local level tsunami programs for facilitating mitigation
- Approaches to risk assessment that account for such things as levels and trends in human populations, economic assets, and critical facilities within tsunami-inundation zones
- Availability of evacuation maps, routes, and structures
- Education and outreach for children, adults, and tourists

The committee will examine a few federal, state, and local mitigation and education activities, including the National Tsunami Hazard Mitigation Program and TsunamiReady Program, to include as specific examples in their overview of the nation's ability to reduce losses of life and property from future tsunamis.

In the final report, the committee will comment on how to optimize instrumental warning with these other elements of tsunami preparedness to serve the needs of end-users. The report will highlight opportunities to improve the nation's tsunami preparedness in the future and identify novel, promising approaches to risk assessment and instrumental warning systems.

Relative Hazards of Near- and Far-field Tsunami Sources

The 2005 National Science and Technology Council report describes the nation's mix of tsunami threats in terms of sources termed "local" and "distant":

> United States coastal communities are threatened by tsunamis generated by both local sources and distant sources. Local tsunamis give residents only a few minutes to seek safety. Tsunamis of distant origins give residents more time to evacuate the threatened coastal areas, but require timely and accurate tsunami forecasts of the hazard to avoid costly false alarms. Of the two, local tsunamis pose a greater threat to life because of the short time between generation and impact. The challenge is to design a tsunami hazard mitigation program to protect life and property from two very different types of tsunami events.[1]

The relative tsunami hazard of local and distant sources varies with the region according to a nationwide assessment prepared a few years ago for the National Tsunami Hazard Mitigation Program (NTHMP).[2] Distant sources account for most of the tsunami hazard in Hawaii, while local sources predominate in Alaska. Washington, Oregon, California, and the Caribbean face a mix of local and distant. So does the U.S. Atlantic seaboard, from its nearby landslides and its exposure to hypothetical tsunamis from the Puerto Rico Trench.[3]

Below are summaries of tsunami hazard studies in Alaska, Oregon, and California that allow direct comparison between local and distant.

ALASKA

Nearby tsunami sources dominate the hazard depicted on tsunami inundation maps of Kodiak, Homer, and Seldovia—communities in the vicinity of the rupture area of the giant 1964 Alaska earthquake.

The Kodiak maps[4] depict seven scenarios: four of them for partial or complete breakage of the 1964 rupture area, one for surface rupture of a thrust fault that extends offshore from Kodiak Island, and two tsunamis of distant origin. The scenarios with the greatest inundation result from repetition of 1964-style earthquakes, and the scenarios with the least inundation result from distant earthquakes off the Aleutians and at Cascadia.

The smallest of the modeled inundations in the Kodiak area corresponds to a distant earthquake on the Aleutian-Alaska subduction zone west of Kodiak Island. The starting assumption here is a break that extends across the so-called Shumagin seismic gap and includes rupture areas of earthquakes in 1938 and 1946. Such a hypothetical earthquake is among the seismic sources adopted in the most recent U.S. Geological Survey seismic-hazard assessment for Alaska.[5] A geophysical speculation has such an earthquake recurring at intervals of 750 years or more.

The Homer and Seldovia maps[6] show two scenarios: a repeat of the 1964 earthquake, and a hypothetical break on a local fault believed inactive in the past 2 million years. Recurrence intervals for 1964-style earthquakes have averaged about 600 years during the past 5,000 years.[7]

OREGON

Far-field tsunamis pose the most expectable source of tsunami hazards in Cannon Beach and Seaside, Oregon. However, as in Alaska, it is the near-field tsunamis that dominate the hazard in terms of tsunami size.

Worst-case inundation extends more than twice as far inland for a near-field (Cascadia) tsunami than for a far-field (eastern Aleutian) tsunami, according to an inundation map prepared by the State of Oregon in 2008 for the tourist town of Cannon Beach.[8] A similar contrast is apparent in maps, of the nearby town of Seaside, that were prepared a few years earlier by a group of federal, state, and local scientists.[9] The Seaside maps show contrasting observations (inundation limits and sedimentary deposits of the 1964 Alaskan tsunami, versus sedimentary deposits of the 1700 Cascadia tsunami), as well as tsunami heights depicted in terms of probabilities that are tied to estimated recurrence intervals.

Far-field tsunamis are the most expectable in Oregon because they can beam toward that state from multiple parts of the Pacific Rim. Thus, in aggregate, they happen more often than do tsunamis from Cascadia sources alone. The Seaside mapping accordingly shows far-field tsunamis as the dominant source of hazard for flooding that would lap onto the edges of town. Only at lower probabilities, commensurate with Cascadia recurrence intervals that average about 500 years, do the waters cover the entire town.

CALIFORNIA

A Cascadia rupture that includes the California part of the subduction zone produces a simulated tsunami that, at Crescent City, runs inland for double the inundation distance of the 1964 Alaskan tsunami. This Cascadia tsunami, moreover, begins with a positive (leading elevation) wave that arrives in less than a half hour.[10]

REFERENCES

1. National Science and Technology Council. 2005. *Tsunami Risk Reduction for the United States: A Framework for Action.* Subcommittee on Disaster Reduction and United States Group on Earth Observations, National Science and Technology Council, Washington, DC.

2. Dunbar, P.K. and C.S. Weaver. 2007. *U.S. States and Territories National Tsunami Hazard Assessment: Historical Record and Sources for Waves.* National Tsunami Hazard Mitigation Program, National Oceanic and Atmospheric Administration, Silver Spring, Maryland.

3. Geist, E.L. and T Parsons. 2009. Assessment of source probabilities for potential tsunamis affecting the U.S. Atlantic coast. *Marine Geology* 264(1-2):98-108.

4. Suleimani, E.N., R.A. Hansen, R.A. Comebellick, G.A. Carver, R.A. Kamphaus, J.C. Newman, and A.J. Venturato. 2002. *Tsunami Hazard Maps of the Kodiak Area, Alaska*. Alaska Division of Geological and Geophysical Surveys and the Geophysical Institute, University of Alaska, Fairbanks, Alaska.

5. Wesson, R.L., O.S. Boyd, C.S. Mueller, and A.D. Frankel. 2008. Challenges in making a seismic hazard map for Alaska and the Aleutians: Active tectonics and seismic potential of Alaska. *Geophysical Monograph* 179:385-397.

6. Suleimani, E.N., R.A. Comebillick, R.A. Hansen, A.J. Venturato, and J.C. Newman. 2005. *Tsunami Hazard Maps of the Homer and Seldovia Areas, Alaska*. State of Alaska Department of Natural Resources, Fairbanks, Alaska.

7. Carver, G. and G. Plafker. 2008. Paleoseismicity and neotectonics of the Aleutian Subduction Zone: An overview. In *Active Tectonics and Seismic Potential of Alaska*, Freymueller, J.T., P.J. Haeussler, R. Wesson, and G. Ekstrom (Eds.). American Geophysical Union, Washington, DC.

8. Priest, G.R., C. Goldfinger, K. Wang, R.C. Witter, Y. Zhang, and A.M. Baptista. 2009. Confidence levels for tsunami-inundation limits in northern Oregon inferred from a 10,000-year history of great earthquakes at the Cascadia subduction zone. *Natural Hazards* 1-47.

9. González, F.I., E. Geist, C.E. Synolakis, D. Arcas, D. Bellomo, D. Carlton, T. Horning, B. Jaffe, J. Johnson, U. Kanoglu, H. Mofjeld, J. Newman, T. Parsons, R. Peters, C. Peterson, G. Priest, V.V. Titov, A. Venturato, J. Weber, F. Wong, A. Yalciner. 2006. *Seaside, Oregon Tsunami Pilot Study—Modernization of FEMA Flood Hazard Maps*. National Oceanic and Atmospheric Administration, U.S. Geological Survey, and Federal Emergency Management Agency, Washington, DC.

10. Uslu, B., J.C. Borrero, L.A. Dengler, and C.E. Synolakis. 2007. Tsunami inundation at Crescent City, California generated by earthquakes along the Cascadia subduction zone. *Geophysical Research Letters* 34:L20601.

Available Tsunami Evacuation Maps

A repository for tsunami evacuation maps does not currently exist in the United States. The following is a list of communities with tsunami evacuation maps, based on information provided by the National Oceanic and Atmospheric Administration (NOAA) Tsunami Program and by an online search of available maps.

COMMONWEALTH OF PUERTO RICO

- Mayagüez (see Figure D.3C), Anasco, and Lajas and current efforts in Carolina and Dorado

Note: This list is based on information provided by the Puerto Rico Seismic Network.

STATE OF ALASKA

- Whittier, Sand Point, King Cove, Cold Bay, and Dutch Harbor
- All TsunamiReady communities have tsunami evacuation maps, including Homer, Kodiak, Seward, Sitka, Valdez (see Figure D.2), and Yakutat.

Note: Information provided by the NOAA Tsunami Program.

STATE OF CALIFORNIA

- A tsunami evacuation zone exists for the entire Californian coastline and will become available on the CalEMA "My Hazards" website (http://www.myhazards.calema.ca.gov/).
- Evacuation maps are being developed as stand-alone brochures (and digital PDFs) for all coastal communities.

Note: Information provided by the California Emergency Management Agency.

STATE OF HAWAII

- A tsunami evacuation zone exists for the entire Hawaiian coastline and is organized around shoreline segments, not specific communities.
- **Hawaii**—North Kohala, Waipio Bay, South Kohala, Laupahoehoe, Hakalau Bay, Wailea, Kailua Bay to Kiholo Bay, Hilo - Part 1 (see Figures D.3D, D.5), Hilo - Part 2, South Hilo / Puna, Keauhou to Kailua, Kealakekua to Kailua, Milolii to Kealakekua, Opihikao to Kau / Puna, Punaluu to Honuapo / Kau

- **Kaua'i**—Kailiu Pt. to Kalihiwai Bay, Kauapea Beach to Anahola Bay, Kealia to Alakukui Pt., Waimea to Barking Sands, Wailua to Kamilo Pt., Nawiliwili, Numila to Waimea, Ninini Pt. to Kamala Pt., Keoniloa Bay to Lawai Bay, Poipu
- **Lāna'i**—Pohakuloa Pt. to Kaiolohia Bay, Kaiolohia Bay to Haua, Haua to Kapoho, Kaumalapau Harbor, Manele Bay to Kapihua Bay
- **Maui**—Kaleia Kealekii Pt. to Keawalua, Hanakaoo to Kaelekii Pt., Kanaha Beach Park to Kuau, Waiehu Pt. to Kanaha Beach Park, Launiupoko Pt. to Wahikuli State Wayside Park, Ukumehame Beach Park to Launiupoko Pt., Hana, Kealia to Ukumehame Beach Park, Kamaole Beach Park to Kealia, Puu Olai to Kamaole Beach Park
- **Moloka'i**—Kalaupapa to Kaupikiawa, Pohakumauliuli, Halawa, Kaalaea to Ualapue, Kapukaulua to Kaunakakai, Halena to Kapukuwahine, Ualapue to Kapukaulua
- **Oahu**—Malaekahana to Sunset Beach, Sunset Beach to Waialua Bay, Hauula to Malaekahana, Kahana Bay to Hauula, Waialua Bay to Mokuleia, Mokuleia to Yokohama Bay, Kaneohe Bay to Kahana Bay, Yokohama Bay to Pokai Bay, Kailua to Kaneohe Bay, Kailua to Kaneohe Bay, Pokai Bay to Kahe Point, Waimanalo to Kailua, Makapuu to Waimanalo, Kahe Point to Ewa Beach, Ewa Beach to Airport, Kahe Point to Ewa Beach, Hanauma Bay to Makapuu, Airport to Waikiki, Waikiki, Wailupe to Hanauma Bay, Waikiki to Wailupe

Note: This listing is based on information at Hawaii State Civil Defense (http://www.scd. state.hi.us/; see Figure D.1A).

STATE OF OREGON

- A tsunami inundation line was developed for the entire Oregon shoreline by the Oregon Department of Geology and Mineral Industries (DOGAMI) to support the implementation of a statewide ordinance (Oregon Revised Statute 455.446-447) limiting the construction of critical facilities in tsunami-prone areas.
- **Clatsop County**—Cannon Beach and Arch Cape (see Figure D.4A; Figure D.4B), Gearhart, Seaside (see Figure D.3B), Warrenton
- **Tillamook County**—Manzanita/Nehalem, Nestucca, Oceanside/Netarts, Rockaway Beach
- **Lincoln County**—Depoe Bay, Lincoln City, Newport, Salishan/Gleneden/Lincoln Beach, Waldport, Yachats
- **Lane County**—Florence
- **Douglas County**—Reedsport, Gardiner, Winchester Bay
- **Coos County**—Charleston, Coos Bay, Bandon
- **Curry County**—Brookings, Gold Beach, Port Orford

Note: This list is based on information from the Oregon Department of Geology and Mineral Industries (http://www.oregongeology.org/sub/earthquakes/Coastal/Tsubrochures. htm; see Figure D.1B).

STATE OF WASHINGTON

- **Grays Harbor County** (Figure D.3A)—Aberdeen and Hoquiam, Cosmopolis and South Aberdeen, Ocean City, Copalis Beach, Pacific Beach, and Moclips, Ocean Shores, Westport, Grayland, and Ocosta
- **Jefferson County**—Hoh Reservation
- **Pacific County**—Bay Center, Long Beach and Ilwaco, North Cove and Tokeland, Ocean Park, Raymond and South Bend
- **Whatcom County**—Bellingham, Lummi Reservation, Point Roberts, Sandy Point
- **Clallam County**—Clallam Bay, La Push, Neah Bay, Port Angeles, Port Townsend, Sequim

Note: This list is based on information from the Washington State Emergency Management (http://www.emd.wa.gov/hazards/haz_tsunami.shtml) and Washington Department of Natural Resources (http://www.dnr.wa.gov/ResearchScience/Topics/GeologyPublicationsLibrary/Pages/tsuevac.aspx)

TERRITORY OF GUAM

Guam, Rota, Saipan, and Tinian.
Note: Information provided by the NOAA Tsunami Program.

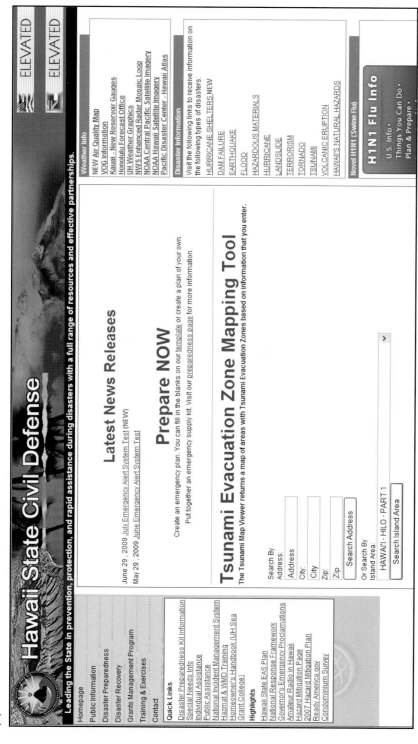

FIGURE D.1 Tsunami evacuation maps available at (A) Hawaii State Civil Defense for Hawaiian communities and (B; next page) Oregon Department of Geology and Mineral Industries for Oregon communities. SOURCE: http://www.scd.state.hi.us; with permission from Hawaii State Civil Defense, and http://www.oregongeology.org/sub/default.htm; image courtesy of DOGAMI, respectively.

(B)

Oregon Department of Geology and Mineral Industries

Home | Maps & Publications | Hazards | Permits | Dept Info | Contact Us | Links | Search

Fossils, Minerals & Gems | Field Offices | News and Events
Mineral Land Regulation and Reclamation | Oil, Gas and Geothermal
Nature of the Northwest Information Center | State of Oregon website
Job listings from the State of Oregon jobs page

Tsunami evacuation maps for selected Oregon coastal communities

IF YOU LIVE AT OR VISIT THE COAST, THE INFORMATION IN THESE BROCHURES MAY SAVE YOUR LIFE. PLEASE TAKE THE TIME TO READ IT AND SHARE WHAT YOU HAVE LEARNED WITH YOUR FAMILY AND FRIENDS.

The evacuation zones on these maps were developed by the Oregon Department of Geology and Mineral Industries in consultation with local officials. It is intended to represent a worst-case scenario for a tsunami caused by an undersea earthquake near the Oregon coast.

Evacuation routes were developed by local officials and reviewed by the Oregon Department of EmergencyManagement.

The map is intended for emergency response and should not be used forsite-specific planning.

These brochures can be viewed and downloaded using Adobe Acrobat PDF Viewer. Some files are 1MB in size and may take a while to open after Acrobat launches.

A tsunami is a series of sea waves usually caused by a displacement of the ocean floor by an undersea earthquake. As tsunamis enter shallow water near land, they increase in height and can cause great loss of life and property damage.

Recent research suggests that tsunamis have struck the Oregon coast on a regular basis. They can occur any time, day or night. Typical wave heights from tsunamis occurring in the Pacific over the last 80 years have been 20–45 feet at the shoreline. A few waves however have been much higher—as much as 100 feet or more—because of local conditions.

To view PDF documents, download the free Adobe Acrobat Reader by clicking on the logo above.

Download a tsunami brochure

- Bandon
- Brookings
- Cannon Beach and Arch Cape
- Charleston
- Coos Bay
- Depoe Bay
- Florence
- Gearhart
- Gold Beach
- Lincoln City
- Manzanita/Nehalem

FIGURE D.1 (B)

Tsunami Evacuation Routes

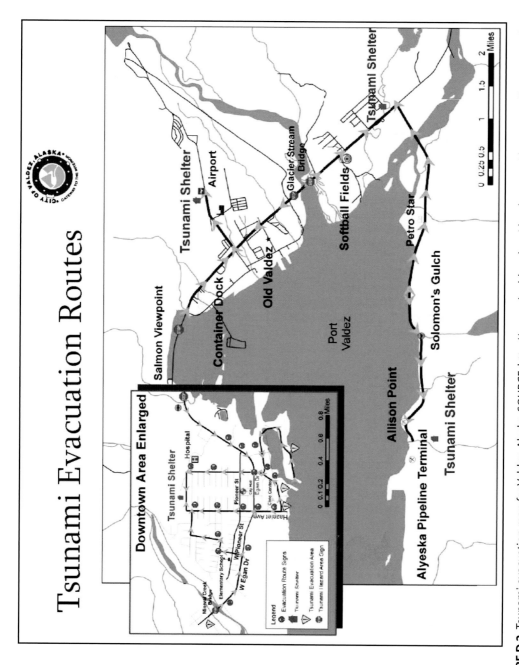

FIGURE D.2 Tsunami evacuation map for Valdez, Alaska. SOURCE: http://www.ci.valdez.ak.us/development/documents/TsunamiBrochureValdez. pdf; with permission from Lisa Von Bargen, City of Valdez.

(A)

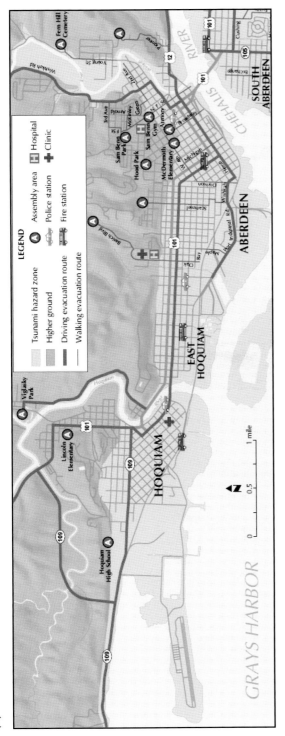

FIGURE D.3 Tsunami evacuation maps for (A) Aberdeen and Hoquiam, Washington, SOURCE: with permission from Washington Department of Natural Resources, Division of Geology and Earth Resources, (B) Seaside, Oregon, SOURCE: http://www.oregongeology.org/sub/default.htm; image courtesy of DOGAMI, (C) Mayaguez, Puerto Rico, SOURCE: http://redsismica.uprm.edu/spanish/Noticias/TsunamiMap_Pub.pdf; with permission from Puerto Rico Seismic Network Tsunami and Educational Outreach Program, and (D) Hilo, Hawaii, SOURCE: http://www.pdc.org/Disaster-Info/Shelters/CDmaps/HawaiiCounty/H-map01.html; with permission from Kevin Richards; Hawaii State Civil Defense.

(B)

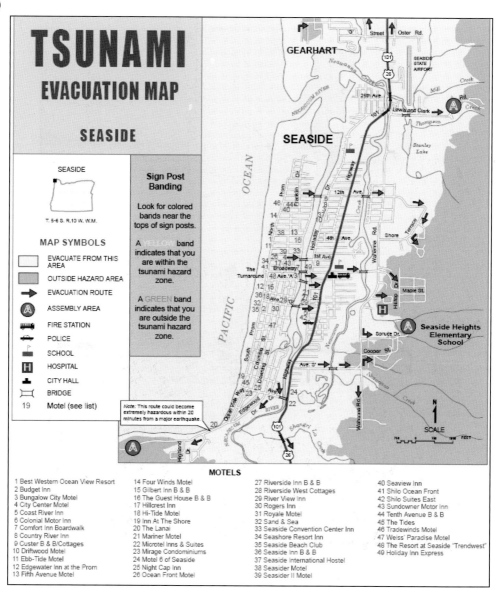

FIGURE D.3 (B)

(C)

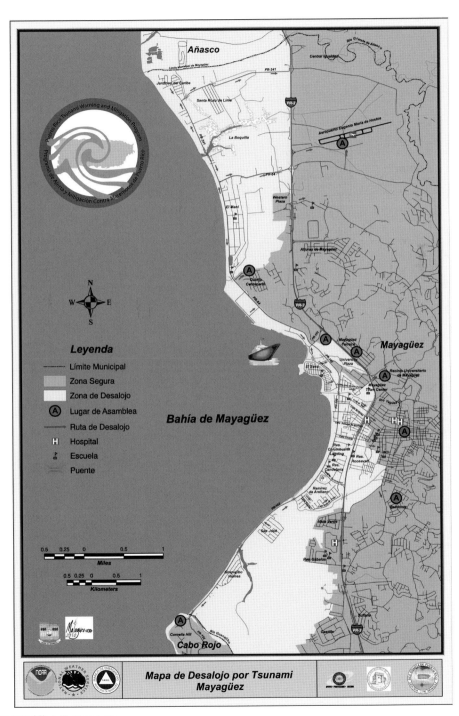

FIGURE D.3 (C)

(D)

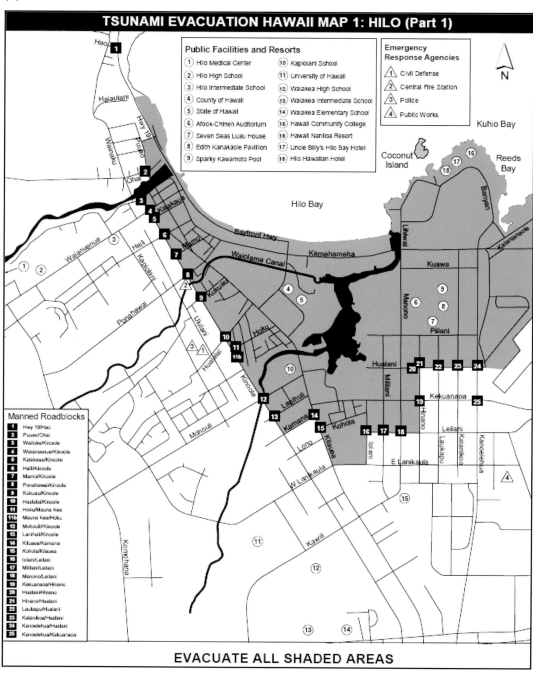

FIGURE D.3 (D)

(A)

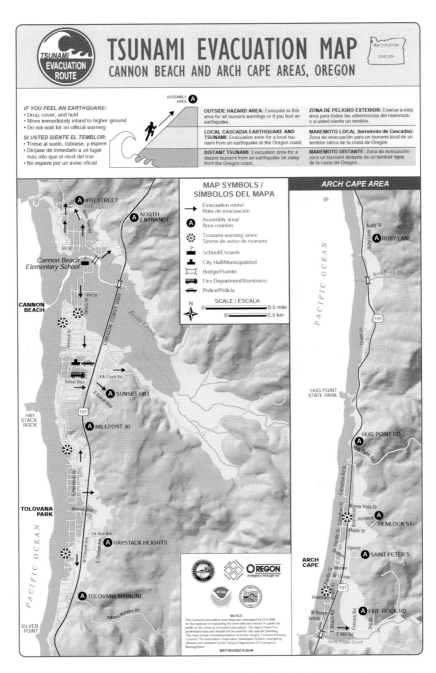

FIGURE D.4 (A) Tsunami evacuation map of Cannon Beach and Arch Cape areas, Oregon and (B) text included in evacuation map brochure for helping individuals to prepare for two types of tsunamis on the Oregon coast. SOURCE: http://www.oregongeology.org/sub/default.htm; image courtesy of DOGAMI.

(B)

BE PREPARED!

Assemble **emergency kits** with at least a 3-day supply for each family member:

- Local map showing safe evacuation routes to high ground.
- First-aid supplies, prescriptions and non-prescription medication.
- Water bottle and filtration or treatment supplies capable of providing 1 gallon per person per day.
- Non-perishable food (ready-to-eat meals, canned food, baby food, energy bars).
- Cooking and eating utensils, can opener, Sterno® or other heat source.
- Matches in water-proof container or lighter.
- Shelter (tent), sleeping bags, blankets.
- Portable radio, NOAA weather radio, flashlight, and extra batteries.
- Rain gear, sturdy footwear, extra clothing.
- Personal hygiene items (toilet paper, soap, toothbrush)
- Tools and supplies (pocket knife, shut-off wrench, duct tape, gloves, whistles, plastic bags
- Cash

How to help with tsunami awareness in your community

- start a tsunami buddy sytem
- make and distribute emergency packs
- initiate or participate in a local preparedness program

WHAT TO KNOW about tsunamis

A **tsunami** is a series of sea waves, usually caused by a displacement of the ocean floor by an undersea earthquake. As tsunamis enter shallow water near land, they increase in height and can cause great loss of life and property damage.

Recent research suggests that tsunamis have struck the Oregon coast on a regular basis. They can occur any time, day or night. Typical wave heights from tsunamis occurring in the Pacific Ocean over the last 500 years have been 20–65 feet at the shoreline. However, because of local conditions a few waves may have been much higher — as much as 100 feet.

We distinguish between a tsunami caused by an undersea earthquake near the Oregon coast (a **local** tsunami) and an undersea earthquake far away from the coast (a **distant** tsunami).

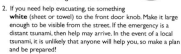

WHAT TO DO for both local and distant tsunamis

1. Evacuate on foot, if at all possible. Follow evacuation signs and arrows.

2. If you need help evacuating, tie something **white** (sheet or towel) to the front door knob. Make it large enough to be visible from the street. If the emergency is a distant tsunami, then help may arrive. In the event of a local tsunami, it is unlikely that anyone will help you, so make a plan and be prepared!

3. Stay away from potentially hazardous areas until you receive an ALL CLEAR from local officals. Tsunamis often follow river channels, and dangerous waves can persist for several hours. Local officials must inspect all flooded or earthquake-damaged structures before anyone can go back into them.

4. After evacuation, check with local emergency officials if you think you have special skills and can help, or if you need assistance locating lost family members.

Look for these hazard zone signs and be ready to leave the area by following evacuation route signs.

Local tsunamis

A **local tsunami** can come onshore within 15 to 20 minutes after the earthquake — before there is time for an official warning from the national warning system. Ground shaking from the earthquake may be the only warning you have. Evacuate quickly!

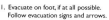

Distant tsunamis

A **distant tsunami** will take 4 hours or more to come ashore. You will feel no earthquake, and the tsunami will generally be smaller than that from a local earthquake. Typically, there is time for an official warning and evacuation to safety.

Evacuation for a distant tsunami will generally be indicated by a **3-minute siren blast** (if your area has sirens) and an announcement over NOAA weather radio that the local area has been put into an official TSUNAMI WARNING. In isolated areas along beaches and bays you may not hear a warning siren. Here, a **sudden change of sea level** should prompt you to move immediately to high ground. If you hear the 3-minute blast or see a sudden sea level change, first evacuate away from shoreline areas, then turn on your local broadcast media or NOAA weather radio for more information.

FIGURE D.4 (B)

FIGURE D.5 Dynamic tsunami evacuation maps for Hawaiian communities served by an online mapping application. SOURCE: http://www.csc. noaa.gov/psc/riskmgmt/tsunami.html; Pacific Services Center, NOAA.

235

Examples of Tsunami Education Efforts

One element in the committee's statement of task was to comment on the status and adequacy of tsunami education efforts, based on existing national compilations or assessments. Such compilations and assessments currently do not exist for tsunami education efforts in the United States, nor do National Tsunami Hazard Mitigation Program (NTHMP) criteria for evaluating the effectiveness of the information content, style, process, and dissemination of tsunami education efforts. It is beyond the scope of this committee to develop such an inventory or evaluative criteria, therefore a thorough assessment is not possible. Instead, in an effort to simply demonstrate the breadth of current tsunami education efforts, the committee compiled the following list based on received information and online searches. This list is not designed to be exhaustive, as NTHMP plans to develop such an inventory and web-based repository in the coming years (as noted in its 2009-2013 strategic plan). This list serves to illustrate the range of education efforts and is organized by passive education (e.g., books, brochures), active education (e.g., workshops, curriculum), education training, and online resources.

PASSIVE EDUCATION DESIGNED FOR AT-RISK POPULATIONS

- **Books:** Several books have been written to teach at-risk populations about tsunamis and how to survive future events. One example is *Tsunami Survival—Lessons from Chile,*[1] which uses past tsunami disasters in Chile and Japan to educate people in the United States about tsunamis related to subduction-zone earthquakes. Another is *Tsunami!,*[2] which documents past tsunami disasters in Hawaii, including first-hand accounts from tsunami survivors. Books designed to educate children include coloring books (e.g., Tommy Tsunami), a tsunami trivia activity sheet developed by NTHMP, and the *Tsunami Warning!* book developed by the Intergovernmental Oceanographic Commission (IOC).
- **Brochures:** Several NTHMP members (e.g., Oregon, Washington, Puerto Rico) use brochures to disseminate tsunami evacuation maps and preparedness information, such as background information about tsunamis, safety tips and instructions on what to do in case of a tsunami, and National Oceanic and Atmospheric Administration (NOAA) Weather Radio (NWR) information. Brochures can be placed where people visit, including visitor centers, hotels/motels, ferry terminals, medical offices, libraries, local businesses, community centers, or even utility offices. Although they are passive educational instruments, brochures do provide a physical cue that may motivate individuals, including tourists that may lack other avenues to learn about tsunamis.
- **Business continuity guides:** The Pacific Tsunami Museum recently published *How to Prepare Your Business for the Next Tsunami,* which is a guidebook for businesses on preparedness and post-disaster continuity planning.

- **Household planning guidelines:** The Washington State Emergency Management Department (WEMD) has published *How the Smart Family Survived a Tsunami,* which discusses tsunami education and preparedness for households in tsunami-prone areas.

- **Paraphernalia:** The NTHMP has supported the development of tsunami education products, such as a heat-sensitive coffee mug, pens, bookmarks, family disaster cards, hazard zone decals, tent cards, trivia cards, videos, and posters. The NTHMP members can order products from a catalog, and several states disseminate products at community fairs and workshops. These products serve to disseminate common tsunami images or messages among NTHMP members.

- **Signage:** Although signage along roadsides or in public places is primarily for identifying tsunami hazard zones and evacuation routes, they may have other educational benefits, such as generating media attention, providing a physical cue to motivate individuals to learn more about tsunamis, injecting tsunamis into hazard mitigation discussions, and disseminating consistent messages across jurisdictions.

- ***TsuInfo* Newsletter:** The Washington Division of Geology and Earth Resources produces *TsuInfo,* a bimonthly newsletter with more than 350 subscribers, to provide current tsunami and research information to local emergency managers, local officials, and others involved in tsunami mitigation efforts.

- **Videos:** Several states have developed tsunami-related videos, including "Run to High Ground" (Washington), "Cascadia," "The Forgotten Danger" (Puerto Rico), and "Tsunami! Surviving the Killer Wave," which can be found in schools and libraries. The Pacific Tsunami Museum worked with others to produce and broadcast three tsunami-safety Public Service Announcements statewide in 2006.

ACTIVE EDUCATION DESIGNED FOR AT-RISK POPULATIONS

- **Community workshops and town hall meetings:** Local emergency managers and state/federal officials organize community workshops and town hall meetings to discuss tsunami hazards, societal vulnerability to these threats, individual and household preparedness, and tsunami warning and evacuation procedures. Local newspapers and other media often advertise these forums and in many cases carry them live over local TV and radio. Workshops provide the public with the opportunity to interact and discuss tsunami topics with scientists, emergency managers, and local officials. Workshops tailored for specific groups (e.g., elected officials, businesses, state agencies, and tribes) can be designed to address educational and preparedness issues specific to their needs. A community-specific tsunami education workshop was held in Tokeland, Washington, in 2008 to address informational needs of a tribe and small unincorporated town. The workshop provided participants with tsunami information specific to their community and led to a review of the community's level of preparedness.

- **Pacific Tsunami Museum:** The Pacific Tsunami Museum (http://www.tsunami.org/) is a nonprofit organization located in Hilo, Hawaii, and is dedicated to promoting public tsunami education for the people of Hawaii and the Pacific Region. The museum combines scientific information and oral histories of tsunami survivors through a series of exhibits and lectures that explain the tsunami phenomenon, tsunami myths and legends, the Pacific Tsunami Warning system, and public safety measures.
- **School curriculum:** Several tsunami education curriculums exist for various school grades. The WEMD has developed tsunami curriculum for grades K-6 and 7-12 that has been distributed both nationally among NTHMP members and internationally via the International Tsunami Information Center (ITIC). In the United States, each state has different laws governing education requirements, and in many cases it is left to the school district or teacher to decide what will be taught in the classroom. A curriculum for northern California schools has also been developed in collaboration with the Redwood Coast Tsunami Work Group. The Alaska Tsunami Education Program (http://www.aktsunami.com/index.html) provides student resources, lesson plans (K-12), multimedia resources, and teacher training workshops. Middle school curriculum has also been developed for schools in Guam. *TsunamiTeacher* is a web-based, distance-learning product (http://ioc3.unesco.org/TsunamiTeacher/) of the ITIC that consolidates tsunami education materials together at the global level and includes education modules, notes for trainers, workshop formats and evaluation materials, media resources, and mitigation resources. The ITIC provides access to the WEMD school curriculum, as well as the ITIC-developed "I Invite You to Know the Earth" education series for pre-school, 2nd to 4th grades, and 5th and 8th grades. Tsunami-related lesson plans are also available at The Bridge, an online teacher's resource provided by NOAA Sea Grant and the National Marine Educators Association (http://www2.vims.edu/bridge/DATA.cfm?Bridge_Location=archive0105.html) and at the Discovery Channel's *Tsunami Teacher's Guide* (http://school.discoveryeducation.com/teachers/tsunami/). School curriculums for tsunamis have been developed also by the Department of Education and Training, Government of Australia (http://www.det.wa.edu.au/education/cmis/eval/curriculum/pathfinders/disasters/tsunamis/).
- **School drills:** In addition to curriculum, many U.S. schools in tsunami hazard zones participate in earthquake/tsunami drills on a yearly basis and invite emergency managers to give talks in the classroom and at Parent-Teacher Association (PTA) meetings.
- ***Stop Disasters* simulation game:** The UN International Strategy for Disaster Reduction (ISDR) developed a tsunami module for its disaster simulation game *Stop Disasters* to teach children how to build communities that are resilient to tsunamis (http://www.stopdisastersgame.org/en/home.html). The interactive, web-based disaster simulation game has players decide how and where to build communities and what additional preparedness and mitigation strategies are needed to protect at-risk populations. A teacher's guide is also included to incorporate the game into lesson plans.
- **State and county fairs:** Fairs can be effective outreach opportunities, especially for individuals who may not live in tsunami-prone areas but are likely to visit these areas.

Products that include tsunami information (e.g., posters, coffee mugs, magnets) can be given away, thereby bringing tsunami reminders into people's homes and businesses. A recent example is the *Tsunami-Safe* fair (http://www.tsunami.org/tsunami_fair.html) held April 2008 in Hilo, Hawaii.

- **Walking guides:** The Pacific Tsunami Museum developed the "East Hawaii walking and driving tour of historical tsunami sites" self-guided tour that includes a brochure and signage at particular sites.

EDUCATION TRAINING

- **Community education guidelines:** The Pacific Tsunami Museum in Hilo, Hawaii, recently published "Tsunami Education: A Blueprint for Coastal Communities," which is designed to help local officials and interested citizens develop tsunami outreach efforts in their communities. The WEMD promotes and trains interested individuals and neighborhood groups on the "Map Your Neighborhood" initiative, which focuses on developing social networks within neighborhoods that ideally lead to greater collaboration in preparing for and responding to disaster.

- **Media outreach and kits:** Radio, television, newspapers, and magazines can be used to educate the public on tsunami preparedness and mitigation efforts; for example, a local cable TV station in Ocean Shores, Washington, hosts a monthly forum with elected officials to discuss tsunami preparedness issues. In response to issues identified during the June 2005 tsunami warning for the West Coast, the WEMD has developed a broadcasters' tsunami emergency guidebook to train broadcasters on the notification process used to send tsunami alerts to the public. The goal of this guidebook is to ensure a consistent message is being sent by all outlets in a tsunami warning process. In Washington, WEMD and the National Weather Service (NWS)-Seattle have visited all broadcasters in Seattle and the majority of the Washington coast to train media staff and provide them the guidebook. The book is updated yearly and is being used as a template for other NTHMP partners.

- **Tourist lodging training:** Several states and territories are working with lodging facilities to provide tourists with tsunami information and what to do in case of tsunamis. The Oregon Department of Geology and Mineral Industries (DOGAMI) has developed a tsunami preparedness guide for Oregon lodging facilities that contains information about tsunamis and their origins, tsunami hazards on the Oregon Coast, preparedness information from the American Red Cross, and evacuation graphics that lodging facilities can provide to guests. A similar disaster response guidebook has been developed by the WEMD for hotels and motels on Washington's Coast. In addition to the guidebook, the WEMD holds workshops with lodging facilities in tsunami hazard zones to discuss the earthquake and tsunami hazard threat, to provide employee training on tsunami warning and evacuation, to set up a NOAA Weather Radio, and to provide education materials to be placed in rooms.

- **"Train the trainers" workshops:** The WEMD and the NTHMP have sponsored workshops designed to train individuals in at-risk communities on how to educate others in their community. Topics include tsunami origins and hazard assessment, societal vulnerability to tsunamis, risk reduction strategies, overviews of the tsunami warning centers, evacuation procedures, and risk communication strategies. The intent is to have local community members spread tsunami-related information through existing social networks to friends, family members, and co-workers.

ONLINE RESOURCES

- **Children of Tsunami:** The Children of Tsunami (http://www.childrenoftsunami.info/) is a multimedia project that tracks Asia's recovery from the 2004 Indian Ocean tsunami disaster through the experiences of eight children and families in India, Indonesia, Sri Lanka, and Thailand. It personalizes the tsunami threat and consequences in other parts of the world by providing at-risk populations the ability to watch real people recover from a tsunami disaster.
- **FEMA for Kids—Tsunamis:** This website (http://www.fema.gov/kids/tsunami.htm) provides background on tsunamis, as well as general, multi-hazard information on disaster supply kits, how to protect your home from disasters, and how to protect pets.
- **Global Education Tsunami Education Toolkit:** The Global Education website (http://www.globaleducation.edna.edu.au/globaled/page1637.html) is funded by the Australian Agency for International Development (AusAID) to support its Global Education Program. The tsunami education toolkit provides background for teachers and students on tsunami issues, as well as larger contextual issues of poverty and development. The website includes worksheets for primary and secondary schoolchildren, teacher activities, case studies, and pet safety plans.
- **International Tsunami Information Center:** The ITIC (http://ioc3.unesco.org/itic/) was established by the IOC of the United Nations Educational, Scientific, and Cultural Organization (UNESCO) and is located in Honolulu, Hawaii. One of its primary missions is to be a clearinghouse for the development of tsunami education and preparedness materials. The ITIC provides a research library, educational materials, media resources, posters, a tsunami glossary, a children's cartoon book, posters, safety flyers, a textbook for 4th to 6th grades, videos, and links to other online resources.
- **NOAA Office of Education—Tsunami Education Resource Kit (TERK):** The TERK (http://www.oesd.noaa.gov/terk_intro.htm) includes brochures, curricula, historical background information, multimedia visualizations, National Science Education (NSE) standards, scientific publications, preparedness strategies, and press releases.
- **NOAA Tsunami Program Education:** This website (http://www.tsunami.noaa.gov/education.html) provides information on the TsunamiReady program, background information on tsunami terms and the warning systems, and links to teacher resources.

REFERENCES

1. Atwater, B.F., S. Musumi-Rokkaku, K. Satake, Y. Tsuji, K. Ueda, and D.K. Yamaguchi. 2005. *The Orphan Tsunami of 1700: Japanese Clues to a Parent Earthquake in North America.* U.S. Geological Survey Professional Paper 1707, U.S. Geological Survey, Reston Virginia.
2. Dudley, W.C. and M. Lee 1998. *TSUNAMI!* University of Hawaii Press, Honolulu, Hawaii.

June 14, 2005: A Case Study in Tsunami Warning and Response

At 7:51 PDT (0251 UTC) on June 14, 2005, an earthquake occurred 90 miles northwest of Eureka, California, that was felt in communities in both California and southern Oregon. The preliminary magnitude of the earthquake was M7.4 and the location was within the Gorda Plate. Within five minutes, the West Coast/Alaska Tsunami Warning Center (WC/ATWC),[1] in conformance with established procedures, issued a tsunami warning for the coastal areas from the California-Mexico border to the northern tip of Vancouver Island, British Columbia. Before the warning was issued, the intensity and duration of the earthquake shaking in Humboldt and Del Norte counties provided the first "unofficial notification" to residents of northern California and southern Oregon of a possible tsunami triggered by an earthquake.

TWO WARNINGS FROM THE TWCS

Although no WC/ATWC staff were in the center facility at the time of the earthquake, within a few minutes after the earthquake TWC staff were at their stations, had assessed the seismic data, and delivered a warning message to its area of responsibility (AOR) regarding possibility for tsunami waves forming in response to the earthquake tremors. As noted above, the earthquake provided the initial notification of the potential for a tsunami to local residents, resulting in local individual actions well before the official warning was received and disseminated by local governments. At 7:59 PDT (0259 UTC), the Pacific Tsunami Warning Center (PTWC) in Hawaii issued a Bulletin for its AOR (including Mexico) stating that there was no tsunami warning in effect (for its AOR). Although the two seemingly contradictory notifications were both correct, state and local officials who received both were left with the impression that the initial notification was canceled. Officials in southern California were faced with the potential of a locally damaging tsunami striking San Diego county that would not impact Tijuana, Baja California, a few miles to the south.

By 8:19 PM, after further analysis, the Gorda earthquake was judged as not likely to be tsunamigenic (the earthquake magnitude was refined to measure M7.2 and was located in the middle of the Gorda Plate, not on the Cascadia subduction zone (CSZ) boundary between the Gorda Plate and the North American Plate). Additional data confirmed that there were no reports of wave inundation along the coast, and the WC/ATWC reported a widespread tsunami unlikely. However, it did not rule out the possibility for a regional tsunami or landslide-

---

[1] Notification of potential tsunami events for the coastal communities of Washington, Oregon, and California are within the Area of Responsibility (AOR) of the West Coast/Alaska Tsunami Warning Center in Palmer, Alaska. Notification of Mexico is within the AOR of the Pacific Tsunami Warning Center in Hawaii.

triggered tsunami. Within an hour of the earthquake, the WC/ATWC canceled the warning for the coast areas of British Columbia, Oregon, Washington, and California.

This tsunami warning came only six months after the horrific tragedies, which took place off the coast of the Indian Ocean during the 2004 tsunami at a time of heightened national fear of tsunami activity. This warning would be the first to be issued for the West Coast for a regional event. Adding to the urgency of decision making action by the TWC was the size and location of the earthquake, the potential for either earthquake- or landslide-generated tsunamis, and the need to issue a statement before potential inundation occurred along the coast (within minutes of the earthquake).

THE EVENT AND RESPONSE IN CALIFORNIA

The June 14, 2005, Gorda Plate earthquake (M7.2) violently shook a wide area of Del Norte and Humboldt Counties, including the cities of Eureka and Crescent City. In assessing the consequent actions at the state and local levels to the tsunami warning it is critical to recognize that the "event" was both a local, widely felt, potentially damaging earthquake, and a potentially damaging near-field tsunami that was the subject of a tsunami warning issued by the WC/ATWC. As noted above, within five minutes of the earthquake, officials at the WC/ATWC issued a tsunami warning to the California State Warning Center (CSWC) both verbally over the National Warning System (NAWAS) and as printed copy over the National Oceanic and Atmospheric Administration/National Weather Service (NOAA/NWS) Weather Wire. These messages were communicated verbally to impacted counties over California's equivalent to NAWAS, the California Warning System (CALWAS) and by the California Law Enforcement Telecommunications System (CLETS), a teletype service connecting local law enforcement agencies to the California Department of Justice. At the same time that official notification to the CSWC and local governments was occurring, the WC/ATWC notification was being automatically transmitted over the Weather Wire to subscribers of email and pager notification services, over the Emergency Managers Weather Information Notification (EMWIN) service, over the California Emergency Digital Information Service (EDIS) to radio and television newsrooms, and over the California Integrated Seismic Network (CISN) Display to local emergency operations centers.

The WC/ATWC warning was transmitted over the California Law Enforcement Telecommunications System to 22 coastal counties (15 counties directly on the coast and 7 counties that would be impacted by flooding in bays and estuaries), California Highway Patrol (CHP), and State Parks and Recreation. Within 10 minutes following the notification over CLETS, CALWAS was used to issue an initial warning to local emergency responders in the surrounding coastal area. Unfortunately at the time in which this took place, the CSWC was not sufficiently staffed, leaving only two employees to handle the surge of emergency calls. The local recipient of the CSWC notifications is the designated Public Safety Access Point (PSAP), usually the county 911 office and/or fire, law, and emergency medical dispatch, and for the CLETS teletype messages, local law enforcement agencies.

CSWC and Office of Emergency Services procedures provide for secondary notification and verification of receipt of warning notifications by local emergency managers. Within minutes of the initial notification of the CSWC by the WC/ATWC, Office of Emergency Services (OES) staff were paged and instructed to verify that the notification was received by the local government emergency managers. This verification process was not able to be completed in the northern most counties (those directly impacted by the earthquake and with the greatest potential for being affected by a tsunami) where telephone service was limited by excessive local use after the earthquake and where hundreds of residents had called 911 dispatch centers to report the earthquake. In Crescent City, Del Norte County, a single 911 dispatcher was overwhelmed by the call volume and was not able to receive calls from state personnel verifying receipt of the tsunami warning.

Given the short time span in Humboldt and Del Norte counties between earthquake shaking, issuance of the tsunami warning by the WC/ATWC, and possible tsunami wave arrival, this demand made it difficult for the CSWC and OES to expedite the warning process, a problem that was exacerbated by overloaded wire and cell telephone systems, inadequate staffing at the state and local government emergency operations centers, and limited training at the local government level.

Despite the confusion that was prompted by the conflicting message sent by the Pacific Tsunami Warning and Mitigation System (PTWS), the CSWC transmitted only the first TWC (from the WC/ATWC) warning using CALWAS. The NWS, following NOAA procedure, activated the Emergency Alert System (EAS) and broadcast the tsunami warning to the potentially impacted counties.

Six minutes following the transmission of the tsunami warning, the CSWC held a conference call between the OES Costal Regional Administrator and earthquake/tsunami specialists to remediate the confusion with personnel from counties that might be impacted. As noted above, overload of the local telephone system precluded OES staff from making contact with local officials in Humboldt and Del Norte counties. The state's satellite telephone system (OASIS) linking OES regional offices and county PSAPs could not be immediately utilized because it could not be accessed from staff residences (the event occurred when OES region staff were at home, as were most local government officials).

Another element of the states' communication procedures with local governments is the convening of conference calls between local government emergency managers, state officials, and appropriate hazard experts at the time of the issuance of alerts to local government. The state maintains multiple 30-port conference bridge lines for this purpose. Attempts to use the conference call procedures at the time of the tsunami warning were unsuccessful because inbound lines with the scientists at the WC/ATWC were overloaded and local government officials could not be contacted.

As noted above, the population of the impacted areas in northern California was "notified" by the earthquake, by broadcasts over the NOAA All Hazards Radio system, and by stations participating in the EAS. In Crescent City, the tsunami sirens were not sounded until 8:30 PDT (0330 UTC) because of the telephone saturation of the 911 dispatch center and the inability of the single staff person on duty to handle conflicting workload priorities. Spontaneous evacuations took place in several communities. In many of the northern California communities, where radio and cable television are not locally controlled, the EAS notification was not picked up and broadcast.

At approximately 9:11 PM PDT, the CSWC received a second message from the WC/ATWC, cancelling the tsunami warning after confirmation that a tsunami had not been generated by the earthquake. The CWSC transmitted the notification of the cancellation via CALWAS to the surrounding communities (noncoastal counties), and the CLETS network notified local government agencies in the potential tsunami zone.

THE EVENT AND RESPONSE IN WASHINGTON STATE

Just as California was affected by the tsunami warning on June 14, the state of Washington was also under alert because the WC/ATWC notified it of its assessment for the generation of possible tsunami waves at approximately 7:55 PM PDT. Upon the arrival of this warning, the Washington State EAS, with input from websites and buoy data, decided to inform local officials of the warning of tsunami activity. Unlike the case in California, officials in Washington found the NOAA message transmitted through the WC/ATWC to be clear and had no problems deciphering the message. Minutes later, the news media picked up the message from the PTWC, seemingly contradicting the initial tsunami warning. This conflicting message in addition to the unofficial sources commenting on the progression of the events to the public caused much confusion among the people of Washington. Upon the next hour, Emergency Operation Center (EOC) officials in Washington contacted local law enforcement divisions and other emergency contacts, while the NWS transmitted the primary tsunami warning via EAS to the counties under alert. The warning was also broadcast on television and AM/FM radio stations in the surrounding area. In the state of Washington, under "Home Rule," local officials are responsible for evacuation orders. Several counties did choose to enact evacuation orders for their communities through local EAS systems. At 8:40, an official State EMD Public Information Officer spoke to several radio and television stations regarding the evacuation suggestions, reminding citizens and officials that the warning of evacuation was only set in place for those living "on the beach" or in "low lying areas." Approximately 20 minutes later, the State EOC received the tsunami cancellation message, which it transmitted to the public via NAWAS 5 minutes later.

Fortunately, no major tsunami occurred and the events of June 14 served as a "stress test" for the notification system technology, the training of warning personnel, and the response of the public to a potential tsunami. The following were important lessons learned:

- The format and content of the information and warning statement from the two TWCs with shared responsibilities for coast lines in the Atlantic and Pacific basin can cause confusion. Tijuana in Baja California was not at risk, but San Diego and Coronado, a few miles to the north in California, were at risk.
- Managing information flow from multiple sources (e.g., media, warning centers, NAWAS, etc.) and reducing confusion requires trained and dedicated staff and the

ability to respond to the surge in demand by a sudden onset event like an earthquake and tsunami warning. Ongoing training of officials, the media, and the public is essential, accompanied by documented procedures, checklists, and predetermined priorities for actions.

- An earthquake will severely disrupt telecommunications and challenge even redundant systems to perform their dissemination responsibilities.

Magnitudes from C. Richter to M$_{wp}$ and the W phase

The process of assigning a magnitude to a seismic event is far from simple, all the more so in the case of the large earthquakes that can generate tsunamis. Originally, Charles Richter[1] used the concept of measuring the maximum ground motion amplitude recorded from local events on a standard Wood-Anderson seismograph. The determination is quite straightforward, involving the measurement of the largest amplitude of the pen during the earthquake in microns—for earthquakes this is generally the shear (or secondary [S]) wave. The amplitude is empirically corrected for distance in southern California, and the observations from several stations are averaged to increase the statistical stability of the measurement. Richter thus defined a regional magnitude scale for earthquakes, which is now referred to as a local magnitude M_L. The definition of M_L was very important because all the subsequent magnitude scales have been tied to this initial algorithm. The concept was soon extended worldwide using a combination of measurements on body and surface waves at teleseismic distances, leading to the definition of two standardized algorithms, a body-wave magnitude m_b measured on short-period P-waves at a target frequency of 1 Hz, and a surface wave magnitude M_s measured at a period of 20 s. These early algorithms were a largely empirical endeavor because of the use of simple models of seismic sources and wave propagation.

In the late 1950s and early 1960s, the work of Vvendeskaya in Russia and Aki in Japan showed that earthquake sources could be described by a relatively complex system of forces, expressed in physical units of dyne*cm or N*m as a "seismic moment" M_0, and directly related to the total amount of slip occurring on the fault plane, integrated over the full surface area of faulting. Unlike seismic magnitudes, seismic moments are directly related to the physical properties of the source. In order to facilitate comparisons with existing catalogues (and also in the process to facilitate communication with the general public), Kanamori[2] and Hanks and Kanamori[3] proposed to recast seismic moment values into a "moment magnitude" scale, M_w, using $M_w = 2/3 (\log_{10} M_0 - 9.1)$, where M_0 is in N*m (or fault area times displacement times material rigidity).

The problem of assessing earthquake size in the context of tsunami warning is several-fold:

(1) The most efficient algorithms for seismic moment inversion require the use of large datasets (in practice, tens to hundreds) of long-period (low-frequency) surface waves, which unfortunately travel slowly and thus delay the warning process.

(2) The conventional magnitude scales, which target relatively short periods (1 and 20 s), are not representative of the low-frequency part of the source spectrum, which controls the excitation of the tsunami. Because of the frequency dependence of the seismic source, m_b and

M_s are theoretically expected and experimentally observed to saturate (around 6.3 and 8.2, respectively) with increasing seismic moment and are essentially useless for estimating the size of the mega-earthquakes capable of producing tsunamis that cause damage in the far field.

(3) If all earthquakes obeyed scaling laws, the measurement of one or another magnitude should in principle be equivalent, and an analyst should be able to predict the low-frequency value of the seismic moment by measuring the source in a different frequency band. However, earthquakes with similar moments produce widely scattered estimates of magnitude, and "tsunami earthquakes" feature anomalous source characteristics. The observational challenge is to somehow identify those in real time.

(4) While the goal of tsunami warning is to quantify (e.g., hypocenter, magnitude, focal mechanism, and fault extent) the earthquake as quickly as possible upon detection, it is also imperative to record the source in its entirety in order to assess its full tsunamigenic potential. Bearing in mind, for example, that the source of the 2004 Sumatra earthquake lasted eight minutes, we realize that assessing its size within five minutes is at best a challenge and at worst an impossible task. Unfortunately, there is really no consensus among seismologists as to the deterministic nature of earthquake rupture, namely whether the early stages of nucleation of a large earthquake carry a fingerprint of the eventual true size of the event. Indeed, several examples of delayed sources (e.g., 2001 Peru and 2006 Kuril Islands, both having generated destructive tsunamis) reveal a sudden increase in seismic moment release as late as one or two minutes into their source process; they constitute another class of events violating scaling laws. In lay terms, at the initiation of a seismic rupture, does Mother Nature really know how large the final product will be? Yet it is that final product that will control the tsunami and that the watchstanders at the Tsunami Warning Centers (TWCs) are charged with estimating, as swiftly and as reliably as possible.

5. Seismic data and sophisticated processing are insufficient to determine the destructive-ness of tsunamis. Guisiakov uses the Soloviev-Imamura tsunami intensity scale based on run-up data to show there is only a tendency of increased tsunamis with an increase in earthquake magnitude. The lack of direct correlation can be attributed in part to secondary mechanisms (submarine slumps and slides) in the generation of tsunamis. This is shown in the findings by Plafker where submarine landslides account for many large and destructive tsunamis.

THE M_{WP} ALGORITHM

The application of geometrical optics to seismology reveals that the earth's ground motion resulting from the passage of P-waves in the far-field is related to the time derivative of the history of the deformation or physical slip at the source. In other words, if a permanent deformation (in the form of a step in displacement) is incurred at the epicenter, the far-field signal will register an impulse (or spike) of short duration, followed by a return to quiescence. Conversely, the deformation at the source should be obtainable by mathematically integrating the ground displacement over time in the far field, and by performing a number of theoretically justifiable corrections, which account, for example, for the path from epicenter to receiver. As most seismic instruments are sensitive to the velocity of ground motion rather than to its displacement or acceleration, in practice a double integration is required. Generally, seismic instrumentation becomes increasingly insensitive at very low frequencies when the earth's noise is increasing and integral estimates of displacement are, perforce, inaccurate.

This is the basis of the so-called M_{wp} algorithm[4] by which initial estimates of the seismic moment of a large earthquake can be inferred from the earliest-arriving seismic signals recorded in the far field. M_{wp} has been implemented and currently constitutes the procedure in use at both the Pacific Tsunami Warning Center (PTWC) and the West Coast/Alaska Tsunami Warning Center (WC/ATWC) for the purpose of estimating earthquake sources for tsunami-genic potential.

SHORTCOMINGS OF THE M_{WP} ALGORITHM

As compelling as the M_{wp} concept may be, it suffers from having never been subjected to the necessary and independent exercise of being tested on synthetic seismograms (i.e., computational renditions of the wavetrains expected to be recorded on a given seismic instrument for a particular scenario of earthquake size, rupture parameters [including for example the case of "tsunami earthquakes"], and receiver geometry). Obviously, the number of great and mega-earthquakes is also highly limited, thereby reducing the opportunities for applications of statistics. The theoretical understanding of seismic sources has reached the point where a number of methods are available to produce these synthetic seismograms. To the best of the committee's knowledge, vetting of the M_{wp} algorithm for tsunami inference was not done prior to its use at the TWCs.

Indeed, it would be expected that the use of M_{wp} will encounter significant and systematic problems both for very large events and for the tsunami earthquakes that demonstrably do not follow scaling laws; that is, precisely those earthquakes carrying enhanced tsunami potential given their size. Tsuboi et al.[4] compared M_{wp} to M_w, but only to sources that had magnitudes less than 8.0 (neither *great* nor *mega*-earthquakes). An M_{wp} algorithm is described in the open source (source forge) website (http://seismic-toolkit.sourceforge.net/) with a comparison of M_{wp} to M_w. For events larger than $M_w = 8.25$, the M_{wp} magnitude significantly underestimates the magnitude and the moment-based M_w measure. As is the case for all other magnitude measures, M_{wp} also saturates as earthquake moment increases. In addition, the development of an efficient algorithm providing automated routine measurements of M_{wp} requires extending a number of its parameters in ways that lead to further problems for larger earthquakes.

There are two categories of problems with the M_{wp} method: systematic problems inherent in the theory underlying the method, and shortcomings of the particular modalities of implementation used at the TWCs. Among the former:

(1) The domain of applicability of the theory, known as geometrical optics, requires in principle the use of wavelengths much shorter than the characteristic dimensions of the structures involved. This principle is violated for large sources such as the Sumatra earthquake whose

fault length (1,200 km) was more than a third of the thickness of the earth's mantle and seismic wavelengths substantially larger than major structural features.

(2) Similarly, the theory assumes a point source and neglects the significant interference effects between the individual elements of an extended fault such as those capable of generating significant tsunamis.

Among the latter:

(3) A practical computation must be limited to a finite time window, the selection of which requires a very subtle compromise between the necessary exclusion of later arriving seismic phases and the use of an interval at least as long as the duration of the seismic rupture at the source.[5] This compromise may not be possible for extremely long sources (eight minutes in the case of the 2004 Sumatra earthquake).

(4) The algorithm implementing M_{wp} at the PTWC uses a distance correction adequate for an infinite homogeneous elastic space rather than a more realistic description.

(5) Similarly, the operational algorithm elects to carry out the integrations in the time domain, which assumes that the response of the seismic sensors is perfectly flat to ground velocity. However, even the best seismic instrumentation (e.g., the Streckeisen STS-1) falls off rapidly in its response at periods greater than 360 s, which leads to a systematic and predictable underestimation of the largest earthquakes, whose true size is expressed only at the longest periods where the instrument begins to decline in response (see below in discussion of W phase).

Such limitations in the operational aspects of M_{wp} were actually recognized by the operators of the centers during the development of their algorithms, through a comparison of their results with published earthquake magnitudes obtained by the Harvard CMT project using geophysical inverse procedures.[6] These authors documented that M_{wp} measurements become increasingly deficient when the magnitude of the earthquake increases, and they proposed to incorporate in the final algorithm (i.e., the one presently used at the TWCs) a linear correction for this effect, defined from an empirical regression of the misfit of the values from their initial dataset with respect to the published reference values.

In summary, the committee expresses concern that the TWCs rely on a single technique applied without sufficient attention to its limitations—both inherent in its concept and resulting from the particular algorithm chosen for its implementation. Furthermore, all magnitude measures will saturate as the seismic moment increases and will not be able to quantify potential tsunami heights for great and mega-earthquakes.

INCORPORATING THE W PHASE

In 1992 during the Nicaragua tsunami earthquake, H. Kanamori identified a "W" phase wave: very-long period energy traveling along the earth's surface by multiple reflections in the mantle. Because of the ray-mode duality in wave propagation, the W phase is best thought of as spheriodal models or Rayleigh waves (combinations of shear and compressional wave propagation) with a group velocity in the range of 4.5 km/s (mantle shear velocities) to 9 km/s (mantle compressional velocities) over a frequency range of 1-10 mHz. The wave comprises the fundamental as well as the first three overtones.

Figure G.1, reproduced from Kanamori and Rivera,[7] shows a synthetic seismogram (red) from the supervision of spheroidal modes mentioned above. Kanamori and Rivera[7] note that at a distance of 50°, the W phase energy is contained within 23 min of the origin time and in time to have a positive impact on tsunami warnings. The W phase arrives following the P-wave and before conventional surface waves, and can be regarded as a very low frequency, fast propagating group of waves. Following a number of investigations, the systematic use of the W phase has now been implemented at the National Earthquake Information Center (NEIC) of the U.S. Geological Survey (USGS) in Golden, Colorado, and it may constitute a significant improvement as a complement to the existing algorithms in use at the TWCs.

In practice, time domain deconvolutions are used to extract the W phase from the data from stations in the Global Seismic Network (GSN). The deconvolution seeks to flatten the velocity spectrum over the broad frequency range above (1-10m Hz) and integrate the result to get displacement. Unfortunately, this is an inherently noisy process, which has to deal with the 360 s (2.8m Hz) corner in the velocity response of the GSN-standard STS-1 seismometer (Figure G.2).

The mass in a seismometer, whether vertical or horizontal, is caused to move by an applied force, which may be associated with the motion of the seismometer frame, or changes in the force of gravity. The linear relationship between force and acceleration causes the mass to move with the applied acceleration. The force and acceleration, of course, disappear following an earthquake so the static acceleration is zero. Most seismometer responses, including the GSN STS-1, are shaped by a judicious choice of poles and zeroes in the feedback electronics to be flat to velocity over some frequency range.

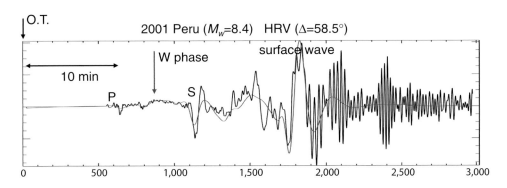

FIGURE G.1 Phase from the 2001 Peruvian earthquake (Mw = 8.4) recorded at Harvard University Seismic Station (HRV), and the synthetic W phase computed by mode summation using the Global Centroid-Moment Tensor (GCMT) solution. SOURCE: Kanamori and Rivera, 2008; with permission from John Wiley and Sons.

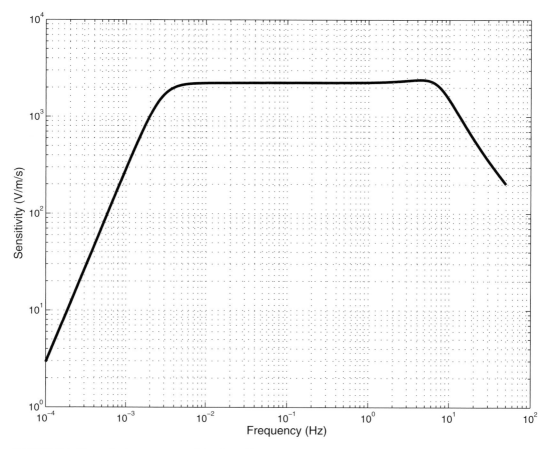

FIGURE G.2 The velocity response of the Streckheisen STS-1 seismometer including the feedback poles and zeroes used to shape the spectrum as discussed in the text. SOURCE: Derived from a standard model of the STS-1 velocity response; with permission from Pete Davis, IRIS/IDA Project.

The response of the GSN standard seismometer has been shaped to be nominally flat to ground velocity between about 2.8 mHz and 5 Hz. For frequencies below 2.8 mHz, the seismometer's output has been shaped so that at zero frequency its output is zero. In the W phase band (1-10 MHz) the response is critically dependent on the exact parameters of the seismometer's feedback circuit.

In a very real sense, the GSN standard seismometer was not designed to support a simple deconvolution of velocity to displacement. Furthermore, many of the STS-1s in the GSN are now more than two decades old and, because the STS-1 is no longer manufactured, spares are not available. More details can be found at http://www.iris.edu/hq/gsn/quality, and a report "The IRIS/GSN Data Quality Initiative: Assessment of and proposed metrics for the GSN dataset" is in draft form. The USGS and the National Science Foundation (NSF) must seriously consider the renovation of the GSN in future years including the requirements for W phase deconvolution.

A new seismometer has been developed and tested, which senses mass position through interferometry using fiber optics for light transmission. Unlike the STS-1 above, a force balance feedback is not used to reduce mass movement to maintain linearity in the displacement sensor. The resultant dynamic range is much greater than a conventional seismometer and is achieved by counting interference rings with a Mickelson interferometer. The output can be shaped computationally as needed and could be used to provide data with high fidelity at low frequencies for measuring the W phase.[8] The optical seismometer has a response that is flat to ground acceleration between DC and about 1 Hz. There is, thus, no need to deconvolve the instrument response for W phase band measurements. Another of the benefits of the optical approach is that with good response at tidal frequencies, absolute calibration against earth tides on a continuous basis is straightforward. The optical seismometer remains under development for horizontal component testing and reducing noise levels at low frequencies—a borehole version is being tested.

To increase the longevity of the STS-1 seismometer, replacement feedback circuitry has been developed to replace the aging electronics (http://www.metrozet.com/). The corner frequency of the STS-1 remains at 1/360 Hz for the new electronics.

PTWC staff indicated that they are in the process of implementing a W phase algorithm, but a careful vetting of the algorithm before it can be reliably applied will be required.

> **Recommendation:** Before implementing the W phase algorithm in TWC operations, the NOAA Tsunami Program should validate the W algorithm to both a sufficient dataset of synthetic seismograms and to waveforms from past great earthquakes, paying particular attention to its performance in "tsunami earthquakes" and to the assessment of a lower-magnitude bound for its domain of applicability.

REFERENCES

1. Richter, C.F. 1935. An instrumental earthquake-magnitude scale. *Bulletin of the Seismological Society of America* 25(1):1-32.
2. Kanamori, H. 1977. The energy release in great earthquakes. *Journal of Geophysical Research* 82(20):2981-2987.
3. Hanks, T.C. and H. Kanamori. 1979. A moment magnitude scale. *Journal of Geophysical Research* 84(B5):2348-2350.
4. Tsuboi, S., K. Abe, K. Takano, and Y. Yamanaka. 1995. Rapid determination of M_{wp} from broadband P waveforms. *Bulletin of the Seismology Society of America* 85(2):606-613.
5. Ishii, M., P.M. Shearer, H. Houston, and J.E. Vidale. 2006. Teleseismic P wave imaging of the 26 December 2004 Sumatra-Andaman and 28 March 2005 Sumatra earthquake ruptures using the Hi-net array. *Journal Geophysical Research* 112:B11307.
6. Whitmore, P.M., T.J. Sokolowski, S. Tsuboi, and B. Hirshorn. 2002. Magnitude-dependent correction for M_{wp}. *Science of Tsunami Hazards* 20(4):187-192.
7. Kanamori, H. and L. Rivera. 2008. Source inversion of W phase: Speeding up seismic tsunami warning. *Geophysical Journal International* 175(1):222-238.
8. Zumberge, M., J. Berger, J. Otero, and E. Weilandt. 2010. An optical seismometer without force feedback. *Bulletin of the Seismological Society of America* 100(2):598-605.

Tsunami Earthquakes

In 1972, H. Kanamori defined a special class of earthquakes, which he called "tsunami earthquakes," whose tsunamis are significantly larger than expected from their seismic magnitudes, especially conventional ones. Such events generally feature an exceptionally slow progression of the seismic rupture along the earthquake fault and can be very treacherous because they lack the high frequencies felt by humans in the near-field, which serve as a natural warning for local populations, while hiding in their enhanced low-frequency spectrum the capability to generate disastrous tsunamis. Examples include the catastrophic events in Sanriku (Japan, 1896) and Unimak (Aleutian Islands, 1946). The real-time identification of tsunami earthquakes remains a challenge in modern tsunami warning, especially because these events are relatively rare; only a dozen have been documented in the past 113 years with only five since the advent of modern digital seismometers.

A case study. On September 2, 1992, an earthquake occurred off-shore Nicaragua with magnitudes $m_b = 5.3$ and $M_s = 7.2$. Note the disparity between the body- and surface-wave magnitudes. The former meant that the earthquake was deprived of the high frequencies typical of ground shaking and felt by humans in the near-field. Indeed, in some coastal communities, the earthquake was not even felt by the population, who thus had no natural warning of the impending disaster. Its higher surface-wave magnitude indicates a "red" source, enriched in low-frequency energy, as was later confirmed by a Global Centroid-Moment-Tensor (CMT) solution equivalent to $M_w = 7.6$, measured at periods of 135 s. The earthquake generated a tsunami that ran up to more than 10 m and killed 170 people on the shores of Nicaragua.[1] Similar scenarios took place in Sanriku, Japan (1896; 27,000 dead), Java (1994, 2006), and Peru (1996); other tsunami earthquakes have been described in the Kuril Arc (1963, 1975), the Aleutians (1946), and Tonga (1982).[2]

A major challenge regarding tsunami earthquakes is to identify them in real time from their seismic records. Once an estimate of the seismic moment is obtained, the earthquake is analyzed for possible extended source duration by computing an estimate of the high-frequency energy carried in its P-waves. The result allows a comparison between the behavior of the source in the bass and treble parts of its spectrum, and if an anomaly is detected, identifies the earthquake as a violator of scaling laws, that is, as a tsunami earthquake, whose tsunami potential is greater than would be expected by its initial seismic waves. This algorithm, which uses the concept of the slowness parameter Θ,[3] has been implemented at the Pacific Tsunami Warning Center (PTWC).[4] It was used to successfully identify in real time the slowness of the Java earthquake of July 17, 2006.

Another, more general challenge is to understand the origin of the anomalous rupture in tsunami earthquakes and in particular in what geological environments they can occur. At least two different (and somewhat contradictory) scenarios have been proposed, involving the activation of splay faults rupturing in sedimentary prisms (Sanriku, Japan; Kurils),[5] or rupture propagating in a jagged mode along poorly coupled interfaces in sediment-starved environments (Nicaragua, Peru).[6] In turn, this raises the questions whether any subduction zone can be the site of a tsunami earthquake, and conversely whether the documented occurrence of such events (most often comparatively small in the magnitude 7.5 range) precludes the occurrence of mega-earthquakes as would be suggested by the available historical record in Nicaragua or Java. Despite active research efforts in this domain, we presently have no firm answers in this respect.

REFERENCES

1. Abe, K., K. Abe, Y. Tsuji, F. Imamura, H. Katao, I. Yohihisa, K. Satake, J. Bourgeois, E. Noguera, and F. Estrada. 1993. Field survey of the Nicaragua earthquake and tsunami of September 2, 1992. *Bulletin of the Earthquake Research Institute University of Tokyo* 68(1):23-70.

2. Polet, J. and H. Kanamori. 2000. Shallow subduction zone earthquakes and their tsunamigenic potential. *Geophysical Journal International* 142(3):684-702.

3. Newman, A.V. and E.A. Okal. 1998. Teleseismic estimates of radiated seismic energy: The E/M_0 discriminant for tsunami earthquakes. *Journal of Geophysical Research* 103(B11):26885-26898.

4. Weinstein, S.A. and E.A. Okal. 2005. The mantle wave magnitude M_m and the slowness parameter THETA: Five years of real-time use in the context of tsunami warning. *Bulletin of the Seismological Society of America* 95(3):779-799.

5. Tanioka, Y., L.J. Ruff, and K. Satake. 1997. What controls the lateral variation of large earthquake occurrence along the Japan trench? *Island Arc* 6(3):261-266.

6. Fukao, Y. 1979. Tsunami earthquakes and subduction processes near deep-sea trenches. *Journal of Geophysical Research* 84(B5):2303-2314.

Samoa Tsunami

The Samoa tsunami on September 29, 2009, had a devastating effect on the islands. The tsunami brought destruction to the small nation only 10-20 minutes after an earthquake about 120 km offshore shook the ground (Figure I.1). However, the tsunami had only a minor effect on Hawaii. The event illustrates (a) how a Tsunami Warning Center (TWC) is unlikely to effectively alert people of a near-field tsunami and (b) how valuable a TWC can be for monitoring a near- and far-field tsunami. It further demonstrates the importance of pre-event education to save lives from near-field tsunamis and the manner in which alert information is disseminated when a tsunami occurs.

The earthquake occurred at 7:48 AM Hawaii Standard Time (HST) and the first Pacific Tsunami Warning Center (PTWC) tsunami warning was sent to the Intergovernmental Oceanographic Commission (IOC) listserv at 8:04 AM HST, 16 minutes after the earthquake[1] (Table I.1). It announced the earthquake's preliminary parameters with an origin time of 7:48 AM HST, coordinates 15.3 South and 171.0 West, located at Samoa Islands Region, and a magnitude of 7.9. The 7.9 magnitude was later increased to 8.3 in subsequent messages.[1] The time of release of the first TWC product gave only 8 minutes of warning to American Samoa and 28 minutes of warning to Samoa, based on tsunami wave arrivals reported at the gauge stations. The delay between the earthquake and the initial message is relatively long. A preliminary earthquake assessment could have been completed as early as 2 minutes, and in the past few years the TWCs have been issuing these messages after less than 8 minutes (TWC presentation to the committee, 2008). Nevertheless, even if the message had been sent 2 minutes after the earthquake, it is questionable whether the message would have alerted the public at risk in time to allow for effective evacuation. In addition, even with a warning message reaching the public prior to wave arrival time, pre-event education would still be required to ensure proper protective action in such a short amount of time.

At 8:05 AM HST on Tuesday, September 29, 2009 the first alert message was issued from the PTWC to the Hawaiian Civil Defense Authority. Tsunami Message Number 1 announced that "A Tsunami Watch is issued for the state of Hawaii effective at 8:05 AM HST." The evaluation included in the text of this message declared that "Based on all available data a tsunami may have been generated…" and gave an estimated time arrival of a possible tsunami to reach Hawaii at 1:11 PM HST.[2] It was understood that the PTWC would send subsequent alerts as its investigation continued.

Tsunami Message Number 2 was issued 52 minutes later, at 8:57 AM HST. This message announced that "A Tsunami Watch continues in effect for the state of Hawaii" and documented an increase in the magnitude of the earthquake from 7.9 to 8.3, but indicated that the time, location, and coordinates remained the same. Additional changes between the initial message and the second message included reports of tsunami wave activity from gauge stations with

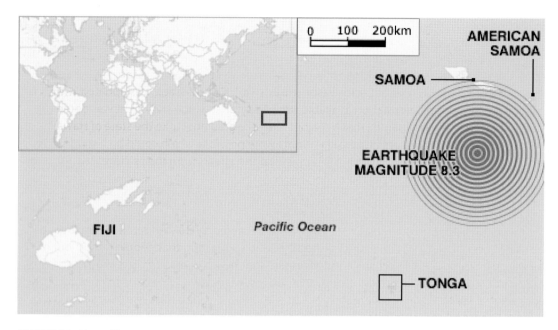

FIGURE I.1 Map of Samoa tsunami source. SOURCE: http://news.bbc.co.uk/2/hi/8281616.stm; with permission from the BBC.

TABLE I.1 Timeline for Key Events During Samoa Tsunami

Arrival time (HR:MIN:SEC) received by USGS QDDS system	Elapsed time (HR:MIN:SEC)	What
17:48:10	0:00:00	Earthquake origin time
17:48:37	0:00:27	P ARRIVAL at AFI (179km)
17:48:58	0:00:48	S ARRIVAL at AFI (179km)
17:49:50	0:01:40	Strong shaking abates at AFI
17:56:24	0:08:14	PTWC M7.1 Hypocenter
17:57:15	0:09:05	WC/ATWC M7.9 Hypocenter
18:03:15	0:15:05	WC/ATWC Tsunami Information Statement (message states 18:02 release)
18:05:10	0:17:00	PTWC Expanding Regional Warning (message states 18:04 release and M7.9)
18:07:15	0:19:05	NEIC M7.9 Hypocenter

SOURCE: Committee staff.

wave amplitude listed, which confirmed the generation of a tsunami. Nevertheless, the official evaluation stated that "Based on available data a tsunami may have been generated...."[2]

After the second message, 86 minutes later, Tsunami Message Number 3 was issued at 10:23 AM HST. The tsunami watch was officially canceled in this announcement for the entire state of Hawaii, and was replaced with a tsunami advisory. This message was updated with a significant amount of information available from additional gauge stations. It was determined, as listed in the evaluation, that a major tsunami would not strike the state of Hawaii, but that a sea level change and stronger currents were possible, which could be hazardous to people along the coastlines. The estimated time that the effects from the sea level change and stronger currents could be expected to begin was 1:00 PM HST. Based on these hazards, it was stated that the tsunami advisory would remain in effect until 7:00 PM HST.[2] Due to the timely delivery of this message, an evacuation was avoided.

An announcement issued by Civil Defense was delivered by police, fire, and lifeguards to the public warning it of potential risks on beaches and suggesting that people remain out of the water.[3,4] Despite Civil Defense's announcement that beach goers should remain out of the water during the time of the possible arrival of the first wave, from 1:11 p.m. and onward, it was reported that Waikiki Beaches were still packed with people during this time.[3] Because Civil Defense knew and communicated that a widespread evacuation was not necessary, people were unresponsive to suggestions that they remain out of the water.

The fourth and final tsunami message, issued at 4:12 PM HST, announced the continuation of the tsunami advisory for the state of Hawaii. This message provided more detail from additional gauge stations with latitude, longitude, time, amplitude, and the wave period for tsunami wave activity. The evaluation in this message declared that "Small tsunami waves from this earthquake are now crossing the Hawaiian Islands. While these waves are not expected to cause any significant coastal flooding they can produce small changes of sea level at the coast and strong or unusual currents that can be hazardous to swimmers."[2]

REFERENCES

1. National Oceanic and Atmospheric Administration. 2009. *Pacific Tsunami Warning Center Tsunami Bulletins to IOC listserv*. Pacific Tsunami Warning Center, National Oceanic and Atmospheric Administration, Ewa Beach, Hawaii.

2. National Oceanic and Atmospheric Administration. 2009. *Pacific Tsunami Warning Center Tsunami Messages to Civil Defense*. Pacific Tsunami Warning Center, National Oceanic and Atmospheric Administration, Ewa Beach, Hawaii.

3. KITV. 2009. *Many Beachgoers in Waikiki Ignore Warnings*. [Online]. Available: http://www.kitv.com/news/21150223/detail.html [2010, February, 3].

4. Hawaii News Now. 2009. *Tsunami Watch Canceled for Hawaii After 8.3 Samoa Quake*. [Online]. Available: http://www.hawaiinewsnow.com/Global/story.asp?S=11225197 [2010, February, 3].

Response to the Chilean-Earthquake Generated Tsunami: The Hawaii Case Study

EVENT TIMELINE

On Friday, February 26, 2010, at 20:34:14 Hawaii Standard Time (HST) (Saturday, February 27, 2010, at 06:34:14Z; Figure J.1) a magnitude 8.8 earthquake occurred offshore of Maule, Chile, in position 35.909°S, 72.733°W, approximately 35 km below the surface of the earth.[1] The Pacific Tsunami Warning Center (PTWC) transmitted its first tsunami information message to the Civil Defense in the State of Hawaii (see Figure J.2) 15 minutes after the earthquake at 20:49 HST (February 27 at 06:49Z)[2] and issued its first tsunami warning bulletin to the Intergovernmental Oceanographic Commission (IOC) at 20:46 HST (February 27 at 06:46Z). The West Coast/Alaska Tsunami Warning Center (WC/ATWC) issued its first tsunami information statement at 20:49 HST (February 27 at 06:49Z). The first tsunami waves arrived at Valparaiso, Chile, approximately 34 minutes after the initial earthquake.[3]

Around 06:00 HST (16:35Z) on February 27, evacuation alarms were sounded in low-lying counties in Hawaii. The first waves reached Hawaii shortly after 11:35 HST (21:35Z), approximately 14.5 hours after the initial earthquake. Life in Hawaii started to return to normal at approximately 14:00 HST (February 28, 00:35Z) when the tsunami warning was lifted.

DESCRIPTION OF TSUNAMI WARNING CENTER MESSAGES TRANSMITTED TO CIVIL DEFENSE IN THE STATE OF HAWAII[2]

The PTWC transmitted its first tsunami message to Civil Defense in Hawaii at 20:49 HST on February 26. The message introduced a tsunami advisory, stated that the PTWC had issued an "expanding regional tsunami warning and watch for parts of the Pacific located closer to the earthquake" and advised that the earliest arrival of any tsunami wave would be 11:19 HST on February 27. Updated messages were released approximately every hour. Measurements and reports of tsunami wave activity were introduced in Message 3 and were continually updated through Message 20.

Message 6 issued a tsunami warning at 00:46 HST (approximately 11.5 hours before the forecast arrival of the tsunami). Message 12 (Box J.1) introduced the first forecasts of combined wave arrival times and amplitudes (crest to trough) for Hawaii. No additional forecasts were provided in Messages 13-16. Updated forecasts were provided in Message 17. Message 18 reported the arrival of the tsunami in Hawaii. Message 20 canceled the tsunami warning and was the final message issued for the tsunami generated by the Chilean earthquake.

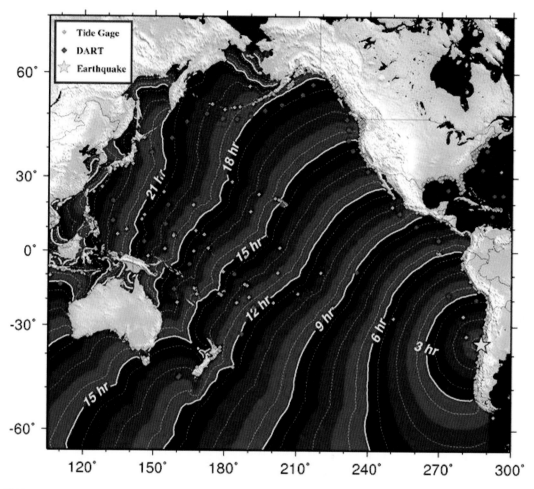

FIGURE J.1 Travel times for the February 27, 2010, Chilean tsunami. SOURCE: http://wcatwc.arh.noaa. gov/previous.events/Chile_02-27-10/Images/traveltime.jpg; West Coast/Alaska Tsunami Warning Center, NOAA.

DECISIONS MADE IN HAWAII

A statewide evacuation was not ordered; evacuation decisions were left to local counties.[4] On September 27, the Oahu Emergency Management Department spokesman was quoted at 03:51 HST in the Honolulu Advertiser as saying, "If you live anywhere in the evacuation zone, you have to evacuate. This is a serious event. We're going to treat this as a destructive-type tsunami" (Box J.2). The Department of Emergency Management issued a Tsunami Evacuation Information notice (no time stamp available) notifying residences in the coastal tsunami evacuation zones on Oahu that they should begin evacuation immediately at 06:00 HST when

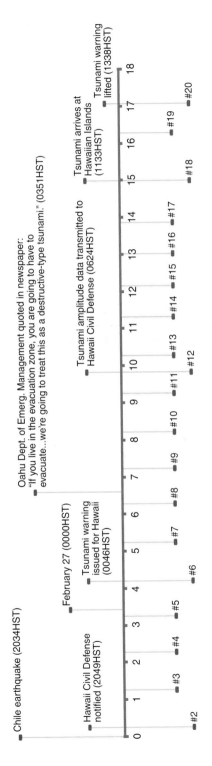

FIGURE J.2 Timeline for key events during the Chilean tsunami. SOURCE: Committee staff.

Time in hours from event. Tsunami Bulletin Numbers are listed below the timeline. Amplifying information is above the timeline. Note that Tsunami Bulletin #1 pertained to a different earthquake; relevant bulletins begin at #2.

BOX J.1

BULLETIN

TSUNAMI MESSAGE NUMBER 12
NWS PACIFIC TSUNAMI WARNING CENTER EWA BEACH HI
624 AM HST SAT FEB 27 2010

TO - CIVIL DEFENSE IN THE STATE OF HAWAII

SUBJECT - TSUNAMI WARNING SUPPLEMENT

A TSUNAMI WARNING CONTINUES IN EFFECT FOR THE STATE OF HAWAII.
AN EARTHQUAKE HAS OCCURRED WITH THESE PRELIMINARY PARAMETERS

  ORIGIN TIME - 0834 PM HST 26 FEB 2010
  COORDINATES - 36.1 SOUTH 72.6 WEST
  LOCATION   - NEAR COAST OF CENTRAL CHILE
  MAGNITUDE - 8.8 MOMENT

MEASUREMENTS OR REPORTS OF TSUNAMI WAVE ACTIVITY

GAUGE LOCATION	LAT	LON	TIME	AMPL	PER
RIKITEA PF	23.1S	134.9W	1536Z	0.16M / 0.5FT	48MIN
DART MARQUESAS 5140	8.5S	125.0W	1531Z	0.18M / 0.6FT	18MIN
QUEPOS CR	0.0N	9.4E	1416Z	0.24M / 0.8FT	52MIN
BALTRA GALAPAGS EC	0.4S	90.3W	1452Z	0.35M / 1.2FT	14MIN
EASTER CL	27.2S	109.5W	1205Z	0.35M / 1.1FT	52MIN
ANCUD CL	41.9S	73.8W	0838Z	0.62M / 2.0FT	84MIN
CALLAO LA-PUNTA PE	12.1S	77.2W	1029Z	0.36M / 1.2FT	30MIN
ARICA CL	18.5S	70.3W	1008Z	0.94M / 3.1FT	42MIN
IQUIQUE CL	20.2S	70.1W	0907Z	0.28M / 0.9FT	68MIN
ANTOFAGASTA CL	23.2S	70.4W	0941Z	0.49M / 1.6FT	52MIN
DART LIMA 32412	18.0S	86.4W	0941Z	0.24M / 0.8FT	36MIN
CALDERA CL	27.1S	70.8W	0843Z	0.45M / 1.5FT	20MIN
TALCAHUANO CL	36.7S	73.4W	0653Z	2.34M / 7.7FT	88MIN
COQUIMBO CL	30.0S	71.3W	0852Z	1.32M / 4.3FT	30MIN
CORRAL CL	39.9S	73.4W	0739Z	0.90M / 2.9FT	16MIN
SAN FELIX CL	26.3S	80.1W	0815Z	0.53M / 1.7FT	08MIN
VALPARAISO CL	33.0S	71.6W	0708Z	1.29M / 4.2FT	20MIN

LAT - LATITUDE (N-NORTH, S-SOUTH)
LON - LONGITUDE (E-EAST, W-WEST)
TIME - TIME OF THE MEASUREMENT (Z IS UTC IS GREENWICH TIME)
AMPL - TSUNAMI AMPLITUDE MEASURED RELATIVE TO NORMAL SEA LEVEL.
    IT IS ...NOT... CREST-TO-TROUGH WAVE HEIGHT.
    VALUES ARE GIVEN IN BOTH METERS(M) AND FEET(FT).
PER - PERIOD OF TIME IN MINUTES(MIN) FROM ONE WAVE TO THE NEXT.

EVALUATION

A TSUNAMI HAS BEEN GENERATED THAT COULD CAUSE DAMAGE ALONG
COASTLINES OF ALL ISLANDS IN THE STATE OF HAWAII. URGENT ACTION
SHOULD BE TAKEN TO PROTECT LIVES AND PROPERTY.

A TSUNAMI IS A SERIES OF LONG OCEAN WAVES. EACH INDIVIDUAL WAVE
CREST CAN LAST 5 TO 15 MINUTES OR MORE AND EXTENSIVELY FLOOD
COASTAL AREAS. THE DANGER CAN CONTINUE FOR MANY HOURS AFTER THE
INITIAL WAVE AS SUBSEQUENT WAVES ARRIVE. TSUNAMI WAVE HEIGHTS
CANNOT BE PREDICTED AND THE FIRST WAVE MAY NOT BE THE LARGEST.
TSUNAMI WAVES EFFICIENTLY WRAP AROUND ISLANDS. ALL SHORES ARE AT
RISK NO MATTER WHICH DIRECTION THEY FACE. THE TROUGH OF A TSUNAMI
WAVE MAY TEMPORARILY EXPOSE THE SEAFLOOR BUT THE AREA WILL
QUICKLY FLOOD AGAIN. EXTREMELY STRONG AND UNUSUAL NEARSHORE
CURRENTS CAN ACCOMPANY A TSUNAMI. DEBRIS PICKED UP AND CARRIED
BY A TSUNAMI AMPLIFIES ITS DESTRUCTIVE POWER. SIMULTANEOUS HIGH
TIDES OR HIGH SURF CAN SIGNIFICANTLY INCREASE THE TSUNAMI HAZARD.

THE ESTIMATED ARRIVAL TIME IN HAWAII OF THE FIRST TSUNAMI WAVE IS

        1119 AM HST SAT 27 FEB 2010

FORECAST WAVE ARRIVAL TIMES AND AMPLITUDES CREST TO TROUGH.

HILO   1105AM HST 2.5 METERS
HONOLULU 1137AM HST 0.5 METERS
KAHULUI   1126AM HST 2.2 METERS
NAWILIWILI 1142AM HST 0.9 METERS
HALEIWA   0.5 METERS
KAWAIHAE   0.6 METERS

THESE ARE ONLY ESTIMATES...ACTUAL VALUES MAY DIFFER.
MESSAGES WILL BE ISSUED HOURLY OR SOONER AS CONDITIONS WARRANT.

outdoor warning sirens sounded. Residents not located in evacuation zones were instructed to stay off the roads so that they would be open for evacuees and first responders.[5] Similar messages were released by Kauai Civil Defense.[6] Roads and businesses in low-lying areas throughout Hawaii were closed through the morning. Low-lying county facilities, including sewage treatment plants and wastewater pump facilities on Oahu and Maui, were closed, later resulting in sewage spills.[7] Private boats and commercial and military vessels evacuated Hawaiian harbors for deeper water.[4] Media sources and residents reported long lines at gas stations and at supermarkets where customers sought water, batteries, generators, and food supplies. Many of these supplies were later returned to the supermarkets.[4]

SUMMARY OF LESSONS LEARNED

Wave Height Forecasts by Tsunami Warning Centers

Availability of Deep-ocean Assessment and Reporting of Tsunamis (DART) buoy data from a buoy near Peru allowed for near-real-time forecasting of the wave heights relatively soon after the initial earthquake. The committee believes that PTWC chose to forecast wave heights using a model developed in-house (Table J.1).

Education of Officials/Citizens

Reports from this event suggest that responsible officials and citizen understanding of tsunamis is insufficient. Specifically, there was a lack of understanding of the different nature of a tsunami wave, which is similar to a massively high tide that cycles on the order of minutes or several hours rather than a common breaking wave, and the potential for strong local currents after the initial waves had passed. One public safety official was quoted as saying that extensive preparations were occurring in Hilo, which was directly exposed to the tsunami, whereas parts of Hawaii that are sheltered by the Big Island would take the brunt of its force.[8] This is untrue; tsunamis can wrap efficiently around islands, and local hydrographic features can significantly magnify the effects of a tsunami (as is the case in Hilo Bay, Hawaii).

It was apparently unclear to many citizens where, beyond higher ground, they should evacuate to. Evacuation shelters were opened in some locations. Others remained closed because they were intended only to shelter displaced residents after a natural disaster, not temporary evacuees.[9] Residents formed long lines at supermarkets to stock up on emergency supplies, many of which were subsequently returned after the tsunami warning was lifted. This indicates an ongoing reluctance of citizens to heed advice to be prepared for emergencies by stocking up in advance.

TABLE J.1 Tsunami Wave Height—Forecast and Observed

Tsunami Wave Height—Forecast and Observed			
PWTC Forecast "Crest to Trough"[a]		PWTC Observed Wave Amplitudes[b]	
0624 HST Forecast[2]		1338 PTWC Message[2]	
Hilo	2.5m	Hilo	~1m[1]
Kahalui	2.2	Kahalui	0.98m
Nawiliwili	0.9m	Nawiliwili	0.28m
Haleiwa	0.5m		
Kawaihae	0.6m	Kawaihae	0.52m
		Barbers Pt	0.19m
		Kaumalapau	0.18m
		[1]Media Reported PTWC Value	
1025 HST Forecast[2]			
Hilo	2.1-2.5m		
Honolulu	0.5-0.7m		
Kahului	1.1-3.0m		
Nawiliwili	0.9-1.4m		
Haleiwia	0.5-1.1m		
Kawaihae	0.6-1.6m		

[a] Reported as "Amplitudes Crest to Trough."
[b] Amplitudes reported relative to normal sea level (approximately one-half of wave height or "amplitude measured crest to trough").
SOURCE: Committee staff.

Response by Decision Makers

The first evidence of evacuation recommendations appears in a 03:51 HST newspaper article in the form of a quote from the spokesman for the Oahu Emergency Management Department ("If you live anywhere in the evacuation zone, you have to evacuate. This is a serious event. We're going to treat this as a destructive-type tsunami;" Box J.2); an official release from the Department of Emergency Management provided additional evacuation information. The official release did not contain a time/date stamp, but refers to the 06:00 HST activation of the evacuation sirens in the future tense, indicating that it was released prior to the siren activation.[6] This suggests that the decision to evacuate low-lying areas was made after a tsunami warning was announced at 00:46 HST, but before the first wave amplitudes (crest to trough) forecasts were received at 06:24 HST.

Actions taken by local officials have been praised, but also criticized as excessive in light of observed tsunami activity. At this time, it is still unclear what information led officials to order the evacuation of low-lying areas and undertake other preparatory measures such as securing power to sewage treatment plants and associated infrastructure.

BOX J.2

honoluluadvertiser.com
Updated at 3:51 a.m., Saturday, February 27, 2010
Residents along shorelines told to evacuate in advance of waves

Advertiser Staff

Oahu officials are urging anyone who lives in a tsunami inundation zone to evacuate in advance of the waves hitting just after 11 a.m. today.

Warning sirens will start sounding at 6 a.m.

"If you live anywhere in the evacuation zone, you have to evacuate," said John Cummings, Oahu Emergency Management Department spokesman. "This is a serious event. We're going to treat this as a destructive-type tsunami."

The last time there were voluntary tsunami evacuations in Hawaii was in 1994.

Cummings said getting out of the inundation zone could be as simple as crossing the street or walking to higher ground.

He and others urged people to stay off the roads as much as possible.

After the warning sirens sound this morning, first responders and Civil Defense volunteers will start going door-to-door in coastal areas to tell people to evacuate.

Cummings also said that there will be city buses going up and down shoreline areas picking up anyone who needs to get out of the inundation zone.

The ride will be free and the special city buses will say "evacuation," Cummings said. The buses will take people to safe areas where they can wait out the waves.

People in need of the ride can flag down the buses, and don't have to wait at bus stops, he said.

The tsunami expected to hit just after 11 a.m. will likely create the biggest problems in enclosed bay areas, including Hilo, Kahului, Haleiwa, where the waves could reach six to eight feet, officials said.

Along other shorelines, the waves are expected to be less than three feet, said Pacific Tsunami Warning Center geophysicist Brian Shiro.

He said the warning sirens that will go off at 6 a.m. and sound regularly as the tsunami gets closer mean that residents in tsunami inundation zones should evacuate.

He said people should not get in their cars to evacuate, but should walk to higher ground.

"All of our predictions and models are suggesting the tsunami in Hawaii is going to be less than three feet. That's not huge," he said. "But in places like Hilo Bay, Kahului, Haleiwa, the tsunami is going to probably get trapped and ... be as high as 6 to 8 feet."

He urged people to stay away from the water.

SOURCE: Honolulu Advertiser. 2010. *Residents Along Shorelines Told to Evacuate in Advance of Waves*. [Online]. Available: http://the.honoluluadvertiser.com/article/2010/Feb/27/br/hawaii100227012.html [2010, June 29].

Findings

- The decisions to order evacuations at the county level on Oahua and Kauai apparently preceded transmission of the first tsunami "amplitude (crest to trough)" model results.
- The Tsunami Warning Centers provided forecasts to Civil Defense in the state of Hawaii that proved to be within the range of observed conditions.

REFERENCES

1. U.S. Geological Survey. 2010. *Magnitude 8.8–Offshore Bio-Bio, Chile 2010 February 27 06:34:14 UTC.* [Online]. Available: http://earthquake.usgs.gov/earthquakes/recenteqsww/Quakes/us2010tfan.php [2010, June 29].

2. National Oceanic and Atmospheric Administration. 2010. *Index of PTWC Messages Hawaii* 2010. [Online]. Available: http://www.prh.noaa.gov/ptwc/messages/hawaii/2010/ [2010, June 29].

3. National Oceanic and Atmospheric Administration. 2010. *Tsunami Event—February 27, 2010 Chile Main Event Page.* [Online]. Available: http://nctr.pmel.noaa.gov/chile20100227/ [2010, June 29].

4. New York Times. 2010. *Hawaii Exhales as Tsunami Warning Is Canceled.* [Online]. Available: http://www.nytimes.com/2010/02/28/us/28warning.html [2010, June 29].

5. City and County of Honolulu. 2010. *Department of Emergency Management Issues Tsunami Evacuation Information.* [Online]. Available: http://www.honolulu.gov/csd/publiccom/honnews10/DEMTsunamiEvacuation.htm [2010, June 29].

6. County of Kauai State of Hawaii. 2010. *Civil Defense Program Description.* [Online]. Available: http://www.kauai.gov/Government/Departments/CivilDefenseAgency/tabid/90/Default.aspx [2010, June 29].

7. Honolulu Star Bulletin. 2010. *Tsunami Response Flaws Draw Review.* [Online]. Available: http://www.starbulletin.com/news/20100228_Tsunami_response_flaws_draw_review.html [2010, June 29].

8. The Wall Street Journal. 2010. *Hawaii Unaffected by Tsunami.* [Online]. Available: http://online.wsj.com/article/SB10001424052748704231304575091624043844744.html [2010, June 29].

9. The Maui News. 2010. *Officials Assess Tsunami Response.* [Online]. Available: http://www.mauinews.com/page/content.detail/id/529308.html

Acronyms

AHAB	All-Hazard Alert Broadcast
ANSI	American National Standards Institute
AOR	Area of Responsibility
ARSC	Arctic Regional Supercomputing Center
ASCE	American Society of Civil Engineers
ATFM	Alaska Tsunami Forecast Model
AusAID	Australian Agency for International Development
AWC	Aviation Weather Center
BODC	British Oceanographic Data Center
BPR	Bottom Pressure Recorder
CalEPA	California Emergency Management Agency
CALWAS	California Warning System
CHP	California Highway Patrol
CISN	California Integrated Seismic Network
CLETS	California Law Enforcement Telecommunications System
CMT	Global Centroid-Moment-Tensor
COMCOT	Cornell Multi-grid Coupled Tsunami Model
CO-OPS	Center for Operational Oceanographic Products and Services
CPPT	Centre Polynesien de Prevention des Tsunamis (Polynesian Tsunami Warning Center)
CPU	Central Processing Unit
CSG	Council of State Governments
CSWC	California State Warning Center
CSZ	Cascadia Subduction Zone
CTBTO	Comprehensive Nuclear Test-Ban Treaty Organization
DART	Deep-ocean Assessment and Reporting of Tsunamis
DEM	Digital Elevation Model
DMS	Data Management System
DOGAMI	Oregon Department of Geology and Mineral Industries
DONET	Dense Ocean-Floor Network System for Earthquakes and Tsunamis
DOS	Department of State
EAS	Emergency Alert System

EDIS	California Emergency Digital Information Service
EERC	Earthquake Engineering and Research Center
EERI	Earthquake Engineering Research Institute
EEZ	Exclusive Economic Zone
EMAP	Emergency Management Accreditation Program
EMWIN	Emergency Managers Weather Information Notification
EOC	Emergency Operation Center
EVO	Engineering Virtual Organization
FEMA	Federal Emergency Management Agency
FGDC	Federal Geographic Data Committee
FORTRAN	Formula Translation/Translator
GAO	Government Accountability Office
GCMT	Global Centroid-Moment Tensor
GEBCO	General Bathymetric Chart of the Oceans
GEO	Group on Earth Observations
GEOSS	Global Earth Observation System of Systems
GIS	Geographic Information System
GITEWS	German Indonesian Tsunami Early Warning System
GLOSS	Global Sea Level Observing System
GMT	Generic Mapping Tool
GOES	Geostationary Operational Environmental Satellites
GOOS	Global Ocean Observing System
GPS	Global Positioning System
GSN	Global Seismograph Network
GTS	Global Telecommunications Service
HAWAS	Hawaii Warning System
HF	High Frequency
HRO	High-Reliability Organization
HRV	Harvard University Seismic Station
HST	Hawaii Standard Time
IAEM	International Association of Emergency Managers
ICC	Incident Command Center
ICG/CARIBE/EWS	Intergovernmental Coordination Group for the Tsunami and Other Coastal Hazards Warning System for the Caribbean and Adjacent Regions
ICS	Incident Command System
IDA	International Deployment of Accelerometers
IMS	Integrated Maritime Surveillance
IOC	Intergovernmental Oceanographic Commission

IOOS	Integrated Ocean Observing System
IOTWS	Indian Ocean Tsunami Warning and Mitigation System
IRIS	Incorporated Research Institutions for Seismology
ISDR	International Strategy for Disaster Reduction
ISEC	Inundation Science and Engineering Cooperative
IT	Information Technology
ITEWS	Indonesia Tsunami Early Warning System
ITIC	International Tsunami Information Center
JAMSTEC	Japan Agency for Marine-Earth Science and Technology
JIMAR	Joint Institute for Marine and Atmospheric Research
JMA	Japan Meteorological Agency
LEO	Low Earth Orbit
LIDAR	Light Detection and Ranging
MARS	Monterey Accelerated Research System
M&ES	Mitigation and Education Subcommittee
MMS	Mapping and Modeling Subcommittee
MOST	Method of Splitting Tsunami
NACSE	Northwest Alliance for Computational Science and Engineering
NAS	National Academy of Sciences
NASA	National Aeronautics and Space Administration
NAWAS	National Warning System
NCDC	National Climate Data Center
NCEP	National Centers for Environmental Prediction
NCG	National Association of Governors
NCSL	National Conference of State Legislatures
NCTR	NOAA Center for Tsunami Research
NDBC	National Data Buoy Center
NEES	Network for Earthquake and Engineering Simulation
NEHRP	National Earthquake Hazards Reduction Program
NEIC	National Earthquake Information Center
NEMA	National Emergency Management Association
NEOWAVE	Non-hydrostatic Evolution of Ocean WAVE
NEPTUNE	North-East Pacific Time-Series Underwater Networked Experiments
NESDIS	National Environmental Satellite, Data, and Information Service
NFPA 1600	Standard on Disaster/Emergency Management and Business Continuity Programs
NGDC	National Geophysical Data Center
NGO	Non-Governmental Organization

NHC   National Hurricane Center
NLC   National League of Cities
NOAA   National Oceanic and Atmospheric Administration
NOMAD   Nonlinear Optimization for Mixed Variables and Derivatives
NOS   National Ocean Service
NRC   National Research Council
NSE   National Science Education
NSF   National Science Foundation
NSSL   National Severe Storms Laboratory
NSTC   National Science and Technology Council
NTF   National Tidal Facility
NTHMP   National Tsunami Hazard Mitigation Program
NWLON   National Water Level Observation Network
NWR   NOAA Weather Radio
NWS   National Weather Service
NWSTG   National Weather Service Telecommunication Gateway

OAR   Office of Oceanic and Atmospheric Research
OEM   Oregon Emergency Management
OES   Office of Emergency Services
OGP   Office of Global Programs
OJP   U.S. Department of Justice Office of Justice Programs
OMAO   Office of Marine and Aviation Offices
OOI   Ocean Observatories Initiative

PC   Personal Computer
P.L. 109-424   Public Law 109-424
PMEL   Pacific Marine Environmental Laboratory
PNG   Papua New Guinea
PRTTRC   Puerto Rico Tsunami Technical Review Committee
PSAP   Public Safety Access Points
PSMSL   Permanent Service for Mean Sea Level
PTA   Parent-Teacher Association
PTWC   Pacific Tsunami Warning Center
PTWS   Pacific Tsunami Warning and Mitigation System

RHM   ROSHYDROMET
RSN   Regional Scale Nodes
RTK-GPS   Real Time Kinematic-Global Positioning System

SAIC   Science Applications International Corporation
SCEC   Southern California Earthquake Center

SELFE	Semi-implicit Eulerian-Langrangian Finite
SHOA	Servicio Hidrográfico y Oceanográfico de la Armada (Chilean Navy Hydrographic and Oceanographic Service)
SIFT	Short-term Inundation Forecasting for Tsunami
SIO	Scripps Institution of Oceanography
SLSMF	Sea Level Station Monitoring Facility
SMS	Simple Messaging System
SOEST	School of Ocean and Earth Science and Technology
SOFAR Channel	Sound Fixing and Ranging Channel
SPC	Storm Prediction Center
STSC	California State Tsunami Steering Committee
TACO	Tsunami Advisory Council for Oregon
TAO	Tropical Atmosphere Ocean Project
TERK	Tsunami Education Resource Kit
TTRC	Hawaii Tsunami Technical Review Committee
TWC	Tsunami Warning Center
TWS	Tsunami Warning System
UAF	University of Alaska, Fairbanks
UHSLC	University of Hawaii Sea Level Center
UNESCO	United Nations Educational, Scientific, and Cultural Organization
USAID	United States Agency for International Development
USACE	U.S. Army Corps of Engineers
USGS	U.S. Geological Survey
USVI	U.S. Virgin Islands
WAS*IS	Weather and Society*Integrated Studies
WC/ATWC	West Coast/Alaska Tsunami Warning Center
WCM	Warning Coordination Meteorologist
WCS	Tsunami Warning Coordination Subcommittee
WEMD	Washington State Emergency Management Department
WHOI	Woods Hole Oceanographic Institution
WMO	World Meteorological Office
XML	Extensible Markup Language
YSTWC	Yuzhno-Sakhalin Tsunami Warning Center

Committee and Staff Biographies

COMMITTEE

John A. Orcutt (Chair) is a distinguished professor of geophysics at the University of California, San Diego. Dr. Orcutt earned a B.S. in mathematics and physics from the U.S. Naval Academy, an M.Sc. in physical chemistry as a Fulbright Scholar from the University of Liverpool, and a Ph.D. in geophysics from the University of California, San Diego, Scripps Institution of Oceanography. His research focuses on cyberinfrastructure and geophysical applications; geophysical studies of ocean seismo-acoustics including rough seafloor scattering, acoustic-elastic interactions, and the use of small arrays; structure of the elastic earth using seismology, synthetic seismograms, and geophysical inverse theory; internal structure of ocean spreading centers; genesis of the oceanic lithosphere; and nuclear test-ban verification methods. Dr. Orcutt is a past president of the American Geophysical Union and a Secretary of the Navy/Chief of Naval Operations Oceanography Chair. He is a member of the American Philosophical Society and served briefly as Interim President of the Ocean Drilling Program in 2000. Dr. Orcutt is a former member of the Ocean Studies Board and has served on numerous National Research Council (NRC) committees, including the Ocean Studies Board's (OSB's) Committee on Exploration of the Seas.

Martha R. Grabowski (Vice-chair) is McDevitt Associate Chair of Information Systems, chair of the Business Administration Department, professor and director of the information systems program at Le Moyne College in Syracuse, New York, as well as research professor in the Department of Industrial and Systems Engineering at Rensselaer Polytechnic Institute. Dr. Grabowski earned an M.B.A., an M.S. in engineering, and a Ph.D. in management and information systems from Rensselaer Polytechnic Institute and a B.S. from the U.S. Merchant Marine Academy. A licensed former merchant officer and retired LCDR in the U.S. Naval Reserve, her research focuses the impact of technology in safety-critical systems, risk analysis and risk mitigation in large-scale systems, the role of human and organizational error in high consequence settings, and financial cybersecurity in complex, heterogeneous data environments. A member of the American Bureau of Shipping, Dr. Grabowski is a former chair of the NRC's Marine Board, a former member of the NRC's standing Committee on Human Factors, and has served on numerous NRC committees. She is currently chairing the Committee on Naval Engineering in the 21st Century and chaired the Committee Evaluating Shipboard Display of Automated Identification Systems. She received the Lifetime National Associate Award from the NRC in 2003.

Brian Atwater is a geologist employed by the U.S. Geological Survey and based at the University of Washington. Dr. Atwater earned a Ph.D. in geology from the University of Delaware.

Having used geology as a long-term advisory of earthquakes and tsunamis in the Cascadia region of western North America, he is now trying to make this strategy helpful to developing countries. He also seeks to mentor scientists in assessing tsunami hazards on the centennial and millennial timescales of great-earthquake recurrence. These overeseas efforts now include a UNESCO project at the Makran subduction zone and a Fulbright in Indonesia. Dr. Atwater is exploring earthquake geology in the British Virgin Islands to help guide the Nuclear Regulatory Commission on tsunami hazards of the U.S. Atlantic coast. Dr. Atwater is a member of the National Academy of Sciences.

Ann Bostrom is professor and associate dean of research at the Daniel J. Evans School of Public Affairs of the University of Washington. Dr. Bostrom earned a B.A. from the University of Washington, an M.B.A. from Western Washington University, and a Ph.D. in policy analysis from Carnegie Mellon University. She joined the Evans School faculty in 2007. Dr. Bostrom has research interests in risk perception, communication and management, and environmental policy and decision making under uncertainty. Her research focuses on mental models of hazardous processes (how people understand and make decisions about risks). Dr. Bostrom co-directed the Decision Risk and Management Science Program at National Science Foundation (NSF) from 1999-2001. Dr. Bostrom is risk communication area or associate editor for *Risk Analysis*, the *Journal of Risk Research*, and *Human and Ecological Risk Assessment*, and she is a Fellow of the Society for Risk Analysis. She has authored or contributed to numerous publications, including *Risk Communication: A Mental Models Approach* and *Risk Assessment, Modeling and Decision Support: Strategic Directions*, as well as NRC and EPA Science Advisory Board and Board of Scientific Counselor reports.

George Crawford is the former Washington State earthquake program manager and has more than 16 years of multi-program design and coordination of local, state, national, and international seismic and geologic programs in collaboration with the U.S. Geological Survey (USGS), National Oceanic and Atmospheric Administration (NOAA), Federal Emergency Management Agency (FEMA) and state, national, international, and private organizations. Mr. Crawford served as the Washington State representative to the U.S. National Tsunami Hazard Program, Western States Seismic Policy Council Tsunami Committee chair, Washington State/Local Tsunami Workgroup chair, a Cascadia Region Earthquake Workgroup director, and the National Emergency Management Association (NEMA) representative to the ANSS National Steering Committee. Mr. Crawford co-developed the All-Hazard Alert Broadcasting (AHAB) Radio that is deployed in U.S. states and internationally, and he advises on tsunami community communication issues globally. He has worked extensively with coastal Native American tribes to link science to tribal oral history, and created the "Run to High Ground" video. He has represented the United States in international forums, and the Washington State Tsunami Program has become a mitigation model nationally and internationally for at-risk tsunami communities. In retirement, he continues to support U.S. Agency for International Development (USAID) and U.N. Development Programme (UNDP) in Indian Ocean country tsunami missions, collaborate with NOAA/Pacific Marine Environmental Laboratory (NOAA/PMEL) to develop the Train-the Trainer Program and

Community-based Education Program in Washington State, assist in the University of Washington Certificate Program in "Tsunami Science & Preparedness," and he is lead contractor in developing a national Tsunami Awareness Course for the National Disaster Preparedness Training Centre at the University of Hawaii. He has written papers on tsunami communication and dissemination, preparedness, and mitigation and continues to collaborate in tsunami research.

Richard Eisner is visiting professor at the Research Center for Disaster Reduction Systems (DRS), Disaster Prevention Research Institute at Kyoto University. Prior to joining DRS, he worked for 23 years as the coastal regional administrator for the California Governor's Office of Emergency Services, where he was responsible for the state's disaster response in the San Francisco Bay Region and north coast counties. He also served as manager for the California Integrated Seismic Network's and the state's Tsunami and Earthquake Programs. Prior to that appointment, Mr. Eisner served as the founding director of the Bay Area Regional Earthquake Preparedness Project, providing planning and technical assistance to promote and support earthquake preparedness and hazard mitigation by local governments and businesses throughout the San Francisco Bay Region. In 2007, he received the Lifetime Achievement Award by the Western States Seismic Policy Council. Mr. Eisner recently served on the NRC Panel on Solid-Earth Hazards, Resources, and Dynamics.

Jian Lin is a senior scientist and Henry Bryant Bigelow Chair for Excellence in Oceanography at the Woods Hole Oceanographic Institution. He earned his Ph.D. in geophysics from Brown University. His research focuses on earth's lithosphere processes that lead to catastrophic events of earthquakes, underwater volcanism, and tsunamis. Dr. Lin has conducted extensive research on earthquake stress interaction and triggering in California and elsewhere in the world. He has led and participated in numerous oceanographic expeditions to the Pacific, Atlantic, and Indian Oceans to investigate mid-ocean ridge volcanism, deep-sea hydrothermal vents, underwater earthquakes, and tsunamis. He is recently conducting two areas of tsunami research: paleo-seismological dating of tsunamis and mega-earthquakes and modeling tsunami sources in the Atlantic. Dr. Lin is a past chairman of the InterRidge International Science Program. He is a fellow of the Geological Society of America and American Association for the Advancement of Science.

Douglas S. Luther is a professor in the Department of Oceanography at the University of Hawaii at Manoa, Honolulu, Hawaii. Dr. Luther received his B.S. in geophysics and electrical engineering from Massachusetts Institute of Technology and a Ph.D. in oceanography from Massachusetts Institute of Technology and Woods Hole Oceanographic Institution Joint Program. His expertise is in circulation variability and dynamics; mesoscale fluctuations; waves in the ocean; and flow-topography interactions. He is a senior fellow, Joint Institute for Marine and Atmospheric Research, National Oceanic and Atmospheric Administration and the University of Hawaii at Manoa. Dr. Luther is also an associate editor for the *Journal of Physical Oceanography*.

Hugh B. Milburn is retired from the National Oceanic and Atmospheric Administration and is now an independent consultant. Mr. Milburn worked at NOAA's Pacific Marine Environmental

Laboratory for 22 years, last as a supervisory ocean engineer. He earned an M.S.E. in ocean engineering from the University of Washington. His interests include the development of moored systems to serve research and operational requirements of global-scale measurements and observation with a focus on materials, sensors, data telemetry, and deployment methodologies. Mr. Milburn is also involved in the development of seafloor observatories utilizing advanced communications for data acquisition and desktop control. He lead the team that developed the NOAA Deep-ocean Assessment and Reporting of Tsunami (DART) system. In 2005, he received the Department of Commerce's Gold Medal for research and development leading to the creation of a tsunami forecasting capability.

Dennis Mileti is a professor emeritus in the Department of Sociology and a research scientist in the Natural Hazards Center at the University of Colorado. Dr. Mileti was director of the Natural Hazards Center at the University of Colorado at Boulder, which is the nation's national repository and clearinghouse for social science research on natural hazards and disasters, and he is currently also a researcher at the Department of Homeland Security's START Center of the University of Maryland at College Park. Dr. Mileti earned a Ph.D. in sociology from the University of Colorado. His research focuses on environmental sociology, complex organizations, research methods, and collective behavior. In 2007, he earned the Outstanding Civilian Service Medal from the U.S. Department of the Army. Dr. Mileti has served on previous NRC committees, including the Committee on Natural Disasters.

Emile Okal is a professor in the Department of Earth and Planetary Sciences at Northwestern University. Dr. Okal earned a Ph.D. in geophysics from the California Institute of Technology. He is currently researching tsunami generation, risk, and warning using post-event field surveys and modeling. Dr. Okal's general research interests include seismology, structure of the Earth's interior, properties of seismic sources, marine geosciences, volcanism, and underwater acoustics.

Costas Synolakis is a professor of civil, environmental, aerospace, and mechanical engineering and the director of the Tsunami Research Center at the University of Southern California. Dr. Synolakis earned a Ph.D. in civil engineering from the California Institute of Technology. His research focuses on tsunami run-up, computer tomography, vibration, isolation of art objects, and earthquake hazards reduction.

Nathan Wood is a research geographer at the U.S. Geological Survey. Dr. Wood earned a Ph.D. in geography from Oregon State University. His research focuses on characterizing and communicating societal vulnerability to natural hazards, with emphasis on tsunamis in the Pacific Northwest. He develops and applies geospatial modeling and mapping techniques, as well as collaborative community-based processes, to better understand how individuals, communities, and coupled human-environment systems are vulnerable to tsunamis.

Harry Yeh is the Miles Lowell and Margaret Watt Edwards Distinguished Chair in Engineering and a professor of coastal and ocean engineering at Oregon State University. Dr. Yeh earned a Ph.D. in civil engineering from the University of California, Berkeley. His research focuses on environmental fluid mechanics, ocean and coastal wave phenomena, flow-structure inter-actions, tsunami induced scour, wind turbulence, structure control (tuned liquid dampers), physical processes in lakes and oceans, and tsunami hazard mitigation.

STAFF

Claudia Mengelt is a senior program officer with the Ocean Studies Board. After complet-ing her B.S. in aquatic biology at the University of California, Santa Barbara, she received her M.S. in biological oceanography from the College of Oceanic and Atmospheric Sciences at Oregon State University. Her master's degree research focused on how chemical and physical parameters in the surface ocean affect Antarctic phytoplankton species composition and con-sequently impact biogeochemical cycles. She obtained her Ph.D. in marine sciences from the University of California, Santa Barbara, where she conducted research on the photophysiology of harmful algal bloom species. She joined the full-time staff of the National Academies' Board on Atmospheric Sciences and Climate (BASC) in the fall of 2005 following a fellowship with the same group in the winter of 2005. While with the Academies, she has worked on studies including the *Analysis of Global Change Assessments* (2007), *Strategic Guidance for the NSF's Sup-port of Atmospheric Sciences* (2007), *Earth Observations from Space: The First 50 Years of Scientific Achievements* (2007), and *Adapting to the Impacts of Climate Change* (2010).

Susan Park was a senior program officer with the Ocean Studies Board until the end of 2009. She received her Ph.D. in oceanography from the University of Delaware in 2004. Dr. Park was a Christine Mirzayan Science and Technology Graduate Policy Fellow with the Ocean Studies Board in 2002 and joined the staff in 2006. She worked on several reports, including *Nonnative Oysters in the Chesapeake Bay*, *Review of Recreational Fisheries Survey Methods*, *Dynamic Changes in Marine Ecosystems*, *A Review of the Ocean Research Priorities Plan and Implementation Strategy*, and *Tackling Marine Debris in the 21st Century*. Prior to joining the Ocean Studies Board, she spent time working on aquatic invasive species management with the Massachusetts Office of Coastal Zone Management and the Northeast Aquatic Nuisance Species Panel. She is currently assistant director for research at Virginia Sea Grant.

Deborah Glickson is a program officer with the Ocean Studies Board. She received an M.S. in geology from Vanderbilt University in 1999 and a Ph.D. in oceanography from the University of Washington in 2007. Her doctoral research focused on magmatic and tectonic contributions to mid-ocean ridge evolution and hydrothermal activity at the Endeavour Segment of the Juan de Fuca Ridge. Deborah was a Dean John A. Knauss Marine Policy Fellow and worked on coastal and ocean policy and legislation in the U.S. Senate. Prior to her Ph.D. work, she was a research associ-ate in physical oceanography at Woods Hole Oceanographic Institution. She joined the National Academies staff in 2008 and has worked on *Oceanography in 2025: Proceedings of a Workshop*, *Science at Sea: Meeting Future Oceanographic Goals with a Robust Academic Research Fleet*, and *Realizing the Energy Potential of Methane Hydrates in the United States* in addition to this report.

Pamela Lewis is the administrative coordinator with the Ocean Studies Board. She joined the Academies in 1997 as a program assistant working in the Division on Engineering and Physical Sciences (DEPS), Board on Army Science and Technology and the Air Force Studies Board. She received a promotion to program associate in DEPS with the Board on Physics and Astronomy, where she was a recipient of a DEPS Staff Award. Ms. Lewis attended Berkeley-Claremont Business School in New York City, majoring in business management. Prior to going to the Academies, she was employed as an office manager at IBM in Poughkeepsie, New York.

Sherrie Forrest is a research associate on the Ocean Studies Board and Board on Science Education at the National Research Council. She graduated from the Institute of Marine and Coastal Sciences at Rutgers University with an M.S. in oceanography in May 2008. Her graduate work employed molecular techniques to explore oceanic nitrogen and iron budgets using quantitative analysis of an essential physiological structure. Ms. Forrest also has a B.A. in English from Pepperdine University. Before transitioning to science, she worked in development and production of feature films and documentaries in both California and New York. More recently, she worked as a freelance science writer for Rutgers University in New Jersey.

Jeremy Justice is a senior program assistant with the Ocean Studies Board. He earned a B.A. in international and area studies from the University of Oklahoma in 2008. Since joining the staff in October 2008, Mr. Justice has worked on *Science at Sea: Meeting Future Oceanographic Goals with a Robust Academic Research Fleet*, *Ecosystem Concepts for Sustainable Bivalve Mariculture*, and *Assessment of Sea-Turtle Status and Trends*, in addition to this report.

Helena Antoun was a Christine Mirzayan Science and Technology Policy Fellow (Fall 2009). She holds a master's degree in marine sciences from the University of Puerto Rico-Mayagüez and a bachelor's degree in environmental sciences and chemistry from the University of Puerto Rico-Rio Piedras. As part of her graduate research, she participated in oceanographic expeditions in the Caribbean, studying the impacts of the Orinoco River plume and mesoscale eddies on phytoplankton dynamics. She has also been involved in different projects ranging from environmental pollution to impact assessment of climate change in the Caribbean and ocean acidification. Ms. Antoun is currently working with the Conservation Trust of Puerto Rico.

Ian Brosnan was a Christine A. Mirzayan Science and Technology Policy Graduate Fellow on the Ocean Studies Board (Winter 2010). A former U.S. Coast Guard officer, his service included command of the Coast Guard Cutter Cobia and liaison to members of the U.S. Congress. He received his master's degree in marine affairs from the University of Washington in 2010 and is presently pursuing a Ph.D. at Cornell University. His research centers around the use of advanced tagging and tracking technologies to address marine conservation concerns in the Pacific.